REAL PLACES

REAL PLACES

An Unconventional Guide to America's Generic Landscape

Grady Clay

The University of Chicago Press *Chicago and London*

TO ELLEN CHURCHILL SEMPLE

Director of The Gazeteer Project, GRADY CLAY has been urban affairs editor of the Louisville *Courier-Journal* and editor of *Landscape Architecture* magazine. He is the author of *Close-Up: How to Read the American City*, also published by the University of Chicago Press.

The University of Chicago Press, Chicago 60637
The University of Chicago Press, Ltd., London
© 1994 by the University of Chicago
All rights reserved. Published 1994
Printed in the United States of America
03 02 01 00 99 98 97 96 95 94   1 2 3 4 5

ISBN (cloth): 0-226-10946-1

Library of Congress Cataloging-in-Publication Data

Clay, Grady.
Real places : an unconventional guide to America's generic landscape / Grady Clay.
p. cm.
Includes bibliographical references and index.
1. Landscape—United States. 2. Human geography—United States. 3. United States—Description and travel. 4. United States—Social life and customs—20th century. I. Title.
E169.04.C57 1994
973—dc20   94-15006
CIP

Title page photos: WINDFARM, see pp. 220–21; KUDZU, see pp. 250–51; FLEA MARKET, see pp. 204–5; HURRICANE PATH, see p. 123; deserted site, see p. 130.

Page v photo: TEMPORARY HOUSING, see page 149.

All photographs by author unless otherwise credited.

♾ The paper used in this publication meets the minimum requirements of the American National Standard for Information Sciences—Permanence of Paper for Printed Library Materials, ANSI Z39.48-1984.

CONTENTS

LIST OF ENTRIES

PREFACE

In 1972 I found myself the possessor of a grant from the Ford Foundation to study American cities, using a cross-section method of investigation. This was a contrivance of my own, derived in part from readings in the history of surgery, and in part from the teachings of Sir Patrick Geddes, the noted Scottish biologist and planner. And, I must add, also from the experience of having been jammed into conference buses, and lectured by exotic guides on endless post-conference tours across American cities. It was enough to convince me I should try my own version.

Two other great exemplars had crossed my route. The first, the founder of *Landscape* magazine, J. B. Jackson, had written in 1957 a seminal essay called "The Stranger's Path."[1] It revealed both Jackson's insights and the generic components of a small Western railroad town of the 1950s unfolding itself sequentially to an observant visitor.

And then there was Gordon Cullen, the English architect-artist whose firm hand and penetrating gaze produced an influential series of articles in the British magazine, *Architectural Review,* later gathered into *Townscape,* in 1961. In this detailed guidebook he analyzed "serial vision" as an organizing device for exploration. It seemed simple enough but in execution required time, patience, and technique.

Kevin Lynch's "Walk around the Block," was an impressive eye-opener, and a predecessor to his memorable work, *The Image of the City.* An evening walk with the incomparable observer Jane Jacobs down Pittsburgh's lively Fifth Avenue, in the early stages of our work on *The Exploding Metropolis,* further opened my mind to the rigorous pursuit of generalizations along a linear path.[2]

To live up to my own 1972 prospectus, I plotted, mapped, and then navigated my own cross-section routes, mostly by auto, across some thirty-four American and Canadian cities from coast to coast. Most were situated in a swathe about 350 miles wide, extending from Santa Monica in Southern California to Ipswich, Mass. Since that time I have "run a section" across many another town, city, and region with varying amounts of time and degrees of rigor. None, I should add, were straight and direct. All dodged about in response to the vagaries of local geography and my own criteria. But all entered one side and exited the other, dealt with The Center, traversed growth and nongrowth areas, slums, and the good address; and extended to the outer limits of locally defined commuting range, usually a forty-minute trip in each direction.

Further to pursue this ever-changing and rapidly urbanizing landscape, I have twice followed the quirky course of "The 450-mile Yard Sale." This is an annual summer event on one three-day August weekend, strung out along both sides of U.S. Highway 127 from the outskirts of Cincinnati, Ohio, through Kentucky, Tennessee, and a bit of Georgia, ending at Gadsden, Ala. Hundreds of families set out their surpluses along U.S. 127. Judging by the traffic, many others share my obsession with this display of horizon-to-horizon yard-sale junk—the end of countless American, Asiatic, and European production lines.

I recorded "a series of jaunts and organized toots, assembled into a cross-section north and south across the geographical grain of North America" in my essay "Crossing the American Grain, or The Happiness of Pursuit," in *Geographical Snapshots of North America.*[3] More recently, I examined the Illinois-to-Ohio stretch of The National Road, U.S. 40 (Cumberland, Md., to St. Louis, Mo.) with Karl Raitz, the editor of a forthcoming book.[4] Earlier, taking part in a television documentary, "Unknown Places," I discovered the disorienting limit to visually absorbing a speeded -up cross-section on film.

All such trips are experiments with continuity. The word "continuity" I see in a special light, and as a tactic for coming to grips with new environments. It is never enough to work with maps alone, or documents alone—although all history depends on both. Nor is it enough to interview Usually Reliable Sources along the route. To come to an understanding of places requires all these: maps, documents, histories, interviews, photography, careful observation. But, above all, understanding depends upon the continuity of moving self-consciously and completely through a place—in-one-side-and-out-the-other—and then repeating that process over time. I have compiled lists (endless lists!) of favorite places which pour out great flows of information.

Such places I have dubbed "epitome districts" in another context. They "carry huge layers of symbols that have the capacity to pack up emotions, energy, or history into a small space." They are "crammed with clues that trigger our awareness of the larger scene. . . ."[5] One must treasure such places; they are worth visiting again and again. Many such places figure in this book.

Year in, year out, I have tested my own hometown cross-section across some forty-odd miles on friends and visitors, noting their reactions, taking (some of) their advice, and keeping track of the changes over time. It is impossible to do this with an open mind and not to be struck by the rhythms of change. Not just one-damn-thing-after-another, but an orderly succession that builds into a structure over time: that oily slick on a front lawn this year; next year the grass is gone, the "lawn" is a parking lot. A couple years down the road the neighborhood is clearly going downhill—or perhaps recovering, as the case may be. Interviews, maps at city hall, all the apparatus of vigorous inquiry, will add to the picture, or change it.

This interest of mine has been reinforced by those editors, publishers, grantsmen, and hosts generous enough to stake my observational and tape-recorded trips through hundreds of American, Canadian, and other urbanizing places. Many of those places are identified in this book. Those on-site identifications have been supplemented by readings, interviews, photography, and feedback—the usual backups of a working journalist.

Meanwhile, a revolution has taken place in what are now called Geographic Information Systems. Thanks to worldwide satellite photography, computerized data, and computer mapping, there are few places left in North America (or the world) inaccessible to electronic viewing and analysis. There are no Secret Places, and those physically inaccessible are diminishing.

But to "know" these place secondhand is shallow knowledge indeed. To hear officials boast of knowing "all that is required" (for taxation or redistricting, for allocation, permissions, highway locations, etc.) by looking at computer printouts is to overhear bureaucracy at its worst, and sometimes at its most dangerous.

But to love, as well as to explore, at a distance is never enough. The cross-section method forces you to deal with middle, rather than vast, distances. This is a term for everything between the microscopic and the global. Modern science has abandoned the middle distance in order to concentrate on "the unimaginably small and the impossibly large" as O. B. Hardison, Jr., puts it in *Disappearing through the Skylight.*[6]

And so I set out to make sense of the outpourings of information that emerge during and after a typical cross-section trip--an endless variety of scenes in the middle distance. In pursuing and compressing these sometimes random encounter zones into a book form, I have given them an artificial (i.e., alphabetical) array–even though we all know that landscape does not array itself alphabetically. Nor do trips give each of us the same set of sequences. If there is order, we must work hard to find it and to give it our own shape.

One set of conditions, however, does shape all cross-section trips–the decrease in density of population and intensity of land uses from The Center to Out There. Even given its endless variations and sometimes imperceptible gradations, that decrease is to be counted upon. Furthermore, all Eastern cities of the United States illustrate a predictable gradation: dry in The Center, and wetter as you move Out There. Urbanization in the Eastern United States has, with few exceptions, consisted of drying out the original landscape. In short, urbanization equals desiccation. The reverse is true across the arid West where cities import or "mine" water to convert themselves into dependent, fragile oases. They live on borrowed time as well as water.

In working my way, so to speak, through this geographic flux, I have imposed a rough kind of cross-section upon the material as it appears in this book—moving from The Center with its DOWNTOWN and THE

SCENE, to Out There with its BOONDOCKS and GHOST TOWNS. By no means, of course, should anyone expect to find landscape arrayed so neatly. I have chosen this as an arbitrary but comprehensible format, as one finds in tourist guidebooks.

It may be helpful to imagine that you, the reader, are taking a long and leisurely trip through your own, or other familiar, locale. The cross-section method can work for a walk through a small neighborhood, as well as for traversing Los Angeles Metro by car. See for yourself. Perhaps you do not have a half-day to spare. So you take a stretch at a time. Stay off too-familiar pathways and deal with whatever unfolds. Your own attention span is an infallible guide, provided you pay attention to it.

Most of us learn avoidance tactics in choosing routes. I recall deciding, in a South Boston suburb during the riotous sixties, that when one sees grown men gesticulating and/or shouting, running down the street, that's a good time to stop or to turn aside.

Another useful rule: when it gets dull, turn off. A memorable example was Savannah's historic Bull Street, which runs south from the Savannah River through or past one historic square after another. While on nearby Whitaker Street, passing Forsyth Park, I grew bored at the predictable array of fine old two-to-three-story Victorian to Colonial houses, at their uniform setback, iron fences, center walkways, and traditional plantings. Off at a right angle, I changed course. Suddenly, I found myself in another world, different and exciting. The houses dropped to one-story, the sidewalk shifted from concrete to brickbats, the verges from clipped grass to loose gravel. The pavement grew potholed, the decor changed from high fashion to motley, and people's faces from white to black. My dictation into a tape recorder speeded up; there was much more to say, and later to be transcribed. When I described this transition to a psychologist, he immediately responded: "Of course. You had entered a strange environment. If I'd had you all wired up, it would show your breathing was shallow, your heartbeat up, and pupils dilated."

This is the essence of the cross-section method: it forces us as travelers to confront changes seldom visible from the Interstate. These don't show up on the Stately Homes Tour, or chamber-of-commerce jaunts for VIPs. And one should not get confused, lost, or disheartened by what is strange along the route. Stick to the same general direction of movement. If you detour, return and pick up where you left off. Don't backtrack or loop about; it's confusing. If you're so minded, keep a logbook, use a tape recorder, mark your course on a map. Snapshots along the way (with place and time noted in the logbook) will reinforce memory. Having a friend or local guide along deepens the experience.

It is quite likely that you will produce documentation, the record of unique insights, never before available for that particular place. Do not keep it to yourself. This is pioneering stuff. And it is equally likely that you will make observations, connections, and arrive at conclusions that may be the beginning of a new kind of geography of your own place. And that is an ending that can make any trip memorable for yourself, and worthwhile for all who can share the experience.

ACKNOWLEDGMENTS

The project that stimulated me to explore this field and write this book owes its inception to an invitation from Professor Lane Marshall, Fellow of the American Society of Landscape Architects, then of Texas A&M University, to conduct a 1984 seminar which I optimistically titled "How Places Work." By the time Texas had finished with me, and vice versa, it was clear that we had ventured into an ever-expanding universe.

In that vast realm, no single book or university course held The Answer; geography alone among traditional academic disciplines sufficiently embraced the matters at hand. And only a handful of specialists (but no literary agents) could we find to tune in on our then narrow-band wavelength. I was obliged "to write the book" to suit its most demanding critic, myself. (Only much later did "Geolinguistics" penetrate our ken.)

In such an enterprise, there's no accounting for all the debts one accumulates, but to acknowledge one's creditors is long overdue. Foremost is my wife, Judith McCandless, whose intellectual support and electronic expertise hoisted this "computer-cottage" venture across shoals and shallows; together with my time-spanning editorial associates, ideal organizers, and in-house critics. First came Ruth Spangler to structure our early efforts; aided and abetted by Elizabeth Van Kleeck, who indexed and shepherded this work through the enthusiastic screenings of editor John Tryneski and his fine-tuned associates at the University of Chicago Press. We flourished in their aura.

Along the way came valued researchers coping with expanding paper and electronic files: Margaret Alam, Jerry Rodgers, James Monohan, Paul Raybon, and especially Chris Caldwell before he was called to a larger ministry. We are also indebted to research librarian Glenda S. Neely of the University of Louisville Library, and to the overworked reference staff of the Louisville Free Public Library.

En route to the ongoing assemblage of some 5,000-odd generic place-names, from which this modest gaggle of 124 emerges, a host of "Topographic Irregulars" contributed one or more place-names to our still-evolving gazetteer. In thus labeling them, I am indebted to *The New York Times*'s columnist William Safire, who adopted from Sherlock Holmes's "Baker Street Irregulars" his own term for linguistic tipsters, touts, and prompters. My own esteemed and certificated "Irregulars" (not all of whom have survived to see this finished work) include the following:

Jim Adams, Bob Adams, Stuart Alexander, Betty Lou Amster, Ann Arensberg, Fred Bair, Jr., Jonathan Barnett, William Belanger, Robert Benson, Robert Bernard, Brian Berry, Raymond Betts, Beverly Beyer,

Sarah Bodine, Joy Bale Boone, George Boone, Tom Bradbury, Lucy Brightman, Sam Brightman, Bob Brown, Stanley Brunn, Jim Brunson, Beth Burns, William Butler, Anne Buttimer, Mary R. Caldwell, Julian Campbell, Jon C. Campbell, Fred C. Cassidy, Charles Castner, Wenonah Chamberlin, Ted Clay, Roger Clemence, Frank Clifford, Ruth Cloudman, James C. Cobb, Michael Conzen, Al Cross, John Cumbler, Marcia Dalton, Gordon Davidson, Neil Dawe, Randall Detro, Robert Dorney, W. C. "Bud" Dutton, Jr., Leonard K. Eaton, David Everett, Nancy Farnsley, Jim Fisher, Ronald Fleming, John Fondersmith, Gordon Garner, Michael Gartner, John Gilderbloom, Michael Greene, John Greenebaum, C. Ray Hall, Dennis Hall, Max Hall, John Fraser Hart, Michael Hayes, Kenneth Helphand, Bob Hill, Tony Hiss, Patrick Horsbrugh, Michael Hough, Lawrence O. Houstoun, Jr., John Howett, Gerald Ingalls, Veronica Jennings, Norman K. Johnson, Kate Jones, Elmer A. Keen, Mike Kennedy, Bill Knack, John Krich, James Krohe, Jr., Lisa Leff, Philip Lewis, Cameron Man, Lane Marshall, Judith Martin, Mary Matheny, Judith McCandless, Gaile McGregor, William Mitsch, William Morgan, Margaret Moseson, Lawrence Muhammad, Cissy Nash, Glenda Neely, Ken Neuhauser, Charles M. Niquette, Pat Noyes, George Oberst, Donald Orth, Risa Palm, Francis Parker, Neal Peirce, Lisa Pennypacker, Larry Peterson, Helaine K. Prentice, Ed Rabey, Susan Rademacher, Karl Raitz, Alek Rapoport, Paul Raybon, Jim Rebmann, Robert B. Riley, Richard Rivers, Bonnie Robinson, John F. Rooney, Jr., Ruth Ryon, Christopher Salter, Shirley Shaw, Robert Shayon, Fern Shen, Clarence H. "Buzzie" Short, Harriet Siler, Anna Steiner, Kathy Sloane, Ruth Spangler, Richard B. Stephens, Marjorie Struckmeyer, Nick Stump, Liz Swearingen, Richard Tewkesbury, James Thomas, William Tishler, Charles Traub, Betsy Tyrrell, Gerald Tyrrell, Jasper D. Ward, Sam Bass Warner, Jr., Ted Wathen, Bill Weyland, Marjorie L. White, LaJuana S. Wilcher, Orme Sandy Wilson, Josie Wiseman, Louise B. Young, and Wilbur Zelinsky.

To all the above go my thanks, and to none of the above, and only to the undersigned, goes the blame for any sins of omission or commission embodied in this piece of work. Further observations/promptings from readers of this book will be appreciated.

An early version of my observations on "ephemeral places" formed the content of *Design Quarterly* magazine, issue no. 143, 1989, edited by Mildred Friedman, and published by the Walker Art Center, Minneapolis, through the MIT Press.

In its evolution since 1991, this book benefited from the exposure of its ideas to the rigors of compression into my weekly public-radio essays, "Crossing the American Grain," thanks to WFPL-FM, Louisville, Ky.

It would be a pleasure, and considerable relief, to give credit to foundations, learned societies and others who might have financially supported this particular work; but that shall have to wait other and perhaps more deserving venues. My thanks do go, nonetheless, to the Nieman, Ford, and Guggenheim Foundations, and to the National Endowment for the Arts, whose fellowships, grants, and support, decades ago, smoothed my early path and steeled my later purpose through these unconventional explorations.

Grady Clay
Louisville, May 18, 1994

See page 108, COMMUNITY BONFIRE.

INTRODUCTION

A lovely tiny stream, flowing from an obscure spring, shines through the woods not far from where I write. In late winter afternoons, sunlight glitters off its riffles. In summer its waters irrigate my garden, and its shaded shallows are aflutter with birds.

But there are deeper levels by which to approach this complex place. Long ago it was an open site beyond the edge of town. At the top of its watershed, an early railroad snaked out along the ridgeline, offering to landowners new access to markets. That access transformed farmland into market gardens, and then into speculation sites. This became the locale of a nineteenth-century commuting suburb with a string of country estates clustered around train stops. Trolley cars reached out to convert it into a growth area, so that by 1900 it had become a streetcar suburb.

The gentle swale where my spring arose became a dump, then a landfill, with a street laid across it. Three generations of infill left the neighborhood on a downhill run. My valley had accumulated working-class cottages on a one-lane street, a trash-pile, and a dogyard. Its tangled slopes became a hillside subdivision gone bankrupt, then a teen-age hangout. In 1974 a smashing tornado converted the valley into a disaster area. It was invaded by a parking lot and its woods became a stumpyard. But gradually the old main drag nearby became a gentrifying area, the valley a rehab site and, in time, a garden spot, with a wooded commons, terraced garden, and stone springhouse. Eventually, it could become, stretching the term a bit, a good address.

These terms—open site, edge of town, dump, hangout, good address, market garden, commuting suburb—arise from that vast pool of familiar generic names used in American discourse about the places where we live, work, and play. We build these man-made inventions-and-conventions out of thin air—hits, runs, and errors—gaudy, gossamer, insubstantial stuff of the mind. They are parts of grammar we inherit and adapt to negotiate changing worlds. This grammar is to be studied with care and disregarded at risk. We test its terms against everyday experience in the marketplace of language. In a democratic society, it is important for image and reality to stick tight to each other. Our meaningful world is what we can describe to each other with a good chance of being understood. Generic place-names are essential lubricants.

Generic names may well be our first infantile "fix" on the world. "Da da" in an infant's speech is believed to precede any distinction between mother and father. It's a generic babytalk term to deal with all familiar grownups. And it suggests, if it does not "prove," that generic place-naming is fundamental to, and probably precedes, any attempt to attach more specific names to people and places around us. "School" precedes "Spring Street Elementary School at Fourteenth Street." "Out There" comes ahead of distinctions such as "four miles beyond the edge of town" or "a ten-minute walk beyond John's house."

So, to understand our man-made neighborhoods and the world beyond, we must deal with classes of places—sorts of places which do not exist "out there" but are products of the human mind. We talk about these generic, man-made places, rising and falling in our esteem, as our places-in-common. As generic places, they have no latitude—no matter how much we take in describing them. They have no longitude. But without them we cannot navigate in today's world.

That great name-giver, Carolus Linnaeus, observed that "the generic name is, as it were, the official currency of our botanic republic," and so it is with the generic names we give our places.[1] Drawing boundaries between these places, dealing with their shifting edges, knowing the difference between here and there—these are not simple chores. To sort out the differences, to choose and use the right names for our surroundings is not "just" semantics, but essential to negotiating the future.

The manufacture of "man-made America"—in which my valley is a tiny fragment—has been a stupefying process. Everywhere we look, we see landscapes poised in the process of conversion. All are adapting, becoming, fermenting, evolving, no matter how slowly. In our spread across the continent, in our rush to capture its riches, we have mastered the art of making two acres do the work of one. Tides of urbanization have overflowed the continent, even though some of its deserts and remote parts are still considered "empty." We live with ever-increasing byproducts of its spread. No place is exempt. Even the most remote fastness is part of someone's market area, or marked as a tourist destination, perhaps offering white water for kayakers, trap lines for its few residents, and ski runs for people at the new winter resort. Generic, man-made places, all of them.

Everywhere, one sees the visible impacts of converting landscape to human usage. We all take part. Millions of us fly over the changing landscape along designated, official flight paths. We view it, travel through it, deal with it. We make (or lose) money off it, pay fees, taxes, or rents on it, and live with the aftereffects of its changes. Some of our byproducts spread for a thousand miles as air pollutants along that route called downwind.

Not only do we live with its byproducts; we *are* its byproducts. We constitute its market. And we continue to incorporate into our houses, our health, and our wealth the bounty of its air, its forests and fields, its mines, fisheries, and pastures. To various ends, by endless means, we produce and modify millions of man-made places, inventing and naming new ones en route, attaching signs, sentiments, and fables as we go. Much of this takes place under such unwieldy titles as "urbanization" and "development."

Little remains as Columbus found it, or as Norse, Spanish, and English settlers viewed it. By now, the original treasure-trove has been explored, spoken for, fought over, surveyed, carved up, named and numbered, settled, and traded on the world's markets.

All these activities expand as the United States enters, grudgingly, into increasingly competitive world trade for its products and capital assets. More than ever, our landscape and its products are components of world commodity flows: Western wheat fields once bankrupted a generation of European farmers by undercutting prices; now they produce grain for hungry Russians. Oregon's forests become Japanese plywood; European investors buy up shopping-center locations. Japan's investors set up new automobile plants through the Midwest along Japan's auto alley, and in order to feed them, hundreds of suppliers cluster along Interstates 65 and 75.

The most powerful tool in this conversion has been the attachment of abstractions—man-made names, values, and prices—to natural places. All places come to be seen as sites—possible locations for new or altered people, crops, products, activities, or structures.

And once Nature—whether recast in the form of real estate, water, lumber, grain, or cattle—was priced and placed in markets, it was no longer Nature.[2] It was converted into a resource, a commodity, an abstraction to be surveyed, recorded, haggled over, traded, speculated with, packaged, and changed beyond all recognition, and usually at a distance.

But never at such speed as since the Civil War. Following the Ordinance of 1785 came other national laws that established the national grid, that giant marketing device that enabled price to do its work. It converted Western prairie into Eastern commodity. It made the United States world-famous for the efficiency by which it destroyed the original landscape and substituted another. Once carved up by a grid of mile-square lots, the West was sold quickly and routinely. Its sections, quarter-sections, tracts, and lots were the new abstraction laid over the natural world. All had their price, quoted in London and New York and now in Tokyo and Berlin. All have their costs—exhausted mines and fields; polluted air and waters; clear-cut forests; run-down, riot-torn cities; and federal and local deficits.

The conversion continues. All locations have been converted into interchangeable parts and parcels that are traded in international markets. They can be described legally, and situated accurately on maps, charts, records, deeds, and in Geographic Information Systems. These "GIS" have become the new currency of our times. Nature is now abstracted for world databanking and trading. All places are now designated, and we are going through the painful process of deciding which shall be saved, which used, which exhausted, and to whose benefits or loss.

Vital to any such abstract system was the attachment of names to places each step of the way—place-names both specific and generic. Place-naming came early, for no place, especially not any out-of-the-way place, could be sold or traded without a name, and often a number, attached somewhere along the way. Thus the Colonies and then the West were littered with place-names by the thousands.

To make sense of these millions of generic man-made places this book explores a small but significant sample, along the way from The Center to Out There. (In order to distinguish those generic places singled out for essays in this book, we capitalize them.) These include ABANDONED FARMS, ANNEXATION AREAS, ARREST HOUSES, BOOMTOWNS, DRUG SCENES, FLEA MARKETS, HOLDOUTS, LANDINGS, NO SMOKING AREAS, SCENES, SETUPS, and WRECK SITES. Every place is DOWNWIND from other places. Some places get their electricity from a WIND-FARM, or heat from a SOLAR FARM. Around or overhead may be FLOODWAYS and FLIGHT PATHS.

In this specialized landscape, our comings and goings are organized by means of places called ARRIVAL ZONES, HUBS, MEETING PLACES, and SHORTCUTS. On all sides are problematic, changeable places of uncertain futures: DECLINING AREAS, GROWTH AREAS, HURRICANE PATHS, and TEMPORARY HOUSING DEVELOPMENTS.

Somewhere—always looking for "the right place"—amid this odd-sounding and perhaps confusing welter of geographic situations, we settle down, work and play, and try to control our surroundings. Even the most solid or stolid of these places may turn out to be in a state of liquefaction. We find them useful, confusing, historic, threatening, pleasureful, or sacred, as the case may be. Given the right (or wrong) circumstances, we will do our utmost to move someplace else—as most Americans do every seven years.

When we change our address, we may focus our thoughts on cost, or on the dwelling itself: how many rooms, how arranged, how close to the doors can we park? But home is also one's address. And that is located in a zip code area, a census tract, a neighborhood. The details of our houses, mortgages, possessions, and incomes have been encoded in electronic dossiers far and wide. Each address includes such place-making elements as front, back, and side yards, a VIEW, one or more hardstands or driveways, a street frontage, and places to put out garbage and trash, locally identified as CURBSIDE. Somewhere nearby there exists that generic place known as THE MAILBOXES.

In this book I attempt to catch these generic names, as well as what happens to man-made places themselves, in transition. Names and places do not stay hitched forever; American English is an especially fluid medium. Its synonyms, metaphors, and coinages continually alter the state of the language, which in turn alters the way we think. As we convert places to new uses, our speech has to reflect those changes.

Otherwise, our world is quickly and hopelessly confused. To tell a friend "Let's meet at the market" when you mean the mall is to waste hours and perhaps lose a friend.

These place-names have been caught in flight, so to speak, as they flow through major American periodicals, TV, radio, books, fugitive reports, and in snatches overheard or usages pursued through interviews and travelogs. ("Facts" caught in passing on TV or radio are among the most ephemeral of traces, to be handled with caution.)

In this flow of language, name is to place as map is to territory. The more inventive among us indulge in a form of coinage to handle place-types not yet current. Some which we examine here may be short-lived: whether LULUS will survive into more placid times is unpredictable. The fad for SAND CASTLES may blow away, DRUG SCENES may shift, LIGHTING DISTRICTS fall victim to high electric bills, and SAFE HOUSES have their covers blown. But for each ephemeral place that disappears, others arise from the endlessly inventive language of everyday life, the quirks and quips of tomorrow.

This book explores our changing universe of places by means of the cross-section examination described in the Preface. Such a viewing process forces one to look at the whole great enterprise, which is the modern urbanized nation. This examination deals with its focal points, cruxes, fluxes, and transition locations. It starts from The Center and proceeds outward through that wide belt of dynamic tension called The Front, and terminates in the great Out There. Such a cross-section view cuts across major segments of life. It takes what comes next. It enables us to rediscover that scarcely anything left Out There has not been, in one way or another, urbanized. And there is no permanence in it.

These components of tomorrow do not offer easy-to-read labels such as "Next Year's Abandoned Filling Station." To track these goings-on requires constant attention. Sharp vision does not make soothsayers of us all. Yet a continuing hard gaze does diminish dreadful surprise. Especially if we stick to it, in darkness and in light, through the four seasons, over the years.

In spite of the mass-production of trademarked places and their urbanizing tackle—motels, drive-ins, gas stations, film kiosks—a thoughtful gaze reminds us that no two locations are identical. We remain ignorant of their singular aspects only if we abandon judgment and fasten our exclusive attention on their mass-produced shapes, signs, symbols, and trade names. We risk loss of insight only when we neglect situation, context, longevity, comparison, and succession. Nothing succeeds like succession, and there is a predictability in what-comes-next that is only beginning to be understood. Each situation, each specific place, is unique. How long it remains so—and when its changes suggest a trend—we can only judge by consulting, or preferably keeping, a record over time.

Keeping track of this is not a game to be played just for the fun of it. It is folly to believe that "anyplace will do"—that "one place is just as good as another" for life to go on; or that we can fully live in a "non-place realm," or fully learn in an electronic universe. The old saying, "Tell me where you sit and I will tell you where you stand" still has the ring of truth. A mis-placed life is a misspent life. "Placement"—that is, identifying the right places for whoever we are, wherever we go, and whatever we do—is a not a matter of frivolous choice, but is necessary to survival.

To see what is right in front of our eyes, as well as what it might become, forces us to look through each place to its possible futures. No place is exclusively as we see it now. Each comes at us from a complex past, en route to a chancy future. Many routes are open.

As we work our way across the grain of this great continent, it is well to remember that our presence, our hopes, and our energies will change it, and possibly for the better.

Overleaf: See page 159, LAST CHANCE/FIRST CHANCE.

BENNIE'S
FIRST CHANCE
BEER
WHISKEY
BEER
WINE
BENNIE'S PLACE

REAL PLACES

SECTION ONE

The Center

T

"The King is dead. Long live the king." That cry from the French Revolutionary mob lives on in the heart of American cities. Once upon a time all things important to a region happened there, at The Center. Pomp and circumstances flourished there. Ambition and connivance expanded there. Everything and everyone of significance converged at midcity.

This fixation on centrality, powerful and persistent, had lasted for thousands of years, ever since the early cities dotted the shores and plains and mountain passes of the Middle East. All roads led to The Center—that is, before the discovery that they led Out There as well—a discovery reinforced in our own times by widespread ownership of automobiles and the cult of mobility.

At the old Center you could find anything you wanted: meals and a show at midnight; dealers in exotic sex, pets, weapons, bonds, transport; inventors, adapters, copyists; loans nobody else would touch; music to put you in touch with the stars. Everything that was traded, everything of value, flowed sooner or later, either in bulk or by symbols, through a city's central marketplaces and storage yards. All roads led to Rome, it was said, the city that was power center of its Empire.

From colonial times, the object of

ATH AND LIFE OF CENTRALITY

American public policy was to establish "centers of population and commerce"—to help them flourish, and to fight over who controlled which. This historic gaze once focused on the frontier West. It had been "conquered" not so much by well-romanticized hunter-trapper-Indian-fighters as by its frontier towns and cities. Towns preceded and then staked woods-clearers; towns were the engines that drove geographic expansion; they drove "a broad wedge of urbanism into Western life," as historian Richard C. Wade put it.[1] Without support from these "spearheads of the frontier," frontiersmen perished. And as regional city-centers of trade and power, towns and cities expanded for a century and a half.

The Great Dispersal

But by the 1960s, power was dispersing away from central cities. In 1967–69 the first President's Task Force on Suburban Problems faced the surge of widespreading cities, their EDGES out of control, spinning off problems into far countrysides. By 1972 it was easy to conclude that "thanks to cheap or free media, to the new interconnected and intercommunicating switched-on society, and to increased personal mobility and movement, the old grip of centrality is loosening. No longer are we certain that, by the grace of God and the laws of geometry, all things of importance are best transacted at the center, all debates concluded at the center, and all great structures located at the center."[2]

The first modern blows at the power of the American center had been struck between 1934 and 1949. In 1935 the New Deal's Rural Electrification Administration set out to extend electricity from The Center to backwoods and BOONDOCKS. In 1949 the Federal Housing Act set off the nation's biggest suburban housing boom. And in 1954, the Interstate Highway system took off to spreadeagle America's cities. It was as though millions of acres of once-remote hinterland had been lifted up and dumped smack at THE EDGE of the city. Suddenly, far corners moved close to town. City and suburb, town and DOWNTOWN would never be the same. The 1990 census confirmed the statistical reality: the majority of Americans had gone suburban.

Thus, in the lifetimes of most adult Americans living in the 1990s, occurred a revolution in places, a turnover in their accessibility, their names, looks, ownership, and functions. Traffic jam flowed into country lanes, mortgages into daily life. TOWN CREEKS were diverted into city sewers, open fields covered with houses, old neighborhoods abandoned, ethnic enclaves populated and then drained in a lifetime. Forests became "the cutover"—stumpyards off to the horizon—and The Environment a cause célebrè. This was a prodigious mass production of new places with names, functions, identities, addresses, access, and populations. Modern publicists used press, radio, and TV to fasten these new places into public consciousness. It was like nothing else in recent history.

No place was transformed more pervasively than The Center—variously called DOWNTOWN, downcity, uptown, center city, inner core, inner city, the central core, Central Business District or "CBD." In Scandinavia the old Latin word was still used: Centrum. In the shorthand of the times, "CBD" fit snugly into American headlines, increasingly followed by the words "In Trouble."

American journalists had long been captives of the belief that everything important happened at The Center. Here was where big-city and small-town newspapers stood their ground. Their traditional "beats" covered city hall, courthouse, police headquarters, farmers market, the financial district, main drag, and convention hall—all immobilized in or close by the old Center. Men who owned newspapers—this was still an all-male domain—spearheaded committees to "save DOWNTOWN." They voiced the accepted wisdom of the 1950s. It said that if new developments could be subsidized—chiefly via federal urban renewal—close to DOWNTOWN, the core would "take care of itself," and the hundred percent location stay put—DOWNTOWN.

News-handlers followed the pattern, giving page one and prime time positions to any change affecting the geographic center of their metropolitan area. "Control the center and you control all," said an old military maxim. It was a lovely concept, slowly losing its anchor to reality. It reflected the day when any good reporter could make the circuit of DOWNTOWN news sources, and quickly get a sense of what's-going-on. They were steeped in the lore of centrality. In 1926 a Manhattan executive had boasted that "from his Times Square skyscraper, he could reach anyone of importance in the business world within fifteen minutes."[3] But that day was either long gone or passing in the 1950s as every new mile of interstate highway siphoned people out to the new EDGE OF TOWN, and enticed many corporate headquarters to suburbs like Tyson's Corners outside Washington, D.C.; Silicon Valley south of San Francisco; to Clayton, Mo., outside St. Louis; and to Princeton, N.J., or Norwalk or Hartford, Conn., on the fringes of New York. Outside—all of them. In twenty years New York's metro area expanded by a thousand square miles to 3,830.

It slowly dawned that increasing proportions of the American people no longer shopped in The Center, no longer worked in The Center, and were spending weeks, months, or even years without going to The Center. To oldtimers in their DOWNTOWN clubs, it seemed like the end-of-civilization-as-we-know-it.

Especially susceptible to the new decentrality were old port cities where DOWNTOWN was geographically down-by-the-waterside. Shipping had made the town. In hundreds of river, bay, harbor, and port cities, to "go DOWNTOWN" was to descend a hundred feet or more, from high ground to the old river-bottom LANDING place. As cities grew outward, DOWNTOWN became more remote; and language, especially in New England, reflected that remoteness. To go DOWNTOWN was to go "downcity" in Rhode Island. But "downcity" as a place-name "has diminished over the years" as department stores, theaters, and other DOWNTOWN magnets closed, "putting an end to the days when friends or family would spend a Saturday 'downcity' shopping and taking in a show."[4]

Meanwhile, forces which had once made centrality inevitable, now made spreading out feasible. Observing all this, top geographers were arriving at similar conclusions: Brian Berry charted the far-reaching 163 urbanized "daily movement systems" to find that all but 5 percent of the U.S. population lived and moved within them. International planner Constantinos Doxiadis identified "daily urban systems" as the new units to watch while they decentralized old cities. Christopher Tunnard identified the emerging Eastern metropolitan corridor; Jean Gottmann named it "Megalopolis," and author Anthony Bailey traversed it for a book, *Through the Great City*. Herman Kahn identified it as "Boswash"—the Boston-Washington corridor; and went on to name "SanSan" (San Francisco–San Diego), and "ChiPitts" (Chicago to Pittsburgh) as new stage settings of interurban life.

It was quite true, as the late urban expert Catherine Bauer wrote in 1962, that "Modern metropolitan trends have destroyed the traditional concept of urban structure." But it is no longer true, thirty years later, that (as she said in the rest of that sentence) "there is no new image to take its place."[5]

That new image was, and is still, being reshaped. It includes a daily range of commuting that would startle the streetcar suburbanites of the early 1900s identified by Sam Bass Warner, Jr.[6] It includes office clusters and shopping centers Out There in endless varieties of array and non-array. It is the stamping ground of urban field runners who operate on a vast scale—growing more exhausted with every new traffic jam.

But still, today . . . if you draw lines on a map that connect all the major power centers of a metropolitan area—its daily commutes, its delivery routes, mail and message flows, electronic money transfers among banks, the offices and home addresses of the Movers and Shakers—one visible fact jumps off the maps. Most lines still converge at The Center. And of course, it's not the same Center as in 1900 before automobiles, or in 1930 before today's model SUBURBIA, or even 1950 before interstate highways helped drain The Center.

Yet, hidden within the traffic swirl, lying at the end of a billion bits of daily electronic flow, and resting in the mental images of millions of commuters, tourists, and image-mongers, there still remains A Center. It still has a SKYLINE going-up-in-the-center, thanks to skyscraper booms of the 1920s and post–World War II. It is more concentrated, less variegated, than the old Center.

Its old walk-in working-class districts nearby have disintegrated, often replaced by festering slums of ethnic minorities. It is more heavily policed, analyzed, and reported-on than ever. Some functions—

retailing, auto sales, wholesale markets—are more spread out, and others quite atrophied. Distant sources of supply continue to dry up, as Chicago learned to its dismay when Upper Midwest–Great Lakes timber was cut out and the lumber boom had begun to taper off by the 1880s.[7]

Today The Center's former monopoly power has waned, its share of the daily commute has shrunk, and its retail sales are only a fraction of yesterday's. Its sidewalks and streets as social centers have been depopulated. But there is still a "There" there. It's different, but its changes we should now view as central to our understanding of tomorrow.

However, even Centers that have lost many functions to the periphery Out There—leaving behind voids of empty land—still retain what art historian Rudolf Arnheim calls "The Power of the Center." Especially if they can be seen from a distance—as can San Francisco, Denver, Seattle, and waterfront cities such as Norfolk, New York, Chicago, Miami, Jacksonville. Any city that can be grasped visually, as a whole, acts as a work of art. As in any manmade composition, "a center, in the dynamic sense of the term, acts as a focus from which energy radiates into the environment."[8]

Being stabilized in finished form and embodied in an urban design, a work of art may continue its radiation unchanged, except as the sensibilities and perceptions of its viewers shift over time. The art of building cities, like the city itself, flourishes or diminishes with its audiences and patrons, as millions of American tourists discover in Europe. In America, especially, it is an art much in need of public support.

New Centering Activities

Within two generations, the heavy hand of government—largely hidden in the nineteenth century—has been openly extended to support The Center. (The irony, of course, is that it was the same heavy hand that promoted decentralization—cheap electricity, federally insured mortgages, and highways.) By the 1990s a host of special-purpose districts had been organized to cover every major Center across the land. Even the term Central Business District had acquired an official sound, and often official boundaries and status.

As American manufacturers moved out of old mining-steel Rust Belts, or offshore, The Center has tried to fill the holes with high-tech and other services. Its struggles end up, one way or another, at the COURTHOUSE DOOR, at city hall, at state and national capitols, and at corporate headquarters—with frantic detours to cope with DRUG SCENES, homeless people, and DECLINING AREAS whose proximity to The Center raises alarm. New purposes get invented, old rationales trotted out. Especially in Eastern and Southern cities, old-family and corporate elites join forces to keep The Center alive. Direct federal aid to cities peaked at $47 billion in 1980.[9] Much of it supported The Center in myriad ways.

So now it boasts new concentrations of space and functions called MULTI-USE COMPLEXES. More of its denizens spend their days, and often nights, spinning around indoors in office clusters, malls, courts, skyways, atriums, and lobbies—conveyed by moving sidewalks and escalators, all under surveillance by TV and such devices. Doorways onto the street are fewer; security is tighter and, in New York's center, outdoor movement on once-bustling sidewalks is at risk, especially at night.

Thus in and around the older Centers have been assembled a new mix of organized and often subsidized goings-on: MEETING PLACES, EVENT SITES, PARTY STREETS, PORNO DISTRICTS, SUPERBLOCKS, SECURITY, corporate headquarters, tourist meccas, convention and sports arenas, disaster management centers, CULTURAL ARTS DISTRICTS—man-made "scenes" of endless invention.

Organized protest had moved onto center stage in many cities during and after the Vietnam War. By the 1990s, Washington's Pennsylvania Avenue and Mall had turned into well-worn venues for protestful assemblies, rallies, sit-ins, marches, demonstrations, and counter-demonstrations. So much so that the District of Columbia's mayors repeatedly sought—with little result—for Congress's help to pay for all the policing. Local media in major cities were hard put to keep track, especially when the scale of riot expanded, as it did in Los Angeles in 1992 after the notoriously televised Rodney King beating.

Beyond the politics of protest, the capacity of the modern city to galvanize crowds is impressive: from Frankfurt, Germany, to San Diego or Dallas, huge portions of the urban pie have been carved out for instant crowd formation. Superdomes such as you find in Indianapolis or New Orleans generate surging crowds that can match the city's population by a few weeks' influx of conventions. Main drags get new "mood lighting" for parades. A mid-sized city can move a half-million conventioneers in and out each year. Boston's Tall Ships festival in May 1980—prototype for Tall Shipments that persist today—tied up every rental helicopter in New England. A night fireworks show called "Thunder over Louisville," 1992, startled police by attracting 700,000 viewers along the Ohio River. Combine a regional version of the Olympic Games with good weather and a strong national economy, and any host city can be swamped.

Light-and-sound districts appear; whole streets are sequestered for festivals, conventions, sporting events, or parades. Spiffy water sports go on display in the midst of once-dry land. Old working ports and waterfronts (Baltimore, Savannah, Sacramento, Manhattan, Norfolk, Jacksonville) have been "renewed" within an inch of their lives—often unrecognizably, as trendy, boutiquey "festival marketplaces." Tourists outnumber local shoppers in many a Center, and tourist DESTINATION has been added to its overlay of functions. But then, DESTINATIONS get overbuilt; fashions in tourism shift, crowds shrink, and vacancies can take over.

Today's Center is ringed or reinforced by a host of management-and-taxation districts. Does the Chicago Loop need subsidy after a damaging underground flood in 1992? Granted. Did Nashville's and scores of other Capitol Hills beg for tax-assisted redevelopment in the sixties? Granted. For many a civic center, the sixties was a period of Ask-and-It-Shall-Be-Given-unto-Thee. For many a financial district, the eighties offered bonanzas. Subsidy holders became key players in the urban game.

Security forces moved in, reinforced by

weapons and tactics sharpened during the Vietnam War and civil riots of the sixties. SECURITY becomes a generic place, its gates, kiosks, TV cameras, and armed sentries standing guard in and around The Center with its banks and command posts.

Where the "action" is today, varies. Most Centers reflect the boom-and-bust cycles of urban history, with spurts of activity (1954, say, to 1973, and again, the 1980s) followed by slack times if not by recessions. In New York, big street parades that had pumped life into Manhattan shops in 1992 found their sponsors thinning out—fewer Coming Attractions for the main drag.

Some cities do better at evening out capitalism's crashes and booms. They nourish a history of strong local planning and enterprises that modify the boom-bust cycles. By fits and starts, a Boston or Philadelphia, Baltimore or San Francisco embarks on a decades-long urban redesign to dramatize a new Center with focal points, grand axis, political *venturi,* and promenades. Some find solace in tourism. Others, as in the Great Plains, suffer from depopulation of the whole region. With each new set of cosmetics or controls applied, The Center continues to reflect basic changes around it. People far distant from it will continue to make it their point of reference, even while viewing it with alarm.

"Hubbing" entered the urban dialogue in the 1980s, after airlines were deregulated. Here was a new centralizing force. Almost overnight, airlines selected HUB cities at which to concentrate their new routes, terminals, and jobs. To become a HUB city, cities such as Atlanta, Denver, Cincinnati, Charlotte, Louisville, and Pittsburgh have sunk huge sums to build new or expand old airports—hoping to attract airlines and airshippers, their jobs, and the new access thus created.

Each new HUB accumulates lines of airpower, while many smaller, backwatered towns lose out. HUB cities issue maps exhibiting air routes radiating out from The Center. Even though the expanded airports might be twenty miles ("only twenty minutes!") from The Center, yet HUB airports come to be seen as working extensions of, vital to the accessibility of, the closest HUB city.

Onset of the World City

Meanwhile, a worldwide shakeout has been taking place among super-cities that sit on locations central to global affairs. Geographer Peter Hall in 1977 identified seven "World Cities": London, Paris, Randstad Holland, Rhine-Ruhr, Moscow, New York, and Tokyo. Today, as the European Community musters power, its capital cities flourish.

Power, derived from aggressively creating accessibility, trade, and new information, enabled these so-called World Cities to arise. Their power derives from their locational access to new worldwide flows of information, travelers, and goods. They absorb or grab new vigor from expanding global markets.

The list of World Cities continues to change. New York attracted billions in capital from frightened investors around the world in its booming sixties, but skates on thinning ice today. In Randstad Holland, Rotterdam tries to leverage its power as the world's Number One port to become the European Community's World City for business—building a mammoth "port city" just across the Maas River. Los Angeles and its "Rimsters" make their moves toward World City status, trying to capture trade around the PACIFIC RIM, where Singapore has given repeated signals that it wants to become the world communications and information center by 2002. Berlin looks to unified Germany as its springboard to European and wider dominance. Such cities often produce an "International Zone" offering perks, tax breaks, and GOOD ADDRESS to an expanding, continent-dropping elite linked into global fax-and-info-nets.

But the future of some aspiring World Cities is at risk. Hong Kong's projected emergence as a World City remains threatened by its 1997 takeover by China. Moscow's centrality functions are being undermined by dissolution of the USSR. Sao Paulo's ambitions for World City status are being thwarted by the burden of its rampant shack-towns. World City London, on the fringes of booming Europe in 1992, has fallen into the doldrums. Other cities wait in the wings, impatiently in the Third World.

Contests continue between Center and Out There, between firsthand and secondhand contact. Ambitious computer freaks insist that the sheer volume of data available to every non-central computer-owner will "win" over personal experience and face-to-face contact. Can one's personal memory of Times Square, of St. Francis Square in San Francisco, or of Bourbon Street, New Orleans, possibly compete with floods of images that can be summoned up on your personal computer? Can one learn personal trust—the core attribute in human relations—secondhand via an "electronic presence" on TV? Can these new forms of social interaction-once-removed take over the functions once corralled person-to-person at The Center?

Beyond these questions, we watch the continued rearrangement of power in and around The Center. In some cities, the lust for metro status has cost city hall its old support of inner-city voters. Yet as it lost to SUBURBIA its claim on family shopping, it picked up other chips in the game: those myriad deals, doings, and derring-do that depend on eye-to-eye-contact, on face-to-face, lapel-grabbing, Body-English-Spoken-Here negotiations. Neither love nor persuasion at a distance has shown a capacity to outdo the immediacy of life-in-proximity at The Center.

Meanwhile, beyond The Center, changes will be widely distributed. Those restless places that distinguish what I call The Front, with its dynamic mix of energies, we will examine later in this book. Our journey ends in the great Out There, where many futures are unfolding.

CHAPTER 1

Back There

Firmly fixed by location, Back There is the "old" city–back where it all began at the original LANDING place, town wharf, or crossroads. But now, Back There is outgrown and passé in the eyes of new majorities out in complex SUBURBIAS. To them, Back There is loaded with locations decrepit, sidetracked, or protected by somebody else.

Back There once dominated its region with a unique capacity for crowd formation. If you saw a queue you knew you were DOWNTOWN. But this, too, is in evolution. Instant crowd formation is spread widely (see THE SETUP and EVENT SITE). Yet many anchors still remain fixed Back There: the old COURTHOUSE SQUARE, financial district, law offices, and corporations that dominate the old "CBD" (Central Business District). There lie historic districts restored to new glory—a Summit Avenue of old First Family mansions, or old water-powered grain and textile mills and the TOWN CREEK that gave them power. Back There was etched, intimately and small-scale, into the original and still-recognizable landscape. But newer settlements on the outskirts ride roughshod over the flattened terrain.

It's a ferment of change. Terms like back country, back forty, backwaters, back-of-the-beyond, backward, the 'way-back, and backwoods originally had a bucolic, country-bumpkin flavor. Those places once were located 'way out in the BOONDOCKS. But centers of population are shifting outward. Millions of suburban folks now think of the city-they-left-behind as Back-There-Someplace in the inner-core city, stuck in a blind or back alley.

But now their daily perceptions have shifted away from inner city. And the folks they left behind Back There often feel rashly abandoned to their own devices—which we explore in the following essays.

THE CAPITAL

Hard at work on the American landscape is a special process of CAPITAL formation. It creates a never-ending array of new CAPITALS, while others fade into obscurity. The traditional CAPITAL is noted for being the location of the state capitol building, and may also be the largest city of a national district or state, as are Washington and Atlanta. Most often these are located at a strategic and historic spot. "The Great Capitals," in Vaughan Cornish's book by that name, are Rome, London, Paris, Athens.[1]

That's one piece of the puzzle. But the puzzle expands with the addition of "dubbed" CAPITALS—those that achieve CAPITAL status chiefly by the ancient practice of "dubbing"—a term used in England as early as A.D. 1085. In the Middle Ages, it meant the conferring of knighthood or other honor by the ceremonial laying of a royal sword upon the neck or shoulder of the candidate, usually a mounted warrior of noble blood. In the absence of royalty, "dubbing" today is everyman's gimmick for attaching status, celebrity, or fame to a place or product, simply by assertion, usually in print.

Most such CAPITALS derive status from a product or service that reaches national or worldwide circulation, and can be promoted for civic benefit. Such products take many forms: chocolate (Hershey, Pa.); artichokes (Castroville, Cal.); or garlic (Gilroy, Cal.). In Louisiana, the town of Breaux Bridge was designated by the state legislature to be the Crawfish CAPITAL of the World. CAPITALS can benefit from food fads, which contributed to the fame of Muscoda, Wis., as the Morel Mushroom CAPITAL. Tarpon Springs, Fla., was long known as the Natural Sponge CAPITAL of the World, a title which circulated internationally, somewhat in response to a global flux of sponge-killing red tides that wiped out Tarpon Springs's competitors.

Animals and wildlife also figure in CAPITAL-formation: thoroughbred horses (Lexington, Ky.); polar bears (Churchill, Manitoba); loons (Mercer, Wis.); clams (Pismo Beach, Cal.); harness horses (Hawkinsville, Ga.); walking horses (Lynchburg, Tenn.); spoonbill, an ancient fish (Osceola, Mo.); white squirrel (Olney, Ill.); and angora goats (Rocksprings, Tex.).

Some CAPITALS are happenstantial. Once antagonistic to bats, Austin, the capital of Texas, has added "Bat CAPITAL of America" to its title. This occurred in 1991 after an introduced flock of nonlocal bats, brought up from McAllen on the Mexican border, took over a local bridge. "As many as 1.5 million Mexican free-tailed bats roost in the concrete crevices of the Congress Avenue Bridge" over Austin's Colorado River. Large crowds of bat-viewers, assembled at dusk, transform the bats' nightly emergence into a civic coming-out party.[2]

Behind many CAPITALS lie local stories of ingenious inventors as well as hard-headed promotion. Brockton, Mass., became the nineteenth-century Shoe CAPITAL (an honor shared with nearby Lynn), thanks to the McKay sewing machine by which uppers and shoe soles could be sewed together, not pegged. With this device Lynn became shoemaker to the Union Army during the Civil War. Dalton, Ga., became the Carpet CAPITAL of the U.S. after local craftsmen experimented for years with jury-rigged looms in garages and backyard shops. This enabled them to convert from chenille bedspreads, sold

The following checklist includes those CAPITALS cited in news reports, periodicals, and books published or broadcast to February 1993. It obviously omits twenty-six Queen Cities or Queen-of-Cities, but does include a scattering of regional nominees.

Akron, Ohio Rubber (to 1980)
Albertville, Ala. Fire Hydrant
Avery Co., N.C. Fraser Fir
Austin, Tex. Bat
Baltimore, Md. Antiques (East Coast)
Battle Creek, Mich. Cereal (World)
Beaver, Okla. Cow Chip
Bellevue, Wash. BMW
Breaux Bridge, La. Crawfish
Brockton, Mass. Shoe (19th-century)
Cadiz, Ky. Crappie (World)
Castroville, Cal. Artichoke
Chicago (South Side of), Ill. Black, Crime
Churchill, Manitoba Polar Bear
Colorado Plateau Western Movie Filming
Corona, Cal. Lemon (formerly)
Crystal City, Tex. Spinach
Cumberland, Wis. Rutabaga
Dalton, Ga. Carpet
Dayton, Ohio Rubber (formerly)
Demotte, Ind. Asparagus
Edinburg, Ind. Veneer
Elkhart, Ind. Mobile Home, Teen-Age Cruising
Fallbrook, Cal. Avocado
Flint, Mich. Unemployment
Fontana, Cal. Fog Belt
Fort Payne, Ala. Sock
Geneva, N.Y. Lake Trout
Gilroy, Cal. Garlic
Glendora, Cal. Ozone

Gloucester, Mass. Whale-Watching (World)
Grand Rapids, Mich. Furniture (formerly)
Grants, N.M. Uranium
Greenfield, Cal. Broccoli
Hawkinsville, Ga. Harness Horse
Henager, Ala. Potato (South)
Hershey, Pa. Chocolate
Hibbing, Minn. Iron Ore
High Point, N.C. Furniture
Hinckley, Ohio Buzzard
Hollywood, Cal. Cinema, Dream
Houston, Tex. Pornography, Manned Space
Illinois, Central Soybean
Indianapolis, Ind. Amateur Sports, Hoosier
Irvine, Cal. Chain Restaurant
Kokomo, Ind. Firsts (mechanical corn-picker, push-button car, radio, etc.)
Kodiak, Alaska Seafood Processing
Lakeview, Ore. Hang Glider
Leamington, Ontario Tomato (World)
Lexington, Ky. Thoroughbred (horse), Bluegrass
Little Rock, Ark. Shadow (formerly)
Los Angeles, Cal. The West, Bank Robbery, Pacific Rim
Louisville, Ky. Volleyball
Lynchburg, Tenn. Walking Horse
Lynn, Mass. Shoe (19th-century)
Malibu, Cal. Glamour (World)
McMinnville, Tenn. Nursery
Mercer, Wis. Loon
Miami, Fla. Bankruptcy, Commercial (Latin America and the Caribbean)
Milwaukee, Wis. Beer
Mission Viejo, Cal. Swim
Muscoda, Wis. Morel Mushroom
Mulberry, Fla. Phosphate

fluttering on the roadsides, to deep-pile carpets sold by the millions of yards at new low prices. Thence came thousands of Carpet City outlet stores. Plant City, Fla., owes its prominence as Strawberry CAPITAL of the Nation to ingenious tractor designs that rototill, pulverize, mound up, and cover the strawberry beds with plastic, which is then punched and hand-planted with strawberry plants—all in one continuous operation that ensures predictable, top-volume production—and CAPITAL status.

CAPITAL formation is as chancy today as when the Roman Empire's CAPITAL shifted from Rome to Constantinople. Dayton, Ohio, has lost its primacy as Rubber CAPITAL; Pittsburgh's status as Steel CAPITAL has been dispersed, as has that of Waterbury, Conn., as Brass CAPITAL. Corona, Cal., was Lemon CAPITAL until houses blotted out its citrus groves. And St. Matthews, Ky., was Potato CAPITAL in the 1920s before the 1937 Ohio River flood prompted subdividers to seek higher ground—and to carve up St. Matthews's potato fields. Prior to the destruction of its supply sources in upper Midwest forests, Grand Rapids, Mich., was dubbed Furniture CAPITAL, a title passed reluctantly to High Point, N.C. The latter city took years to capitalize on its supply of underemployed skilled woodworkers, local hardwoods, and promotion. Finally—with the backing of an Atlanta bank—it supplanted Grand Rapids, only to find itself challenged by Dallas, Tex., with that city's new furniture showrooms in the Trinity River industrial district.

Many CAPITAL titles are peddled into prominence by local promoters puffing hometown assets and compiling scrapbooks to prove local claims to capitalhood. Santa Rosa County in the Florida panhandle was designated the Canoe CAPITAL in 1981 by the Florida legislature. In 1991, *Wisconsin Tourism Development* magazine listed Sauk Prairie as the Cow Chip Throwing CAPITAL. Battle Creek, Mich., long enjoyed the title of Cereal CAPITAL of the World. Springfield, Mass., contested it in 1969 by making an end run, producing the "longest breakfast table" serving pancakes to 33,869 people as proof-to-title. No contest.

The production of figments has become a national obsession. Thus the watering down of CAPITAL stocks is under way. Occasional others are randomly coined, bestowed, or dubbed by media reporters in passing. Elkhart, Ind., was described by visiting researcher Susan Orlean as the (teenage) Cruising CAPITAL of the U.S., but had already achieved more substantial manufacturing status as the Mobile Home CAPITAL.[3] A visiting reporter dubbed Little Rock, Ark., as the Shadow CAPITAL of the U.S. during the interregnum between Governor Bill Clinton's election and his ascent to the presidency in 1993.

James T. Yenckel reported in the *Washington Post* in 1985 that "tourist officials list more than 40 golf courses within a 20-mile radius of Palm Springs, CA, and have thus dubbed the city 'The Winter Golf Capital of the World.'"[4] Not to be pinned down to one product, a well-known New Jersey city, in its manufacturing heyday, merely asserted, on a huge sign visible to then-passing railroad passengers, that "Trenton Makes, The World Takes."

Few CAPITALS get such intense scientific

scrutiny as Parkfield, Cal., population 34, "now the Earthquake Capital of the World. Hundreds of scientists monitor it as thoroughly as a patient in intensive care."[5]

Finally, a variety of cities resist would-be titles bestowed, in passing or otherwise, by rivals, by critics, or by reporters scrabbling for a news-peg. Thus: Washington, D.C., has been called Murder CAPITAL; Calvert City, Ky., headlined as "Cancer City"; Roxbury, Mass., as Crack CAPITAL; Newark, N.J., as Auto Theft CAPITAL; Los Angeles, Bank Robbery CAPITAL; Houston, Tex., as Pornography CAPITAL; Times Square, N.Y., "unofficial CAPITAL of Raunch"; and besmogged Glendora, Cal., as Ozone CAPITAL.

Thus has the word descended from its original honorific function. For every Grand Rapids or High Point that achieves CAPITAL status by years of hard work and ingenuity, others make do with flatulent claims, accumulating no capital, no machinery, and only intangible products. They follow Gresham's Law of money: bad CAPITALS drive out the good. Most would eventually disappear; many have done so. No doubt, some gas remains in the list of asserted CAPITALS listed here. Caveat emptor.

See also HANGOUT, in Chapter 4: Ephemera.

THE COURTHOUSE DOOR/YARD/SQUARE

THE COURTHOUSE DOOR is a vital component of public life in every county seat where it performs its various functions as the center, as an entry, entrance, stoop, doorway, podium, vantage point, prospect, demonstration site, promontory, setting, auction site, and inauguration platform. Many state laws require public auctions of foreclosed properties, etc. Legal notice of such auctions was required to be posted "upon THE COURTHOUSE DOOR," which in simpler days was an important bottleneck for public life. So THE COURTHOUSE DOOR(S) always served a myriad of political and social functions. In more unruly jurisdictions they did bloody duty as dueling grounds and more often as THE SCENE of public speeches, ceremonies, swearings-in and swearings-at, debates and exhibitions. In political campaigns THE COURTHOUSE DOOR offers endless if repetitious PHOTO OPPORTUNITIES for candidates and the media. A photographer, standing at THE COURTHOUSE DOOR can encompass in one shot passersby, local traffic, curiosity-seekers and bystanders, giving the appearance of action even on a dull news day. In the 1960s, during the coming-of-age of what was loosely called "participatory journalism," THE COURTHOUSE DOOR was a favorite locale for TV photographers to encourage protesters to "get a little action going." Following such advice, there occasionally ensued a modest but photogenic fracas. Later, in unfriendly courtrooms, this would be described as "inciting to riot."

The DOOR has great symbolic value, as when Alabama's Governor George Wallace stood at the statehouse door to prevent federal intervention in school segregation policies of Alabama. The courthouse is the historic center of political and, in time of insurrection, military control. When THE

Murray, Ky. Snuff
New York, N.Y. Imperial
New York, N.Y., South Bronx Homeless
New York, N.Y., Times Square Raunch
Newark, N.J. Auto Theft
Newport, R.I. Vacation-Land
Niagara Falls, N.Y. Honeymoon
Olney, Ill. White Squirrel
Omaha, Nebr. Agricultural
Orange Co., Cal. Fraud
Orlando, Fla. Theme Park (World)
Osceola, Mo. Spoonbill (formerly)
Palm Springs, Cal. Winter Golf
Parkfield, Cal. Earthquake (World)
Pismo Beach, Cal. Clam
Pittsburgh, Pa. Steel (formerly)
Plant City, Fla. Strawberry
Prescott, Ariz. Cowboy
Quantico, Va. Crime (staged)
Ridgway, Ill. Popcorn
Rockford, Ill. Screw
Rocksprings, Tex. Angora Goat
Roxbury, Mass. Crack (Northeast)
St. Matthews, Ky. Potato (formerly)
Salinas, Cal. Lettuce
San Saba, Tex. Pecan
Santa Monica, Cal. Hip
Santa Rosa Co., Fla. Canoe (Florida)
Seattle, Wash. Coffee (World)
Tarpon Springs, Fla. Natural Sponge (World)
Troy, N.Y. Collar
Tulsa, Okla. Oil
Uvalde, Tex. Honey
Walnut Creek, Cal. Walnut (1920s)
Washington, D.C. Murder, Imperial
Waterbury, Conn. Brass (formerly)
Wilmington, Del. Chemicals
Wichita, Kans. Air (World)

THE COURTHOUSE DOOR/ YARD/SQUARE

If the law says it happens here, it happens here: at, in, or around the local COURTHOUSE DOOR and its formal surroundings—a familiar venue in counties across the land. The focal point was once the COURTHOUSE DOOR and its steps—for legal notices, formal arrivals-and-departures, stand-offs, duels, speeches. But pieces of the action have gone inside, to be air-conditioned and televised; or suburban, into multi-service centers. Sketch by Annette Cable.

COURTHOUSE DOOR is taken over by the state or federal military or by local police it always tests the locus of control.

The emerging use of THE COURTHOUSE YARD as a substitute for the SQUARE marks a watershed in the history of the county seat—the heart of traditional local government in the United States. It marks the gradual disappearance of THE COURTHOUSE SQUARE as a social center, where old-timers and patriarchs congregated. Now the onetime local movers-and-shakers' voices as well as their numbers grow thin. "Even in midsummer, when the shade trees are full, there are scarcely enough loafers to start an argument in the once-crowded Hopkins County courthouse yard in Madisonville, and at others like it across Kentucky. So gradual has been their decline that hardly anyone noticed they were vanishing."[1]

It is still possible, in hundreds of county seats, to stand at THE COURTHOUSE DOOR and survey what remains of that community's financial district and lawyers' row—and of the Civic Center, laid out amid ambitious stirrings of the City Beautiful movement after the Chicago World's Fair in 1893. Over there, the new Public Safety Building (renamed Hall of Justice) shot up after the riots of the 1960s. Across the land, billions of dollars were poured into civic center construction—$300 million in the Borough Hall section of Brooklyn, N.Y.[2] Chicago's new Civic Center Building, 631½ feet high, was designed as "the new dominating feature of Chicago's skyline."[3] Dearborn, Mich., got a new one on fifty-four acres donated by Ford Motor Company.[4] And so it went—broadening, blocking the view from THE COURTHOUSE DOOR.

Amid all this, the local power structure once had its focus. Here, within a ten-minute walk one could "see everybody that counted" in an all-male world. That structure began to crumble after the suburban rush of the 1950s. Courthouse Annexes appeared far from DOWNTOWN. Auto licenses, hunting and fishing licenses went suburban. Tax payments and voting registration took to the mails or went electronic. Nearby men's clubs were opened (often by court order) to women. And federal courts took over countless contentions that once were tried behind the (county) COURTHOUSE DOOR.

By the 1980s new federal laws further decreed that the typical COURTHOUSE DOOR—approachable only through Greek columns at the top of imposing stone steps—denied access to the handicapped. Thus countless remodelings, side doors and new ramps were added. It was a long way from the Acropolis in Athens, Greece—still the prototype for all who look at THE COURTHOUSE DOOR in historic terms.

DATELINE

Ostensibly there is a "here" here, since DATELINE labels the location, usually a city, in which a newsworthy "event" occurs, takes place, or originates. But exactly where this particular "here" can be pinpointed geographically is another matter. For DATELINES are fictions suitable for manipulation by editors. The origin of the news or feature story, and thus the DATELINE preceding it, is a matter of editorial decision making. It does not arise from a location fixed inevitably at a named locale of given latitude and longitude. The latitude belongs to the editor. Thus where news happens is wherever a news editor decides that the newsworthy situation took place, or where a news-peg could be found. Any complex article can be datelined to respond to any one of several needs of the editor or publisher.

Further, the date itself gradually disappeared from most DATELINES during the 1980s as newspaper editors, fearing TV's omnipresence, attempted to suggest that each news article is up-to-the-moment, unconstrained or uncontaminated by yesterday's date and data. Meanwhile, TV was successfully touting yesterday's event as "today's story." The *New York Times* almost uniquely persisted (1993) in publishing not only the place-name, but the date each nonlocal story was filed. That particular note of time-wise realism is largely eliminated from TV news, as well as from newspapers when they reprint news reports from wire services, or from other papers (up to a fortnight past the original publication date). It is rare indeed for newspaper readers to encounter a report labeled, "From our Correspondent in Timbuktu, filed June 29, and delayed en route; with additions [in italics] from Associated

DATELINE

Exactly (or even approximately) "where" a news event occurs has become obscured as U.S. newspapers increasingly abandon the "DATELINE" that includes both the place of origin and date of the news article. The *New York Times* is one of few still recognizing, by using DATELINES, that most news events occur on a given date and at a geographic locale, rather than in an editorial office or TV studio. (Headlines from June 27, 1992 *New York Times*.)

A New Tide of Immigration Brings Hostility to the Surface, Poll Finds

By SETH

Special to The

LOS ANGELES, June 26 — With both legal and illegal immigration into the United States approaching historic highs, a public reaction

U.S. IS TAKING AIM AT FARM CHEMICALS IN THE FOOD SUPPLY

EMPHASIS IS ON CHILDREN

Policies of Three Agencies and a New Study Signal a Shift in Government's Stance

By MARIAN BURROS

Special to The New York Times

WASHINGTON, June 26 — The Federal Government has decided to reduce the use of chemicals in the

Problem With Cl Creatures Devo

By LINDSE

Special to The

NEW YORK, June 26 — Mariners used to joke that New York City was "clean harbor" — a quick way of sa ing that the water pollution was severe that it killed any organi attached to the hulls of their ships now the tables have turned: the w around New York are so much c voracious mollusks and

On Syrian Streets, They Dare

By WILLIAM E. SCHMIDT

Special to The New York Times

DAMASCUS, Syria, June 25 — In the ancient souks and the new office buildings of this city, growing numbers of people are now saying openly what was once unspeakable: peace with Israel is coming, and it is time to think about Syria's role in the Middle East after a

His brother Amir agreed, adding, "We must be prepared for Israel," a reference, he acknowledged, not to Israeli tanks and soldiers, but rather an invasion someday of merchants and business people from Tel Aviv and Haifa.

Press correspondent Flange Gardo, Moscow."

The national newspaper *USA Today*—cracked up to be "the first American hamburger manufactured from newsprint"—created yet another variation of ubiquity by manipulating by-lines. Five out of six front-page articles in a typical (May 2, 1988) issue identify "the source" by printing the reporter's by-line followed by *USA Today*. Thus the newspaper inserts itself into the DATELINE'S place—a case of ubiquity replacing geography.

DATELINES serve many purposes. Addressed to nobody in particular but everybody in general—that eighteenth-century invention called The Public—they aim at a universal anonymous You. And so "where" news happens can be wherever a news editor decides that the news can be hung on its proper peg. Datelining a news story "Hogeye, Arkansas," or "Monkey's Eyebrow, Kentucky," or "Pumpkin Center, Indiana," or "Scrabble, Virginia," gives it instant recognition as a uniquely backwoodsy, down-homey, folk-knowledgeable sort of place, with those qualities shared by the writer.

A hard-news event, such as a plane crash in downtown San Diego, permits news-handlers little choice in datelining anyplace but San Diego. But in managing a situation or "trend" story, the choice of DATELINE is up for grabs. Thus a roundup of plane-crash news could combine reports from a Washington reporter badgering the Federal Aviation Agency, fill-ins from a stringer at the crash scene, additions peeled off television news—all whipped together by a desk man in New York, with no DATELINE at all beyond the ambiguous phrase "from AP and overseas dispatches." By manipulating locales in such fashion, drought news in the summer of 1980 could "originate" wherever a national news editor found a suitable place-name to fit both the occasion and the need to attract readers. The DATELINE could be Sauk City, Iowa, because this is where a reporter conducts the worst-case interview with a bankrupt farming family. Or it could be an obscure county seat (preferably named "Cactus Junction") where sits a local banker with a quotable array of droughty horror stories that will suitably dramatize the situation.

Exactly when the shift took place is unknown. Former newspaper publisher Barry Bingham, Jr., thinks it grew out of the widespread practice of holding news stories several days while awaiting more newspaper space. Such stories become news on Thursdays, which is a heavy advertising day producing a big "news-hole" for held-over stories.

Manipulation of DATELINES to serve imperial needs is common practice in all media, although the Associated Press rule book requires specific time-and-place identification of all its stories. But a local editor's decision creates headline value. In order to beat his small-town or suburban competitors, a metropolitan news editor can attach to an otherwise routine story the county-seat-name of an adjoining county. This stratagem is designed to prove his paper's PRESENCE in remote locations. PRESENCE, being undefinable, is useful to such claim-makers.

A Los Angeles editor, deciding how to play a PACIFIC RIM story, has the choice of using (and thus boosting) his home base as a DATELINE city, since the story leads off with a roundup of quotes from Los Angeles bankers negotiating loans in Taiwan, Australia, and Japan. Or, given suitable quotes, the same "story" could be datelined Taiwan, leading off with examples of local industrialists anxious to penetrate mainland U.S. markets. Similarly a Hong Kong DATELINE will dramatize its bankers' readiness to compete with mainland (U.S.) banks before China takes over, and will serve as warning to U.S. readers of coming PACIFIC RIM competition.

"DATELINE cities" come and go with the flow of world news. At the right moment (i.e., during the 1989 riots and put-downs), the world hung on any scrap of news from Beijing. The continued flow of financial power into a handful of World Cities guarantees their DATELINE value, perhaps for generations. Meanwhile, lesser cities attempt to become DATELINE cities with engorged promotional budgets to trumpet their wares and catchy labels ("World Class City," "Gateway to the Midwest," etc.) to balloon their affairs.

All the foregoing is well known and used by public relations counsel as they jockey to boost clients' names into important news slots, either in print or broadcast. A hospital corporate spokesman, learning that a competitor will soon announce a new medical-science "discovery," rushes into print with his own forestalling announcement, confident that his hometown press or TV stations—themselves anxious to be recognized as doing business in a "DATELINE city"—will give him front-page or prime-time coverage.

When earthquake struck the San Francisco Bay Area (Santa Cruz) on October 17, 1989, the ubiquitous, non-datelined Dan Rather flew in from New York, to give on-site coverage. He set himself up alongside the broken-down ramps to the Bay Bridge, and held court for a succession of local sources anxious to get their quake story out on TV. Said Rather: The epicenter of the quake may have been down around Santa Cruz, but the "perceived epicenter" was right here!

Thus DATELINES do not bestow themselves upon deserving locations. Rather, as man-made artifacts, they long since have become tools in interregional and international struggles in and for the media.

DECLINING AREA

Even in the booming years of the late twentieth century the number of counties with declining population increased. "Since World War II, the faster our national population has grown, the faster and more extensively our small communities have declined."[1]

On the face of it, any DECLINING AREA is an affront to prevailing Progressive sentiments. According to myths widely believed in good times, all Real American villages are destined to become towns, towns to grow into cities, and cities to expand ad infinitum. When growth turns into decline, somebody is to be blamed: the mayor, president, Congress, Russians (to 1990), Japanese auto makers, or Arabs.

Cape Cod in 1990 was suffering an intense lack of interest in its glut of unsold new houses. Would-be sellers were dropping sales prices, and most houses and condos had "been on the market unusually long. [It was] another painfully slow year."[2]

Yet local media are usually reluctant to pin the donkey's tail on a market area that may be glutted, but is also gifted with local boosters who resist use of the term "declining." As the flow of speculative money in a DECLINING AREA shrinks, so does scholarly interest. Academic as well as popular researchers have tended to analyze BOOMTOWNS rather than DECLINING AREAS. Before a depression becomes obvious to all in a region, outsiders are hard put to find much about it in print. This began to change in the 1980s under the impact of post-Marxian studies by scholars radicalized during the 1960s.

Despite prevailing sentiment to the contrary, some areas of decline—based on a resource exhausted, or old time-and-distance advantages lost—can do nothing about it in the short run. Their mobile citizens, eager to live in a GROWTH AREA, do well to cut their losses and bail out while they can. Short of such desertion, large segments of any local population will devote great labors to spotting or encouraging any signs of a revival. And once an area's bankruptcies have been written off, its assets cheapened, and a new generation of lean and hungry survivors become active, it may attract a new migration of speculators and others willing to make a new start. And so the cycle continues.

See also TOADS, in Chapter 4: Ephemera; DEPRESSED AREA/REGION and GHOST TOWN, in Chapter 7: Power Vacuum.

DOWNTOWN

The story began down at the LANDING, down by the docks, down at the water's edge, or down at tidewater. That was where most early American settlers arrived by boat. Their descendants may still celebrate Pioneers' Day down at the Old Town LANDING. In New England, they still speak of going "downcity."

DOWNTOWN remains a centrally located business district of an American city. It is still erroneously believed to be a product of unique local get-up-and-go, independent of outside systems and forces. This belief was fostered when any small town of the nineteenth century could become the echo of one determined man or gang who could whip a settlement into town shape. But another reality took over when capital began to flow via long-distance and satellite phones and fax. Such outside pressure altered the ways DOWNTOWN worked and looked. The game expanded when footloose capital from frightened overseas owners poured into New York and other DOWNTOWNS in the 1970s, and then began seeking noncentral outlets as well.

By the 1990s DOWNTOWN was no longer the only game in town for investors, speculators, power-seekers. Some DOWNTOWNS were only one-among-many retail centers. A few found themselves outflanked by competing suburban office clusters (optimistically labeled "edge cities"). But the great majority still hosted the local (often only) cultural arts center, multi-use SUPERBLOCKS, civic center, convention complex, and 100 percent location (with highest per-square-foot land price). Most were still the biggest single DESTINATION for daily metropolitan trips and usually had one or more high-turnover bus and/or transit stations. The roots of ancient monopolies—geography, status,

DOWNTOWN

Navigators-by-map of metropolitan Toronto, Ontario, get a view of generic places not often identified on city maps, such as motel ROW, Chinatown, and Harbourfront (i.e., THE LANDING). As could be expected, "Lakeshore Blvd." [arrows] is sometimes blocks back from today's much-extended lakeshore. Map by Unique Media, Inc., Ontario, Canada.

and control—still ran deep under many a DOWNTOWN.

The biggest casualty of the new DOWNTOWN—dreamed in the forties, financed in the sixties, evident in the 1990s—was street life. In the name of "security" millions of pedestrians from Miami to Montreal were shunted into underground tunnels, or over-street skyways. Flagship buildings were designed as citadels with one-doorway-per-block. Small shops were out. Pedestrians got sucked off the streets and sidewalks. It was potentially a form of depopulation.

What was lacking was DOWNTOWN'S onetime certitude of mastery and control. For the most part, its old family banks and corporations had gone under or gone national. Retail trade was dominated by national superstores (increasingly suburban). New "Save Downtown" campaigners came and went. Each tried anew to tap federal and state subsidies—for expressways, mass transit (a recurring dream), historic preservation. The basic thrust remained: "Save Downtown and you Save the City." Local media still reported nightclub life, art gallery openings, and social gossip as ingredients of "the downtown scene."

But when Detroit, Washington, and in 1992 Los Angeles erupted with underclass riots close to, but not immediately threatening, DOWNTOWN, a new era had begun. Before the 1992 presidential election, so-called urban unrest was beginning to produce many a DOWNTOWN tightened up and battened down as the region's tightest-security zone.

THE GOOD ADDRESS

The manufacture of GOOD ADDRESSES involves history, luck, and hype. Thus the keywords used to indicate a GOOD ADDRESS (real, fancied, or merely hoped-for) can be judiciously mixed and matched for marketing purposes, using this grid of enticements normally found in real estate advertising.

THE GOOD ADDRESS

Like the Good Old Days, THE GOOD ADDRESS is hardly what it used to be. Nostalgia, and the redistribution of wealth and population have put crimps in what Used to Be. Yet THE GOOD ADDRESS still shows on the mental maps of potent minorities. In hundreds of towns and cities, it magnetizes new money and old—and what it buys.

Manipulating this valuable and tangible property continues to obsess ambitious real estate agents, movie/TV stars, and hosts/hostesses of lavish parties where "everybody" (who-is-anybody) must be seen. Hundreds of firms now specialize in screening zip-code data for clues to THE GOOD ADDRESS and the spending habits of its indiginees.

In the process, key neighborhoods come to be defined, located, and bounded—and then hounded by inquisitive outsiders. Places recognized to be THE GOOD ADDRESS command high, sometimes astounding, prices. They also command extravaganzas of language which an English observer terms "the verbal paraphernalia of the pseudo-baronial."[1]

With few exceptions (i.e., Manhattan skyscraper apartments not yet gone condominium), THE GOOD ADDRESS consists of private property. Its size is often measured in acres, and described in semiprivate language. Old-family owners talk down: they own "a little place in the country"—which may in fact cover two hundred acres.

The Los Feliz area, fifteen minutes from DOWNTOWN, "represents to Los Angeles what Gramercy Park does to New York or Nob Hill to San Francisco. Los Feliz mansions suggest old money, good cloth coat, not fur money, their elegance so understated it's a murmur."[2]

New upcomers and newly rich tend toward specifics: "This was the Fred Astaire / Buck Rogers / Irene Dunne Estate and cost a cool million when it was built."

The Good Address

Buzzwords for the manufacture of a Good Address, as found in contemporary real estate advertisements. This mix-and-match grid may be useful in assembling a suitable array of enticements for merchandising purposes.

0	Luxurious	waterfront	estate	(mint condition)
1	Spacious	prestigious	location	(close to . . .)
2	Exclusive	country	horse farm	(great view)
3	Impressive	waterfront	seat	(unique)
4	Commodious	antique	penthouse	(panoramic vista)
5	Unique	restored	place	(rare find)
6	Elegant	English	landmark	(historic ambience)
7	Ultimate	historic	establishment	(all amenities)
8	Gracious	colonial	retreat	(14-foot ceilings)
9	Stately	classic	enclave	(servants' quarters)

Examples: *0962* = Luxurious classic landmark (great view)
7111 = Ultimate prestigious location (close to . . .)
6709 = Elegant historic estate (servants' quarters)

Sales agents employ special lingo, notably in print: "This secluded, wooded property occupies a Prestige Location in exclusive Maryland Hunt Country. For three generations, it was the country seat of a distinguished Eastern Shore landowning family. This is a first-time offering for a discriminating, horse-loving family." And so on into the millions.

Aside from such exotics, THE GOOD ADDRESS in most American cities goes by local, often unmapped names: Pill Hill, where wealthy doctors live; or the better-known Nob Hill, a.k.a. Nobby Hill or Snob Hill, the topographic landmark in San Francisco. Developers may take over the old name long after First Families depart. Society Hill in Philadelphia was headed toward high-turnover decline until architectural restorers and urban-renewal townhouses attracted higher status newcomers after the 1970s.

Not easily does THE GOOD ADDRESS make it Out There to the far suburbs, unless one considers locations that once were old wealthy resorts, or railroad commuting suburbs of the nineteenth century. Or "Blue Blood Estates," so identified by Michael J. Weiss, such as 960-acre Gibson Island off the northwest Maryland shore of Chesapeake Bay, "a paradise few outsiders can afford."[3]

By 1959, however, social observer Vance Packard was tracing THE GOOD ADDRESS as it moved outward to New York's Long Island. Hence, a builder asserted "Your address . . . in Manhasset is more than an address, it is a symbol of tradition and prestige—a supreme achievement in luxurious suburban living."[4]

And even as they consider moving, most occupants of THE GOOD ADDRESS cherish their anchors Back There—to The Bank, The Club, The Church, The Trust(ed) Officers, charities, and arts anchored to the old DOWNTOWN.

The explosive anger of jobless minorities in the riotous sixties and seventies and again in the nineties brought on installation of new defensive screens around THE GOOD ADDRESS. There arose a fear of injury, and of property damage ranging from burglary to arson, from trespass to trashing. By the 1990s, blatant display of THE GOOD ADDRESS and its landed trappings had become somewhat muted. Yet, so long as differences between rich and poor, pioneers and newcomers, the established old-moneyed and the nouveaux riches continued to exist, THE GOOD ADDRESS would magnetize competitive social capital.

See also SECURITY, in Chapter 3: Perks.

HOLDOUT

Once moved from its verbal usage, HOLDOUT as a noun indicates an isolated property, building, structure, tract, farm, or other location surrounded, overshadowed, or threatened by newer, usually bigger, and obviously different, structures or activities. It could be a single corner house left standing alone, while all around it falls to demolition. Or an odd-shaped lot, its owners refusing to sell for a MULTI-USE COMPLEX.

HOLDOUTS attain their highest visibility, and often notoriety, in center-city locations. But they may occur as the cornfield at a four-way stop on the EDGE OF TOWN, the other three corners occupied by shop-

HOLDOUT

Indianapolis, Ind.: This old, white, center-hall motel on a cramped site northwest of Memorial Circle stood pat during the 1980s' building boom that produced multistory parking decks all around it. Litigation to clear title prevented its sale, the boom passed, and the long-vacant property was still on the market when photographed in 1993.

ping center, filling station, and drive-in bank.

The term HOLDOUT is used admiringly by those who favor old-timers vs. newcomers, indiginees over refugees from away, the historic over the contemporary. But the versatile term works both ways, also serving as a propaganda weapon for supporters of "progressive new development" as against old-fashioned stick-in-the-mud inhabitants who persist in "standing in the way of progress" by refusing to sell out.

The sight of a historical structure that becomes an isolated HOLDOUT excites varied reactions: (1) local preservationists tend to "rally 'round" to save it and succeed, fail, or compromise; (2) no significant body gives a damn, and another landmark bites the dust; or (3) the structure is skillfully incorporated into the new scene and remains an example of "adaptive reuse."

HOLDOUTS take many forms. Midtown Manhattan long remained "the last holdout against the big discount chain stores that sprouted in the malls." Then Kiddie City and Toys R Us arrived in 1988.[1]

City officials often are anxious to help purchasers of historic HOLDOUTS to demolish rather than preserve, and to build anew on the old site. They pursue a tried-and-true scenario: they permit themselves to be persuaded to look the other way as the property is allowed to degenerate into squalor. This will induce neighbors to demand the eyesore be removed.

Varied techniques are used to thwart HOLDOUTS. In Washington, D.C., U.S. Senator Francis Newlands "got Congress to allow his Chevy Chase Land Company to run a streetcar along Connecticut Avenue into Maryland. He and his friends then secretly bought up the property along the projected route, punishing holdouts by altering the route."[2]

For illustrations of city properties whose owners persist in holding out against takeover, and against all odds, see the highly prejudiced accounts in *Holdouts!* by architects Andrew Alpern and Seymour Durst, who champion the cause of large-scale development as against HOLDOUTS.[3] In *The Experience of Place,* Tony Hiss explores the old Klein family farm in northeast Queens, N.Y.[4] Only two acres remain of the original 140, and these are surrounded by the huge Fresh Meadows housing development. The Kleins hold out by serving a huge, hankering public of vegetable buyers. And also by ingenious use of other farms as a source of veggies.

HOLDOUTS come from many backgrounds. By 1984, the National Inholders Association, composed of property owners within areas of federal land control, had grown to a membership of nine thousand, its members dedicated to holding out against federal takeovers, and to protecting members' rights.

Almost by definition, "X People"—so named by social observer Paul Fussell in his book *Class*—constitute a purported new social elite: "X people move away when they, not their bosses, feel they should. . . . Their houses, which are never positioned in 'developments,' tend to be sited oddly—on the side of mountains, say, or planted stubbornly between skyscrapers."[5] (Aha: HOLDOUTS!)

On the other hand a HOLDOUT area is often a likely candidate for becoming a registered historic place or neighborhood, simply by continuing to represent a distant period in local history. Some HOLDOUTS merely melt into the local scenery and become an unnoticed part of the city fabric, only to strike outsiders with amazement.

THE LANDING

Until its recent association with roulette and craps, the term LANDING had been used in England since the fifteenth century to mean a place for disembarking passengers or unloading goods from vessels. This usage expanded to include "foothold," with suggestions of "beachhead"—any meeting point of land and water which has been taken over by a stranger, intruder, foreigner, or enemy.

Its early U.S. usage was the town LANDING, notably at Colonial ports along the Atlantic and Gulf; at St. Louis on the Mississippi; Cincinnati, Louisville, and Pittsburgh on the Ohio; Manhattan, Kans., on the Kaw; Lafferty's Landing (later Gadsden, Ala.) on the Coosa; at Fernandina, Fla.; and at other waterway cities. Tourists continue to be amazed to find town LANDING traces or markers alongside shallow, out-of-the-way streams that once actually handled boat and barge traffic in the early nineteenth century. But on the whole, the town LANDING is badly marked or not easily found by tourists.

For hundreds of places, the original LANDING occurred where shallow harbors, tidal meanders, or streams barely deep

THE LANDING
***The Mississippi Queen:* Nuzzling into the Ohio River shore landing at Henderson, Ky., this giant river-cruising sternwheeler uses the historic tactic for boarding and offloading: a bow ramp adjustable to the height of the bank or dock. Photo by Mary Alan Woodward.**

enough to float a potato chip in dry season could at wetter times support explorers' canoes, and settlers' barges pushed far enough inland to start a trading settlement or pioneer village.

Such LANDINGS today are often discoverable only via historical markers, though sometimes they are expanded into historic districts. A well-located or lucky pioneer village at the head of navigation often grew into a port city. But as ships drew more water, and as natural harbors silted up (as at old Port Tobacco, Md.), heads of navigation moved downstream, and original ports disappeared or dwindled into local obscurity. Along navigable waterways a public LANDING still performs public duties, especially as a ferry LANDING or as a site for contrived ceremonies—such as reenactments of the pioneers' arrival—to attract tourists.

Such low-key goings-on began to tune up in the 1990s, when the gambling industry entered its greatest boom, becoming a thirty-billion-dollar "entertainment" enterprise by 1993. Enroute to their new multimillionaire status, promoters of floating gambling casinos saw old town LANDINGS as prime sites for their new operations. Some cities, seeking new income from decrepit waterfronts, welcomed gamblers with open arms and relaxed laws. What was now called the casino LANDING became a hot property. "Casinos are cropping up everywhere from mountain saloons to Mississippi riverboats," headlined the *New York Times,* October 25, 1992. By that time there were three casino boats operating out of Davenport and Dubuque, Ia., and Rock Island, Ill., on the Mississippi. At Metropolis, Ill., some five hundred "happy souls . . . boarded the hulking three-story Players Riverboat Casino" on February 22, 1993, to usher in legalized gambling on the Ohio River.[1] "Splash Casino" arrived at the Mississippi River waterfront of Tunica, Miss., in time "to transform the economy of a Mississippi county that was the nation's fourth poorest in 1990." The city, with a total annual budget of $2.6 million, hoped to net $1 million a year from a percentage of cover charges.[2]

Four companies with gambling boats on the Mississippi "have seen their stock prices more than double in recent months." By mid-February 1993 the Mississippi state Gaming Commission took in twenty-nine applications for casino licenses.[3] By the end of 1993 there were "more than two dozen casino boats working the waters and betting handles, and 40 more authorized in six states . . . in cut-throat regional competition, lately aboard land-locked 'riverboats.'"[4]

To land or not to land is the question. At some LANDINGS, it is argued that gambling is legal only if done while afloat in the historic aura of the "Mississippi riverboat gambler tradition." Others argued: So what's the difference ashore or afloat? It's still gambling and to hell with it. States along the Mississippi, Missouri, and Ohio Rivers combed their boundary laws to prevent or permit gambling on "our waters."

Small-town Jeffersonville, Ind., found itself entertaining three competing proposals in late 1993 for LANDINGS along its Ohio riverfront. These offered such bait as two 250-foot-or-longer sternwheeler riverboats modeled on the old "Robert E. Lee," a 59,000-square-foot "gambling pavilion," six thousand parking spaces, a twelve-screen theater, eight-story hotel, and 1,400 gaming positions.[5]

Nearby New Albany, where the original

"Robert E. Lee" was built, used its redevelopment commission to ask casino operators for proposals. Several nearby towns and a county joined up to get their pro-rata shares of future gamblers' take. With millions of tax dollars at stake, Kentucky and Indiana jockeyed over ownership of the proposed LANDINGS, since Kentucky claims the Ohio River up close to adjacent states' shores. All three proposals were killed, for at least two years, when voters rejected legalized gambling in November 1993. No one could know how soon the market for gambling at and on the many new LANDINGS would be saturated.

Meanwhile, heavy industry on old waterfronts has shrunk or moved; much of the heavy work along inland and outer coasts is shifting overseas, or to new marine industrial districts following years of arguments and local bond issues to deepen local moorings. Or THE SITE may develop as a MULTI-USE COMPLEX such as "Jacksonville LANDING," a "$42 million retail and restaurant COMPLEX" along downtown Jacksonville's St. John's River frontage. Or "Catalina LANDING," a $50 million, 11.3 acre project completed in 1986 at Long Beach, Cal., with three four-story office buildings, a 1,500-car garage, shops, and reconstructed ferry LANDING—"a place where you can sit in a boat and watch the offices go by, or sit in an office and watch the boats go by."[6]

Thus, in quiet coastal backwaters, the term LANDINGS has come to suggest land on or near waters that can be declared "navigable," purchased by a developer and reinforced with houses, condominiums, a boat dock, boardwalk, and, by the application of publicity, converted into a VIEW property and rechristened LANDING.

Such transformations occurred notably during the much-advertised "return to the waterfronts" of expanding, boat-oriented older cities; and along any waterway that could be converted to a resort, or adapted to the newly promoted lifestyle of "waterfront living." By the 1990s the "working waterfront" had found new bosses. THE LANDING had become a generic term to signal the introduction of a new consumer system based on THE VIEW.

For some, an important ingredient was access to cheap, stoutly built old brick warehouses, ripe for boutiquery. LaClede's Landing on St. Louis's cobblestoned Mississippi riverbank, Cleveland's Flats and Cuyahoga Riverbank ("on America's North Coast"), and both banks of downtown Jacksonville's St. John's River and Philadelphia's Delaware became sites typical of many such redeveloped MULTI-USE COMPLEXES.

Casino LANDINGS, for all their local pull and gaudy publicity, were fated in the longer run to be overshadowed by mechanized container port complexes being put in place around U.S. coastlines to pursue international trade. Tampa has an $873,000 "juice berth" just for piping aboard semi-frozen citrus slush. Some thirty-five thousand jobs are built into the mechanized Port of Houston. When chips are down and stakes are high, the old town LANDING and its modest scale is being backwatered and overshadowed by the throw of the dice, the spin of the wheel, and the thrust of the giant propellers of international shipping.

See also VIEWSHED/THE VIEW, in Chapter 4: Ephemera.

THE LIGHT

THE LIGHT has penetrated beyond its role as an essential commodity, bought, possessed, and maintained at great cost. It has become not only a localized condition of a place—believed to be a fundamental human entitlement—but is spoken of as though it were a place itself: "Let's go out into THE LIGHT." It occupies such an important position in urban life that an unlit DOWNTOWN is considered unsafe and unmanageable.

When properly lighted, DOWNTOWN with its nighttime glow is expected to send far and wide the signal, "Here's where the action is!" The Great White Way—Manhattan's Broadway—was a distinctively center-city phenomenon, and its brightly lit stores magnetized customers as it amazed its first admirers.

City Lights was not merely the title of a Charlie Chaplin movie, but came to symbolize romance, excitement, the hustle and bustle typified by a bright-as-day Times Square. By the 1980s no U.S. DOWNTOWN considered worth saving failed to promote its Christmas LIGHTING DISTRICT and events.

It was not always thus. Few living Americans can recall the excitement generated when man-made illumination first cast its gassy glow over nineteenth-century cities. By 1900, an entire culture in lighting had developed, for the most part in central cities, and has continued outward ever since. THE LIGHT has become more than a desirable place to go. All urban skies have brightened up: "It's a golden roof over Manhattan," bragged New York columnist Jimmy Breslin.[1]

THE LIGHT had early spiritual connotations: THE LIGHT was central to the expression of Christian doctrine, and its shining on the Sun's Day contributed to an early sense of joy and revival. A fifth-century Father Maximus of Turin said Christians reverenced the Lord's Day "because on it the Savior of the world, like the rising sun, dispelled the darkness of hell, shone with the light of resurrection."[2]

Modern law recognizes "solar access zones" as places of entitlement: nobody can legally deprive you of your rightful share of sunlight. "Rightful share" and legal access have been repeatedly defined in court decisions.

"Sun-worship" arrived in Western society during the 1920s when sunbathing on chic French beaches became fashionable. Within a half-century, sunburn balm and the bikini extended the fashion until the penalty of skin cancer became evident in the 1970s. Then came television and portable phones to preempt more than five hours per day of American families' time budgets. Inexorably, millions of Americans have moved slowly into a perpetual TWILIGHT ZONE—working, communicating, and watching TV indoors, traveling, trucking, or tractoring in enclosed vehicles, many with tinted windows; turning down indoors lights to save energy, wearing sunglasses, and—for those still at work DOWNTOWN—walking shadowed by tall buildings. Increasingly, THE LIGHT as a locale has come to resemble a reserve for specialists: for sun freaks, garden nuts, outdoor types, and "mad dogs and Englishmen [who] stay out in the noonday sun."

THE LIGHT also represents the triumph of the continuous. As Fernand Braudel said of the transition when sixteenth-century European traders shifted from seasonal fairs to regular shops, trade became "a continuous stream, replacing intermittent encounters."[3] Today, some places are never allowed to become THE DARK. Thus what Rudolf Arnheim once called "a gift of life-generating power that flows toward you" has become one of life's commonplaces.[4]

In everyday talk, THE LIGHT may become a command, a possession, a condition, or a place. On the road, one "gets the green light," and responds to its code-of-the-road by driving off. One possesses it in order to "throw light" on a complex subject. As a condition, it gets assigned to locations naturally or artificially light enough for normally anticipated human visibilities, and for activities that depend upon light from whatever source. As a fashion, lights may be attached to clothing; showoffs may sloganize their clothing with neon signs; dancers, like glowworms, may glow in THE DARK.

During a curfew (originally "cover-the-fire!" in French) lights go out. But the unexplained absence of light sets off many reactions: an eclipse, in pidgin English, comes out "Him kerosine blong Jesus Christ all bugger up done finish." When it goes to work as an adjective, LIGHT is the object of repeated efforts by advertisers to deprive it of the "gh," add an "e," and work it for their own use: Lite beer is an example of cutting away "the undergrowth of silent consonants."[5]

THE LIGHT is also a place, a DESTINATION. You go outdoors into THE LIGHT of day the better to see; you turn on THE LIGHT indoors or out to the same effect. THE LIGHT'S sinister counterpart is THE DARK. From primitive peoples' viewpoint, THE LIGHT arrives just in time to turn night into day—the magic of the sun thus freed from the forces of darkness to do its work. Later voices took it to be "the angel who rolls away the darkness."[6] No longer listening to primitive voices, we now see THE LIGHT not only as a place, but as a complex emission system that encourages plants to grow, and human skin and psyche to absorb necessary stimuli from solar rays. Psychic disorders flourish and are well documented in Scandinavian countries after prolonged exposure to short days, dull sun, wintry weather—and cramped housing. By the 1970s this had acquired a name: "Seasonal Affective Disorder" or SAD. Under the influence of late dawns and early twilights, SAD's victims—not limited to Scandinavians—"sleep fitfully and lose energy and libido in fall and winter [and] binge on carbohydrates." Research by neuroscientist George Brainard, Jefferson Medical College, Philadelphia, and others connects THE DARK'S effect on the pineal gland and its capacity to produce melatonin, a sleep-inducing hormone.[7]

Possession of THE LIGHT brings with it power and prestige. Reporting on the Alaska earthquake of 1963, the author was told that a number of new community leaders arose from the crisis; mostly men with home generators, who produced THE LIGHT visible throughout the confusing darkness. Neighbors flocked to THE LIGHT, expecting its possessor to give orders. Under pressure from such expectations, new community leaders emerged. Meanwhile, across much of the United States, millions of farmyards have emerged from THE DARK. Two generations of electricity salesmen have introduced barn lights which automatically turn on at sunset, so that once-dark countrysides are lit from horizon to horizon.

"Crime lighting," also known as "security lighting," became a fad in the 1970s when many neighborhoods lobbied their city halls to get lights installed on minor streets and alleys, and even in mid-block "to discourage robbers and muggers." Ordinary lights were replaced by yellow lights that were thought not to cast sharp shadows, making it easier to catch a thief. Evidence that THE LIGHT made much difference is widespread. "Twelve times as many crimes of violence are committed at night as in the daytime," and in the darkest months—December to March—it went up 12 percent over the rest in a 1961 study.[8] Meanwhile, key intersections of the interstate highway system are bright-lighted far out to populated suburbs.

Baseball has been transformed by THE LIGHT. "The first night game was held on May 24, 1935, in Cincinnati's Crosley Field; by 1948 every ballpark but Wrigley [Field, Chicago] had lights. . . . Night baseball plus the automobile have equaled disaster for both urban ballparks and their neighborhoods." It discourages walk-in customers and wipes out ballpark neighbors with parking lots, argue Philip Bess and Howard Decker.[9] Public gatherings are transformed differently: TV cameramen will flood a public gathering with portable light—then suddenly switch it off once the quick-quoted "sound bite" has been taped—TV's self-serving signal that "It's all over, folks."

To escape altogether from THE LIGHT has become a long-distance enterprise. This was discovered by millions of amateur sky-watchers who flocked to city outskirts in 1986, looking for THE DARK. But they found it almost too bright for a clear view of the barely visible Halley's Comet. And as true darkness on the edge of the city virtually disappeared by the 1980s, the term "light pollution" found its way into public debates. Federal Aviation Agencies have long exercised some controls over distracting light sources around airports. In San Diego County, Cal., a 1985 law banned all nonessential light in unincorporated areas. This was intended to help darken skies over the Mount Palomar and Mount Laguna observatories. Tucson and "at least two dozen other Arizona communities have approved some portion of a 'dark sky' ordinance proposed by the Kitt Peak National Observatory" near Tucson.[10]

The most extreme example occurred early in World War II, recalled by John McPhee in "La Place de la Concorde Suisse," the *New Yorker,* November 7, 1984: "as the rest of Europe blacked out, Switzerland still sparkled with light and served as a beacon to [Allied] aircraft. The Germans were particularly uncomfortable with this flashlight at their side. They told Switzerland to shut off the light or Germany would come and do it for them. Switzerland went dark."

Manufacturing THE LIGHT is, however, by no means confined to gas-and-light companies and the makers of portable generators and flashlights. Daylight saving time is another device for extending THE LIGHT. France was advised to adopt it by U.S. Ambassador Benjamin Franklin in 1784. A major move was made by London contractor William Willett, who observed his workers suffered more accidents in winter's short days than in summer. His solution, recommended in 1907, was: daylight saving time (DST). It was adopted during

World War I by England, Germany, France, et al., and later in Argentina, Spain, and the Soviet Union. It is still widely adopted but locally debated in the United States, where the tip of Maine predictably moves into THE LIGHT hours before it is replaced by THE DARK on THE BEACHES of Hawaii.

Indiana is an oddly divided HOLDOUT with seventy-seven of its ninety-two counties on twelve-month Eastern standard time (EST), and five counties on EST for five months and DST for seven, while ten remain on central standard time (CST) five months and DST for seven. In winter "when the clocks of Hudson Lake say it is 5 p.m., the clocks of New Carlisle, just two miles east, say it is 6 p.m. The situation is even more confusing in Vevay, where the time can be 1 p.m. in one pharmacy and high noon in another. In Vevay, it is possible to keep a 10 a.m. appointment with the optometrist and then drive to the doctor's office for a 10 a.m. checkup."[11]

Meanwhile, the nighttime hustle in suburban malls and the pinpointed bustle in downtown cultural centers are reminders that, before THE LIGHT, nighttime was almost universally a time of indoors repose, recovery, and quietude while cities at large remained in THE DARK.

See also THE DARK, in Chapter 7: Power Vacuum.

MEETING PLACE

Seldom acknowledged on a map, MEETING PLACES abound in local folklore and among knowledgeable frequenters, meeters, and greeters. For generations of East Coast collegians around World War II, "under the clock at The Biltmore" (Hotel) in Manhattan, New York City, was a widely known and crowded weekend MEETING PLACE. During the "hippie takeover" of

MEETING PLACE
Old dangers, distant views, and human densities make Baltimore's Inner Harbor a favored meeting place for millions of tourists. The dangers are mostly past—this once was a rough-and-ready fisherman's wharf. The views keep changing—the SKYLINE has altered with each addition of an aquarium, offices, etc. The density of tourists has brought on a continuing boom to such MEETING PLACES as this dockside restaurant.

Key West in the 1950s, the western town dock at sunset was a favorite place to meet, sing, smoke, and cheer when the sun disappeared. For thousands of like-minded groups, "meeting" was an evanescent, catch-as-catch-can affair. To learn where "everybody" was likely to be located became a widely shared obsession.

Among the new MEETING PLACES arising in the 1980s were the so-called fern bars with their fake or living greenery, macrame hangings, yard-sale decor, all in a purplish glow. Writer Joel Garreau dismissed them as "clone bars." New dimensions were added at such high-velocity points-of-contact as Boston's Copley Plaza. Its new owners built a ground-floor sit-and-stand bar with many different levels. Hundreds of patrons could thus drink and sit or stand at comfortable eye-contact-level with strangers-but-not-for-long.

Beyond such face-to-face contacts, MEETING PLACE began to be routinized. By the 1980s the manufacture of look-alike convention centers, civic centers, auditoria, resorts, et al. had prompted many a developer to set up formal and well-designated MEETING PLACES to help occupants to keep bearings and appointments, and expand their address books. Resorts took to signboarding every nook and cranny that could be featured in its brochures: "Meet your friends for tropical drinks on the Dolphin Deck!" Rio de Janeiro made its declaration with an English-language sign, "Meeting Place," in its international airport's arrival lobby.

Inevitably, this geographic migrant ended up as the name of many newspapers' classified advertising sections, under a headline, "Meet The Love of Your Life in The MEETING PLACE." In this format, a lonely "inimitable, well-grounded, adroit and loving, 34-year-old 5′6″ SWF" advertises for friends in the space adjacent to a "SWM CPA" (single white male who is a certified public accountant) "looking for energetic female for companionship and possible long term relationship."[1] The MEETING PLACE ads became syndicated by a firm in Minneapolis, Minn., with an 800 number.

The folklore of MEETING PLACE includes such anecdotes as from "Amos 'n Andy," a radio comedy team of mid-twentieth century: "I'll meet you at the corner. If you get there first, YOU make a chalk mark on the light pole. If I get there first, I'll rub it out."

PHOTO OPPORTUNITY

A derisive term, this designates any location—often "in the center of things"—chosen by a political candidate so as to manipulate press and TV photographers into shooting a particular scene/episode. The term was popularized by the skill of President Ronald Reagan's press aides in contriving such opportunities in the 1984 presidential campaign.

PHOTO OPPORTUNITY became the only venue open to reporters or photographers attempting to question an evasive, manipulative president over matters of state. Its typical location was in Washington, D.C., on the White House lawn, where the president walked from house to helicopter, answering shouted questions over his shoulder and the noise of the helicopter.

But the opportunity grew. The biography of Harrison E. Salisbury, long-time overseas correspondent for the *New York*

Times, was reviewed as "a book about the way it was before reporting turned into what now seems to be one great, long, unending photo opportunity." By 1988, national political conventions were described derisively by *Planning* magazine as "The Biggest Photo Opportunity of All"—"elaborately staged" and "designed for television."[1]

Long before Reagan, however, the "photo-op," as it came to be called, had acquired its own rules. At a national convention in Washington, D.C., 1958, a sign was placed near the speaker's platform advising press photographers: "Do not photograph the speakers while they are addressing the audience. Shoot them as they approach the platform."[2]

Thus rule-driven and politicized was a practice that had long been aimed at tourists—the posting of signs at central locations, scenic overlooks, viewpoints, vistas—advising camera-toting tourists where to stand to "get a good shot." So instructed were tourists at World's Fairs such as Knoxville; and at sightseers' DESTINATIONS such as San Francisco's Golden Gate Bridge. The resulting photos inevitably look alike; they reproduce the elements of scenic composition made famous by Renaissance painters who discovered perspective as a geometric device for organizing a scene.

PHOTO OPPORTUNITY and its picture-takings implies theft, stolen glimpses, sneaked shots—as though once an exposure is fixed upon photographic film, that's-all-there-is. It implies momentary titillation, with the picture-taker unmoved by the experience. This is far from the art of seeing, of the "pointed mindfulness" envisioned by the teachers of Zen and practitioners of the art of *feng shui*. This is the ancient Chinese art of sacred placement—in response to mysterious earth forces—that guides the feet of pilgrims. Tourists at many Japanese shrines are instructed where to stand or walk to be in phase with cosmic power. So far as is known, *feng shui* has not been adopted by American presidential aides in staging PHOTO OPPORTUNITIES.

PORNO DISTRICT

Variants: adult entertainment district, combat zone, fleabag area, sex center.

Within walking distance of old downtown hotels, within easy taxi distance from big convention hotels, and readily described to, and found by, strangers, PORNO DISTRICT is a leftover from the days when explicit sex among strangers had not yet moved into home videos. Here, often adjacent to the former red light district of most older cities, sexually predatory males could find partners for cash.

The same police unconcern that once hovered over red light district came to benefit PORNO DISTRICT, especially as scores of cities catered to conventions to prop up local economies deserted by heavy industry. PORNO DISTRICT became an unpublished enticement to conventioneers, mostly male.

They came in many geographic variations: a notorious "chicken ranch" of mobile-home prostitution retreats outside Las Vegas, Nev., and strings of adult book stores along commercial strips. New York's Times Square area became a world-class PORNO DISTRICT, attracting hustlers, buy-

PORNO DISTRICT

For those unsatisfied by the rising diet of pornographic films on TV, PORNO DISTRICTS, such as Boston's "Combat Zone" still flourish, in spite of expanding neighbors: Chinatown and Tufts Medical Center. A milestone year for this "combat zone" was 1973 when new property owners felt secure enough to erect large, expensive rooftop signs.

ers, and sellers of international ilk. Boston's notorious Scollay Square was redeveloped out of DOWNTOWN to the EDGE of Chinatown and Tufts Medical Center, where by 1973 it had become well-established enough for its businesses to invest in expensive electric signs. Other notorious "blocks," such as Baltimore's, survived total demolition but were persecuted and/or purified to one degree or another. In five years, the number of pornographic-film houses across the nation had dropped from around eight hundred to four hundred in 1988.[1] A sharp boost to local anti-porno forces came in 1977 when a noted case, Miller vs. California, shifted the burden of proof in porno-prosecutions to local standards, as determined by local juries. This offered the growth machine new Puritanical partners among religious, fundamentalist, parent-teacher, and other groups anxious to kill PORNO DISTRICT, or to render it invisible.

Thus over the years, PORNO DISTRICT had became a SINK into which powerful elements in a community could dump or delegate disapproved or proscribed goings-on. But PORNO DISTRICTS were also subject to waves of reform, of local Puritanical dominance, or to economic booms when the local growth machine saw the old PORNO DISTRICT as a redevelopment site for new, usually corporate, purposes. At such times, property owners in PORNO DISTRICT become second-class citizens, subjected to mild-to-flagrant-to-illegal takings of their properties. The stronger the Puritan strain locally, the slimmer the chance of PORNO DISTRICT'S survival.

THE ROW

Variants: advertising ROW, antique ROW, auto ROW, bondsmen's ROW/block, boathouse ROW, book(seller's) ROW, boutique ROW, cannery ROW, cathouse (or whorehouse) ROW, the diamond block, fraternity ROW, furniture ROW, jewelers' ROW, lawyers' ROW/block, newspaper ROW, officers' ROW, pawnshop ROW, press or printers' ROW, restaurant ROW, shoe avenue, skid ROW (nee road), Tory ROW, wreckers' ROW.

THE ROW'S history involves a medieval congregation of pitchmen, noted by Fernand Braudel when he examined the sixteenth-century and earlier European roots of modern capitalism. These were tradesmen-merchants who foregathered first outdoors at fairs and later indoors in shops, up and down the same street or alley. They offered the essence of proximity; if the customer didn't like it here, try next door, and on down THE ROW. It marked the appearance of a revolutionary affair: the comparison-shopper's market—with employees, credit, warehoused goods, a shop open all year round at the same address.

THE ROW is all pre-strip, pre-automobile. Its proximities are born of foot-slogging porters toting goods up and down muddy streets, and weary country folk haggling, buying, and selling. In such a close-coupled context, "next door" was close at hand.

Until the Great Fire of 1666, St. Paul's churchyard was London's main center for chockablock booksellers. By the nineteenth century, London was dotted with other ROWS: printers, used-clothing merchants, tradesmen, guildsmen—ROWS, straggles, and close-in versions of later strips. Fourteen clustered booksellers still survived into the 1980s within a short walk north of St. James's Park.

Here as elsewhere, book ROW is in transition. "In the 1950s it took Paul A. Solano [a book lover] a week to stroll the six blocks from Union Square to Astor Place in Manhattan, a corridor of three dozen shops selling used books."[1] But by 1988 the last bookstores on Manhattan's Book ROW had dispersed, the typical surviving store bigger, computerized, awash with mail orders. In many cities the ROW had disappeared, bookstores scattered, many into drive-in malls.

In Cambridge, Mass., one of the continent's greatest book ROWS is dispersing along the Harvard-MIT axis of Brattle Street, but Harvard-Brattle Squares still magnetize the greatest cluster, some two dozen stores, several expanded in the 1980s. Here as in other bookish precincts—outside Sather Gate along Telegraph Avenue, Berkeley, Cal.; around DuPont Circle, Washington, D.C.; along

THE ROW

The backs of these buildings on warehouse ROW, downtown Main Street, Louisville, Ky., once gave access to the lower-level Ohio River wharf, when this was a busy Ohio River port city. The ROW, its unified cast-iron facades protected as historic artifacts, undergoes gentrification, while a new Riverport expands ten miles downstream. Composite photo from Lawrence P. Melillo and Associates, and Louisville Central Area, Inc.

east-west streets through the University of Chicago area—bookstore-cafes sometimes add to the menu.

The old affinity for proximity persisted in ROWS along most American big-city streets. Many concentrations such as Manhattan's Diamond Block (District) or its Gallery Row, Louisville's shoe avenue, the boutique ROW on Vero Beach, Fla. (with those "hug-a-bunny-names"), and the oil-drillers' parts-and-repair district of Houston—all came to depend on a national or international clientele to reinforce the local walk-in trade.

In the art boom of the 1990s, many local art and antique galleries struggled to find the proper mix—modest rents, high visibility, and locations along the route of target customers—to start a ROW. In another sense, many a ROW was started along cathouse (or whorehouse) ROW, with which it existed in cheek-by-jowl proximity in many a nineteenth-century town or city. By 1880 few such ROWS had survived amateur and PORNO DISTRICT competition. Across from Cincinnati, the "notorious Monmouth Street strip" of Newport, Ky., once sported seventeen sex-joints. By 1986 it was down to ten, straight businesses taking over some sites. Las Vegas, in true twentieth-century fashion, had dispersed (1977) its legalized houses of prostitution along State Highway 93 north of town. Its signs advertised "Kitten Creek," "Shari's Ranch," and "Judy's Coyote Spring Ranch."[2]

Politics had a hand in naming ROWS such as Tory Row along Brattle Street, Cambridge, Mass. Here wealthy Tory families built colonial mansions which, come the American Revolution, acquired the then-opprobrious name Tory Row. But now, with that Revolution safely historicized, Cantabridgians relish the archaic

and innocent name still fastened thereto by guidebooks and by local histories.[3]

The Millionaire's Mile (and a half) of Fifth Avenue, New York, was an odd mix of self-merchandising, architectural display, and intense social-cum-financial competition, not so much for customers but for admirers. Between 1890 and World War I, millionaires, new and old, built extravagant mansions from East Sixtieth to Ninetieth Street. "The Whitneys imported a seventeenth-century ballroom from Bordeaux. . . . Mrs. John Jacob Astor had a two-ton bathtub cut out of a solid block of marble."[4] On Fifth Avenue and in adjacent Central Park, the new rich promenaded or coached in deliberate full view of the lesser orders. Today Millionaire's Row, though much diminished, is a hodgepodge display of converted mansions, new offices, and other outer-directed institutions such as the Guggenheim Museum.

Another strung-out Manhattan variant was Ladies' Mile. In 1900 it offered frontage to two dozen department stores along a circuitous carriage and walking route including Sixth Avenue and Broadway. When the Seigel-Cooper and Co. store opened in 1896 it attracted a crowd of 150,000 pedestrians.[5] But it was proposed in 1986 to declare a Ladies Mile Historic District, with some nineteen million square feet of old stores and warehouses. This set owners to debating whether being officially "historic" would help or hurt business.[6] Oil millionaires in Tulsa, Okla., found Black Gold Row the nickname attached to their concentration of mansions on Madison Avenue and on nearby streets. This short, European-style boulevard carries an elegant melange of Italian, French, Tudor, Georgian reminiscences.

Madison Avenue in Manhattan, known as Advertising Row, could once boast that

its ad agencies handled one-half of American industry's advertising budget.[7] Manhattan's ROWS of fashion-art-shopping were, for generations, caught in the recurring glitter of that city's self-promotional media. Madison Square, originally a potter's field, became a parade ground, then the focus of Fifth Avenue Mansions (Millionaires' Row) "the center of elegant New York in the 1860s."[8] From World War II into the 1990s it slowly evolved from a street of small-but-select shops into a mix of skyscrapers (housing hundreds of advertising agencies) and high-fashion shops, with Givenchy's corner boutique at Seventy-Fifth Street once described as the cornerstone of his American fashion business.

Another form, boutique ROW, added a piquant mix of fancy shops, festooned along streets of fashion and their imitators. Melrose Avenue, Los Angeles, once "a lackluster strip of boarded-up storefronts and furniture refinishers" by 1988 had been turned into "a funky, neon-lit retailing mecca" a five-mile kaleidoscope of trendy fashions, gift and interior design boutiques, and nouvelle cuisine restaurants—much of its glitter emerging from punk rock culture.[9] As boutiquers and their lingo move into ROWS, they work their metaphors overtime, peddling ice cream at Toute Sweets, recordings at Flip Side, and snacks at Purple Onions, Blue Heavens, or Red Tomatoes.

Antique ROWS abound in scores of cities. Baltimore, which advertises itself as the Antiques CAPITAL of the East Coast, lists "25 shops offering Old World charm" clustered along the 800 block of North Howard Street; Louisville's antique ROW is along Bardstown Road in the Highlands. Atlanta's once was arrayed along Peachtree Street but is now dispersed; Manhattan's is still anchored to Madison Avenue, and New Orleans's is around the French Quarter axis of Bourbon Street.

Facing many an old COURTHOUSE SQUARE, the lawyers' ROW or block has persisted—visible, along with bail-bondsmen's shops, in such disparate purlieus as a scatter northeast of Albuquerque's courts complex and a historic redbrick row across the street from the Harrodsburg, Ky., courthouse.

Pawnshop ROW has its locational druthers—along a major street between a large factory and a working-class neighborhood; or strung along the highway outside a big Army/Navy base. Florists line themselves up across from cemetery entrances; office and quick-copy suppliers gravitate to the DROP (off) ZONE just beyond financial district—all capitalizing on proximity and serendipity—assets that survive few journeys beyond the city's EDGE.

Cannery Row in Monterey, Cal., immortalized in John Steinbeck's same-named novel, is of another breed. This was a workingman's ROW, a noisy, noisome, toilsome place close to the docks, and occupied in the early 1900s by Italian sardine fishermen and later by Chinese squid fishermen. That was before Steinbeck wrote and the tourists arrived. Cannery Row rose and fell with the supply of fish off the West Coast. Over-fishing during and after World War II killed off THE FISHERY. Cannery Row's shacks and sheds rotted until tourists, toting Steinbeck's book, transformed scenic Monterey into a trendy, pricey, boutiquey resort. By 1988 the tourist flow had been reinforced by a giant aquarium. Only one shack remained on Cannery Row.

Beyond tourists, other ROWS also invite

surveillance: fraternity ROW and whorehouse ROW share to some degree the recurring interest of authorities; pawnshop ROW is watched by local police; gallery ROW is a standard port of call for pulse-takers from national art journals trying (usually without success) to dig out current real prices paid for well-known works. And fraternity ROW faces occasional close-downs—seldom the whole ROW, but usually a chapter house notorious for liquor, drug, or sexual outrages.

Skid ROW'S dependence on curb-sitting homeless men for its identity is a West Coast (U.S.) phenomenon, and a corruption of the earlier skid road, a muddy or greased slope for moving logs. The term is used cautiously or freely by local media, depending on whether skid ROW is still its old self, or being upgraded by aspiring developers. In the latter case it is referred to as a "resurgent neighborhood," or former skid ROW.

Why do some ROWS flourish and other wither? Is there a universal rule at work? One answer lies in location theory, much used and abused by marketers. Geography usually holds a key: Bucks County, Pa., for instance, is halfway between New York and Philadelphia, "the crossroads of the main antiques trading routes between North and South."[10] But the ROW here is a huge scatteration rather than a single main drag. In New Orleans's French Quarter, the seven blocks of Royal Street between Iberville and Dumaine is more like it: "more quality antique jewelry and furnishings than any comparably sized area in the United States."[11] Braggadocio of this sort surrounds many a ROW as its denizens strive for the big time.

London's Regent Street is a noted European predecessor of U.S. ROWS, built mostly between 1817 and 1823 by John Nash. Its great colonnaded curve of fine houses and shops was located precisely between the raffish purlieus of SoHo and the royally frequented St. James's district. This bipolar situation yielded the classic mix of high- and lowlife that appealed to buyers and sellers, to strollers, procurers, and adventurers. It remains a great shopping street into the 1990s.

All ROWS carry loads of local history, but few are published and some get wiped out. In Oklahoma, hundreds of poor families swept out of the Dust Bowl in the 1930s migrated first to Oklahoma City where they sold their meager or other belongings to scrap up cash for the migration to California. Two blocks of Reno Street became their depository: antique ROW. (Local folks disdained the vernacular but opprobrious term Okie Row.) The owners of small pyramidal-roofed cottages conjoined their sagging front porches to form a block-long

THE ROW

When thousands of poor families ("Okies") were swept off their farms in the Dust Bowl in the 1930s, they sold their belongings here in Oklahoma City, Okla., for cash to get to California. These cottages along Reno Street, their sagging verandahs joined together, formed a picturesque furniture ROW. It later became a mecca for popular-culture hounds, tourists, and souvenir-hunters. This lasted until its 1970s demolition made way for a grandiose redevelopment called The Myriad.

THE START

A multiple-access, longitudinal mechanical device, the starting gate was invented so that high-ticket horse-racing could avoid competitive shenanigans among trainers and jockeys awaiting (and sometimes jumping) the starter's gun. The device is wheeled to the designated starting-place for each race. Once all horses are installed —officially "at the starting post"— the race can begin. At the Kentucky Derby (right), "They're Off!" leaving handlers still standing at each stall. The starter and starting post are at right. Photo copyright 1988, the Louisville *Courier-Journal*. Reprinted with permission.

display of secondhand furniture. It persisted—a mecca for collectors, tourists, and popular cultists until its site was wiped out for a grandiose redevelopment called The Myriad in the 1980s.

Many an old auto ROW has moved, often at great expense, chasing customers as churches follow parishioners, from center city to suburb. There it reappeared as auto center or auto mall, still located along a main drag, but now occupying a quarter-mile frontage, and up to a hundred acres of display, parking, showcase buildings, and repair shops. Dealers in California paved the way in the sixties near Riverside, Cal., with night-lighted spreads, several dealers in one huge lot. By the 1990s, shoppers could choose from a dozen U.S., Japanese, and German makes in one center. The San Francisco Auto Center carried (1987) two dealers with fifteen lines from seven countries. At Ontario, Cal., eight dealers bought up available sites on forty-eight acres, the first phase of a ninety-six-acre auto center in 1986.[12] To boost its fading economy, Downey, Cal., assembled land, using its redevelopment power, for an auto center on its 3.2-mile Firestone Boulevard corridor.[13] In Albuquerque, a new (1989) auto center west of the Rio Grande River penetrates the regional skies with its bright floodlights. By the late 1980s more than half of all U.S. dealers were selling both domestic cars and imports, and doing so increasingly at new auto centers.

"Megadealers" the big newcomers were called—about 275 of the nation's 25,000 auto dealers. One of them, owned by giant Penske Corp. of Red Bank, N.J., had a center in El Monte, Cal., so big its five-thousand-foot "showroom" was wholly devoted to chairs and tables for closing sales—a computer for each salesman. Most of its 18,000-plus cars sold there in 1986 were parked—outside.

Back of such mobile scenes lies fear—small one-make dealers afraid that megadealers will dominate the business—and manufacturers afraid of losing control over "their dealers" no longer specializing in one make of vehicle.

Many ROWS become strung-out communities of interest. Most ROWS have their cast of characters equipped with egos, jealousies, rumors, and tall tales. Sometimes they cooperate with joint ads, maps, promotions, THE ROW then taking on some aspects of theater. In such places, what is being exhibited is an affinity that gathers likes to like, cheek-by-jowl with others in the same line of work, offering similar wares, working the same racket, racking up the same scores, scoring off the same marks, marking up the same merchandise, or, more simply, merchandising themselves along with their goods.

THE START

Precisely where it all began is a matter endlessly debated among historians and revisionists, participants in a street fight, courtroom witnesses, and patriotic zealots. Did World War I really start at the street assassination of the Archduke Ferdinand of Serbia and his wife in Sarajevo, capital of Bosnia-Herzegovina? Was not the true beginning of the American Civil War long before the bombardment of Fort Sumter, S.C.? To what point in time, to what place, will surviving historians refer when sorting

out the start of World War III? And where were you when you saw the defendant start shooting?

In the everyday world of sporting events, THE START is preordained and well-marked for marathons, parades, and other goings-on that use public rights-of-way as their venue. Often THE START is situated at a prominent civic location—in front of city hall, or the parking lot of a motel or business anxious for publicity. THE START of a protest march may be carefully pin-pointed at the door of a fallen martyr, or THE SCENE of a historic riot.

Once upon a time THE START of a race was a simple affair—perhaps a white string held taut by teammates at a high school hundred-yard dash. But gradually it grew complex, acquiring more space, official observers, stands for photographers, and special provisions for that minority of fans who prefer being there at THE START rather than at the finish line.

As beginnings became formalized they proliferated. The running or racing START at the Indy 500 on Memorial Day evolved into a complex ritual distinguished by "Gentlemen, start your engines!" Race cars are organized into a predetermined starting grid formation, all led by a pace car (donated for the publicity by its manufacturer and/or distributor), all crossing THE START in a pecking order that immediately turns into a free-for-all down the stretch.

A longitudinal mechanical device, THE STARTing gate was invented for high-ticket horse-racing. This complex and mobile contraption is wheeled into place at THE (STARTing) post athwart the racetrack, having a gated and padded enclosure for each horse. Action begins when the front gates of each slot snap open, the crowd shouts "They're off!" and horse-and-rider fling themselves into the pell-mell, racing for the first turn, the inside track, the lead position. THE STARTing gate is retired off-track until the next race.

TOWN CREEK

Variants: town branch, town run.

Seldom to be found surviving at ground level in any major city of the 1990s, most TOWN CREEKS have long since been shoved underground and paved over by settlers, speculators, and town developers anxious to "get this town out of the mud," and away from the banks of oft-flooding creeks. The same mind-set persists among civil engineers and developers who see all springs and streams as threats to their plans for high-and-dry streets, rather than as the protected nuclei for natural areas. One of the more celebrated is Minetta Brook or Creek, once a bucolic stream flowing off hills around Twenty-Third Street through lower Manhattan into the Hudson River. It was covered up during the nineteenth century, but its course is still identified by the slant wall of a house at 45 West Twelfth Street that conforms to the old brookside bank.[1] The Minetta's persistent flow deep underground was tapped by an apartment building west of Washington Square to feed an indoor fountain.[2] Beargrass Creek, which emptied into the Ohio River at downtown Louisville, Ky., was filled and rerouted a mile upstream, and now courses through much of the city in a concrete culvert. (A 1991 redevelopment plan for the waterfront sought to recapture the long-gone water's presence by a new recreational harbor around the former mouth of Beargrass Creek.) Denver's rambunctious Cherry Creek, rising in eastern suburbs, has been dealt a similar, concreted form through most of the city. Hangman's Creek in downtown Placerville, Cal., now courses under a new name, Placer Creek, in culverts beneath much of that town's central area. Town Branch and its commons, once the center of a Bluegrass pioneer settlement that became Lexington, Ky., had degenerated into a civic eyesore by the 1940s and was pushed underground by the 1970s, persisting but polluted where it surfaces blocks downstream. Silver Creek

TOWN CREEK
Originally, here was a bucolic, open, spring-fed stream valley, the center of a frontier settlement. Vine Street, now a major downtown artery in Lexington, Ky., follows the course of de-watered Water Street, a former railroad, and old Town Branch, the latter now culverted underground. An office boom of the 1980s flowed downhill to fill the valley and cover the remnants of the TOWN CREEK with a new thoroughfare.

through Charlotte, N.C., had become an open sewer by the 1960s, and portions put underground.

Other streams found stronger supporters. Strawberry Creek and its canyon through the University of California, Berkeley, campus and its thirty-acre Botanical Garden has staunch defenders including Faculty Club members who lunch along its redwooded, shady and protected banks. Jones Falls Valley, into downtown Baltimore, Md., has been intermittently protected and still has semi-preserved stretches. Boulder Canyon Creek through downtown Boulder, Colo., now (1992) gets guaranteed freshets of reservoir water to maintain trout in season and whitewater kayakers in mid-town. And Rock Creek Park through Washington, D.C., enjoys the assiduous protection of wealthy and powerful neighboring property owners. It remains a scenic asset to Georgetowners who resist plans to enlarge the park's roadway.

History buffs can often navigate through strange cities by finding old place names that reveal what once was, and has now been wiped away: Dock Streets blocks away from the filled-in waterfronts (as in Philadelphia); Mill Streets where once stood a grist mill, its stream now gone; as well as many a de-watered Spring, Lake, Front, and Pond Streets, their former waters visible only on old maps.

Not until the 1960s did TOWN CREEKS begin to get official recognition in urban renewal and other plans, as havens of refuge in overbuilt cities. In Texas, the downtown San Antonio River acquired protectors, cleaner and more permanent water sources, and a national reputation as a busy thoroughfare for sightseers' boats and promenades amidst convention hotels. The capital city of Austin, Tex., and its university, protected little Waller Creek by means of a Hike-and-Bike trail policed by student joggers—although too late to prevent the Lyndon B. Johnson Library from encroaching onto its banks.

See also PRESIDENTIAL SITE, in Chapter 3: Perks.

CHAPTER 2

Patches

City Centers and their environs are a patchwork of start-ups, projects half-finished, ventures under way, and others still holes-in-the-ground, lying dormant, then starting up anew. Each COMPLEX or SUPERBLOCK is conceived as a free-standing "signature" structure, with a much-touted corporate logo. Each is designed to separate itself from its neighbors, so far as the city plan can be altered to allow it to do so. If you know the local power structure, you understand how this visible patchwork came to be.

On every block, down every street, are expedient projects built on ruins of old dreams. THE VIEW constantly changes. Rarely does a block stay solid, seldom does a familiar SKYLINE last a generation. Some Patches expand and become more self-sufficient. Others act as spawning-grounds for new enterprises. In spite of City Beautiful plans of the 1930s that came and went, and the waxing-waning powers of federal-aid urban renewal; and beyond the reach of repeated efforts to create a memorable urban design, the old Center remains a patchwork quilt fashioned by many hands.

ARREST HOUSE

Could it be a sign of things to come—when "arrest" becomes an electronic condition, and a new breed of "The Man in/of the House" is electronically monitored by keepers who keep their distance?

The new ARREST HOUSE of the 1990s can be anyhouse. No signs visible from the street announce its constraints, no neighbors are visibly alerted to the news that a convicted felon, rapist, mugger, or drug dealer is now a probationer, a parolee, a detainee in residence, back home, just down the street. They would perhaps gain scant comfort from the fact that, in many communities, and under most state laws, ARREST HOUSE is available only to persons accused of nonviolent crimes.

Standing outside ARREST HOUSE, no passerby could or need easily know that the detainee inside may be electronically linked (via nonremovable bracelet or anklet) to a computer at Police Headquarters where his/her movements can be traced. Or that the detainee must report periodically by telephone to a distant keeper—his or her own personal voice-print electronically scanned for verification down at headquarters.

ARREST HOUSE was seen as a society's relief valve for overcrowded jails—a rented, nonpublic place to put prisoners, a structure that did not require a construction bond issued by the already over-bonded local fisc. It was an adjunct to other new expedients of the 1980s, those privately run jailhouses that operate under contract, some occupying disused hotels, apartments, sanitaria, hospitals, or tents.

Will we decant the denizens of high crime district into house arrest district? Will a strange hush descend over such neighborhoods where the man-on-the-street has become the man-in-the-house, an indoorsman no longer available for outside work and jobs, for errands and deliveries—house-bound, tied like an infant to the electronic bedpost? Will headquarters also monitor and record the detainee's every phone call, and to whom? How will the arrestee identify himself/herself at the door to unsuspecting visitors? How is one sprung from this not-so-big house? How is one's free state signaled or announced to the world?

Or will ARREST HOUSE be signposted? Will lights flash and alarms sound if the arrestee leaves the premises (furnishing a new market for gadgets of control)? Or is this knowledge accessible only DOWNTOWN in electronic files at police headquarters? Such questions abounded during ARREST HOUSE'S early years in the 1980s. All this costs less than building new prisons—but still money: Michigan's first three-year monitoring program, covering 1,300 prisoners, probationers, and parolees, cost $1.8 million.

Such new forms of surveillance area can be traced to New Mexico Judge Jack Love, who, in 1979, saw a comic in which the villain attached a gadget to Spiderman's wrist to track his goings and comings. Love persuaded a Honeywell corporate executive, Michael Gross, to develop on his own such a gadget. It was later sold to B.I., Inc., based in Boulder, Colo., "which now has about 50 percent of the monitoring contracts in the country."[1]

Was this only another step toward constant television monitoring of wired-up ARREST HOUSE inmates? Even though it reeks of Big Brotherhood, it's a step beyond Jeremy Bentham's eighteenth-century Synopticon. That too was the essence of rationality for its day: a circular jailhouse, all its prisoners stashed in pie-shaped wards, under constant surveillance by all-seeing wardens peering outward from the central hub.

Not without precedent, ARREST HOUSE has served historic personages. A one-time owner of famed Longfellow House, built in 1759 at 105 Brattle Street, Cambridge, Mass., was Andrew Craigie. After declaring bankruptcy, Craigie was confined for the rest of his life to house arrest—except for Sundays "when he had immunity from arrest." A Cambridge street bears his name; the *Pocket Guide to Harvard Square* (1973) tells the tale.

The word *arrest,* separate from house,

has come a long journey from its Roman origins. To arrest in itself means to halt, to prevent, to slow down, to hinder, to force to a standstill, to focus intention, to seize, to capture. But its later meaning implies the presence of police, and the imminence of jail and punishment, in this case being stuck in an ARREST HOUSE.

ARREST HOUSE is not to be confused with SAFE HOUSE, which is any house so innocently located, outfitted, used, and frequented as to harbor and protect secret agents, refugees, or criminals in hiding. They are alike only in their deliberate obscurity, which was dramatized by a craftily obscured SAFE HOUSE in John Le Carré's novel, *Tinker, Tailor, Soldier, Spy;* and the hard-to-reach Mafia house of *The Godfather,* shrewdly and unapproachably situated by author Mario Puzo.

See also TENT CITY, in Chapter 7: Power Vacuum.

CURBSIDE

Officials have long been telling the public what to do at CURBSIDE: from "Curb Your Dog," to the latest (1990) advisory to trash-pickers: "Steal a Can, Go to Jail." The curb or curbing itself makes a sudden vertical declaratory judgment in the form of an abrupt four-to-six-inch rise that marks the limit of the street and THE EDGE of the non-street realm. It also marks the final, irreducible, outer limits of most building sites, as distinct from the public right-of-way. As architectural critic Reyner Banham once observed of an intrusive building in Britain: "One foot closer to the road and that building will constitute a parking violation."

Along the way, CURBSIDE has become the omnium-gatherum of modern, mobile life, a transaction zone performing multiple duties. They all hang out at CURBSIDE: drive-by motorists picking up cash at the CURBSIDE bank, or negotiating with hitchhikers or hawkers; drug-dealers, or pimps and prostitutes haggling or flirting with mobile customers. Carhops serve up fast snacks; volunteers solicit contributions to local charities, pitchmen cajole motorists for gifts, lifts, or sales.

Here garbage/trash haulers do what needs to be done with things everybody throws away, notably on streets out front, but in older parts of cities, along the alleys out back. In some older neighborhoods, built-in garbage racks on the alley are the only spot where neighbors regularly confront/greet each other in predictable and generally neutral settings. One Midwest informant, veteran of a half-dozen SUBURBIAS, reports that "meeting around the dumpster" is the start of neighboring.

Most CURBSIDE gutters flow with trash and debris after heavy rains. But in Logan, Utah, street gutters glitter while doing double duty carrying irrigation water from the up-mountain side of town to the valley cropfields below. Many a CURBSIDE becomes the single-track walkway for broom-wielding cleaner-uppers and on designated days becomes a no-parking zone cleared for mechanical street-sweeping and garbage-hauling vehicles. Between CURBSIDE and trafficway is likely to be the newly white-lined route for bikers or joggers.

It was in the 1980s that CURBSIDE began

CURBSIDE

In response to impatient customers' demands for auto-accessible convenience, newspaper vendors (no longer human) cluster at this CURBSIDE in downtown San Antonio, Tex. Some attempts by cities to regulate such jam-packed curbsiding collide with freedom-of-the press defenders. Meanwhile, CURBSIDE pickups and deliveries of every sort are extending far into SUBURBIA.

its slow conversion into a legally defined outpost of municipal power to regulate waste-recycling. By legal fits and contentious starts it became the way-station between householders separating (often under protest) their trash into glass, plastic, paper, aluminum, and organic stuffs, and municipal collectors operating under varieties of duress, monopoly, or skills. Some cities claimed this cut the volume of land-filled garbage by 25 percent, meanwhile bringing revenue to the city.

Thus CURBSIDE became the focus of intense debate and lawsuits during a decades-long process of learning to recycle household wastes. As hundreds of towns and cities ran out of space to dump generalized garbage/trash, they began to force households and landlords to segregate "recyclables" into separate bins at CURBSIDE. These attracted mobile scavengers, who creamed-off whatever items—especially aluminum—would yield highest prices at resale, leaving city hall to collect stuff of less value. City halls responded by declaring trash left at CURBSIDE to be public property. Hence the warning: "Steal a Can, Go to Jail."

Legal questions dogged the rise of CURBSIDE. Private garbage haulers contended cities were—in the absence of just or other compensation—taking their property (i.e., the exclusive right to collect along the public way at CURBSIDE). Amateur scavengers, accustomed to treasure-hunting along CURBSIDES, were outraged to find their trove-sites put off-limits by city hall.

As in earlier battles over fluoridation of city water in the 1950s, civic garbage-intervention was attacked in the 1990s as dictatorial, communist, etc. In many a dispute over CURBSIDE, old politicians warned: "Don't try to pass the (curbsiding) ordinance in an election year!"

See also DRUG SCENE, in this chapter, below.

DISASTER AREA

Variant disaster site.

Here is where widespread nonmilitary destruction and/or death has just occurred. Communication is disrupted, electricity cut in its course if not at its source. There is wreckage and debris in the air, on the ground, deposited along stream banks or tornado path, and/or embedded to ill effect in the collective human body. Medical aid is in short supply if not immediately exhausted. If bridges are down and tunnels clogged, movement has been discouraged or blocked. Looting has begun—at first innocent pickers among public debris, then organized forays into rich precincts still unguarded.

DISASTER AREA (Quake Zone) Large-type poster circulated by The City and County of San Francisco in 1982, seven years before the 7.1-magnitude quake in October 1989 struck that region. It was centered around Loma Prieta, Cal. Despite repeated warnings, damage was widespread, especially to structures built with no quake-reinforcement.

Quickly enough, local police, National Guard, and U.S. Engineers are called in, rehabilitated in the public eye as the case may be, and now viewed as saviors, universal order-givers, helpers, and problem-solvers. Heroes unheralded emerge—assorted unknowns from all walks of life who, confronted with disaster, stay cool, give directions and first aid, help establish, and keep order. On occasion, old habits and attitudes are so disrupted that the DISASTER AREA is transformed. From it emerge new businesses, leaders, attitudes, goals. History also reminds us of other communities for whom disaster was, indeed, disastrous.

When disaster strikes, at the first quiver, blast or crush, the media flock to THE SCENE, themselves becoming, or supplying, major objects of public attention. In cities such as New York—always at the brink of disaster—crisis managers in large corporations have been rehearsing for this day of transition. Their game plans are on public display, their careers on the line at the next great storm or quake. They stash cots (ConEd keeps nine hundred in its Manhattan headquarters), equip company cars with two-way radio and first aid kits, and acquire first-priority warehouse and dormitory space. Other contingency-managers in the Manhattan financial district have been remoting their data bases to the suburbs, spending millions on back-up electronic systems for the next blackout. By 1988 the bulk of that city's computer power was "no longer in the city at all but in vast processing centers" in the suburbs or further away.[1]

Prior to the event itself, DISASTER AREA was believed by optimists to be a locale untouched by the law of averages except when an Act of God or other "natural disaster" occurs. Realists, on the other hand—always a tiny minority in such matters—have insisted that this particular locale of vulnerabilities will inevitably be THE SCENE of the next flood, hurricane, earthquake, or out-of-control fire.

We reserve special terms for events we wish would leave us alone. Natural and inevitable occurrences such as seasonal high waters we call "floods" when they do damage. Periodic and predictable droughts are said to cause "famines." Hurricanes turn into "disasters" only in the presence of human settlements. Any increase or spread of population into high-risk areas equals more "disasters." The average annual number of identified "disasters" increased worldwide from fifty-four in the 1960s to eighty-one in the 1970s. More than six times as many people died in disasters in the 1970s as in the 1960s, according to Earthscan, Washington, D.C.[2]

Scientists predict th
major earthquake c
sometime within th
s the time to prepare you
iome for such an event.
our responsibility to be
uake will tax local gov
esponse agencies, prob
y. It is imperative that
or such an event.

"Disaster" and similar terms are widely used to exonerate the human stupidity that fuels continuing overpopulation of high-risk locations.

DISASTER AREAS appear inevitably to accompany the continued swarming of people to such exposed sites as shorelines and barrier islands, or into floodways of major rivers and quake zones along major faults, as well as the settlement of semi-humid farmland subject to repeated drought. They also mark the arrival of "foreign" newsmen who perceive Disasters where natives may have seen only the inexorable and familiar working of Fate.

One of the key phenomena of the present generation is that it has watched disasters and DISASTER AREAS emerge from the closet to become the object of intense media attention (few events can command larger TV audiences than a major hurricane), as well as official and scholarly study. Since the first atomic explosions, the scope and reach of disaster sites, and of disaster studies, has expanded. Suddenly, to survivors and to spectators, the world itself seems to have turned malevolent. Disasters have become familiar demons.

Before this turn-of-mind, disasters were widely presumed to occur elsewhere. When they did occur Here, their sites were unlikely to be well marked or memorialized locally. Occasionally a historical marker reveals the high-water mark of, say, the Great Flood of 1900. But local boosters still work to ensure that historians do not dwell on it. Victims forget, or try to forget. They suffer versions of what once was called "shell shock," and exhibit guilt for having survived. They become anxious, lose sleep. "In a major catastrophe like the Mexican earthquake or the mud slides in Colombia, experts say, the psychological world collapsed just as resoundingly as the physical one."[3]

Meanwhile, as media-hype became a more standard TV practice in the 1980s, "Disaster" slid down the scale of events so that announcers could refer breathlessly to the site of a slightly out-of-the-ordinary wreck as DISASTER AREA. Gresham's Law for currency—bad money driving out the good—was clearly at work in the media. This downward mobility had long been aided and abetted by federal legislation which made the official declaration of a "disaster area" the signal for federal subsidies. Consequently, under pressure to get local relief, local high waters were redefined as "disasters" so that residents who had moved into known-to-be-floodable areas might obtain cash for their indiscretions. Mayors generally leap for the cash by claiming Disaster.

Eventually, THE SITE itself may be declared a historic spot—such as ground zero in Hiroshima, Japan, where the first atomic bomb struck; or the spot where the legendary "Mrs. O'Leary's cow" was said to have started the Chicago Fire of October 8, 1871, "one of the quintessential American disasters."[4] What happens after the great disaster may be The End of local greatness, or the start of a new era.

See also EMERGENCY CENTER, in this chapter, below.

DRUG SCENE

From far and near flow its ingredients: from cocoa plantations of Central America, from gang-plagued, low-income neighborhoods, from segregated blacks and other minorities with high unemployment and widespread despair; from distribution rings set up by criminal gangs; from white addicts in from the suburbs—all these working themselves into public view at the DRUG SCENE.

Usually it's a city sidewalk, alley, or street corner (crack corner), located in a populous, pockmarked old neighborhood where vacant apartments, storefronts, and shops can be converted for illegal drug sales and consumption, indoors or out. It's got easy access. Some dealers, pushers, and gangs paint on wall and sidewalk coded graphics, or graffiti, to advertise their turf. If raided, they reopen shop nearby to keep old customers juiced up.

DRUG SCENES offer many natural meeting points, as observed in West Spanish Harlem, New York City, by researcher-author Terry Williams: "'drug-copping zones' (drug-selling locations)" where one may see "'steerers' bringing customers to dealers, 'spotters' watching for and warning of police presence, and 'pitchers,' or street dealers, openly beckoning passersby to purchase their wares. Some buildings have been invaded by crack spots, and users sit on stoops and hang out in stairwells waiting to buy at all hours, every day."[1]

The SCENE is composed of fleeting encounters, moving targets, hit-and-run sellers cruising—afoot, in cars, or on bikes. There's the cashier, who takes cash in exchange for bits of paper with cryptic signals. The buyer then goes to a nearby storefront, or parked-vehicle narcotic outlet to exchange the chit for drugs. Its seller may keep a stash in nearby shrubs, flowerpots, garbage, or trashed apartments. Many sellers transact through holes in the walls in boarded-up apartment houses, commandeered and turned into flophouses, "rock houses," or crack complexes with crack pads, lookouts, and escape routes.

"This is also a talking neighborhood. The public phones are always in use. People stop to talk on street corners or in the middle of the block, carry on conversations out of apartment doorways, atop well-worn stoops, leaning from car windows."[2]

DRUG SCENE
Seldom do all ingredients of a DRUG SCENE array themselves for photographic purposes. Pip Pullen's sketch captures elements of the action: a backfiring drug bust in progress, a beat-up building (with multiple escape routes outback), a reclusive wino with a drink-in-a-sack, and a sidewalk mugging being carried out. Sketch by Pip Pullen.

U.S. Customs agents in Tucson, Ariz., worried in 1989 that, "a crackdown on 'Cocaine Alley' will just force smugglers to go somewhere else."[3] Violence is close at hand; specialists identified in 1989 the "drug-and-murder belt" of Washington, D.C.

The economic impact has not been insignificant. "Some 400,000 people live in the New Haven hinterland, and most of the drug users there tend to do their drug shopping in the city, so the cocaine boom has given New Haven a marked economic boost."[4]

Nor was the social cost less marked: "Most of the guys I hung out with are either in prison, dead, drug zombies, or nickle-and-dime street hustlers. Some are racing full-throttle toward self-destruction. . . . Many of my former running pals are insane . . . ," wrote reporter Nathan McCall in "Dispatches from a Dying Generation."[5]

Around the sales, indoors and out, neighbors may cringe, look fearfully the other way, move out, but sometimes organize to cut the flow. Some try to reduce drug sales by having their streets blocked to through-travel, or organizing via local churches to confront the gangs with Christian patience and presence.

Across the street at an upstairs screened window, police cameramen may be filming passing episodes, sometimes for months, to catch actionable sales, and produce recognizable identities in court. National and local governments have been spending billions to "get control." Far beyond the SCENE, in Central and South American farms, the raw ingredients have become the fuel that powers drug cartels with their own planes, police, and foreign policies. The motivating forces behind the SCENE in the 1990s seem to possess staying power far beyond what most people anticipated back in the 1970s. That was when "making THE SCENE" still implied all the innocence of teen-age youth merely looking for fun and friends.

EMERGENCY CENTER

Variants: disaster/emergency information and relief center or headquarters; first aid/civil defense station.

"It can't happen here." Against such ancient folk belief the promoters of EMERGENCY CENTERS fight uphill battles. In spite of the fact that the so-called hundred-year flood comes more often; that hurricanes strike and chemical plants explode, popular aversion to government planning hampers preparation for disasters. More often than not, funds to fight the next disaster must be concealed in official budgets. There is little patience for emergency planning to prepare for the ever-increasing impact of floods, fires, quakes, spills, and explosions in urbanized areas. It took the rise of domestic riots in the 1960s to give disaster planning new life.

As one result of the downgrading of emergency planning—and also to qualify for federal disaster funds—local "emergencies" are increasingly labeled "disasters." Training for emergency has been co-opted by a new sub-profession of crisis managers. In these regulatory times, the locale of a wreck, flood, storm, quake, etc., does not qualify for federal relief until it is officially and presidentially declared to be a DISASTER site or AREA.

Thousands of "Safe Places," Air Raid Shelters and Civil Defense Centers were designated and placarded during the Cold War, many of them serving as EMERGENCY CENTERS. But by the 1990s they were gradually disappearing, as prospects of a nuclear war receded (temporarily?) into the background.

Meanwhile, as local radio and TV coverage grew more competitive in the late twentieth century, "disasters" became an important stock-in-trade for editors eager to improve their ratings. In the absence of war, riot, famine, or such cataclysm, any local multi-car wreck, large fire, or a multiple murder could be labeled a "disaster" by local media. Thus acquired was a repertoire in which small collisions could be upgraded to front-page/prime-time material to attract wide audiences.

When incidents do approach true disaster scale, the local mayor or county executive applies to the Federal Emergency Management Agency, Washington, D.C., for a "declaration"—i.e., a presidential declaration that this place is now formally a DISASTER AREA.

At once, a train of events is set in motion. Headquarters is designated, often at city hall or the central police station. In some emergencies, a special trailer, station wagon, tent, kiosk, or other impromptu shelter can be set up, usually on-site, to dispense facts and dispel rumors that inevitably accompany disasters. Since the mass media play such a central role in covering (if not creating) disasters, James L. Holton, special consultant to the Federal Emergency Management Agency, Emmetsburg, Md., advocated that a formally designated disaster site information center be set up to accommodate hordes of media reporters and cameramen converging to cover the aftermath of such natural events. Such a media center "could be anywhere—a motel, school gymnasium, city hall auditorium, even under a large tent in an open field."[1]

A large motel, with plenty of meeting rooms, would be an ideal location, provided it were intact and not needed to temporarily shelter disaster victims. Wars in Vietnam and the Persian Gulf trained both national and local troops in procedures adaptable to domestic emergencies. Thus current EMERGENCY CENTERS take on military trappings and routines. Such a CENTER can be airlifted, as need be, in the form of a prefabricated, mobile command post, a TENT CITY, or headquarters, to such places as WRECK SITE, fire, crime or riot scene, or DISASTER AREA.

HUB

HUB predates the invention of the wheel, which its name implies, for it is a response to early human calculations. As a device for strategic management-at-a-distance, HUB served from ancient times as a gathering place based on the minimum time-and-energy required to get there from someplace else.

It was the Roman colonial port; it was the ancient Danish council-ring of stones on which elders came together at a central place to decide about next year's crops. It was Boston, the Hub City, the first colonial port encountered by ships from the mother country England. When groups decide to foregather, deciding on the HUB comes quickly next.

From the HUB radiated early settlers' paths, pioneer roads, and later highways. Often they bore the names of those remote settlements at a day's hard plodding or trekking through the forests. Such DESTINATIONS gave their names to roads spreading like spokes from the HUB. And thus they remain today—easy signals to travelers who might otherwise be confused or lost.

Not at all common in everyday U.S. speech, HUB swelled in utility and popularity after the U.S. government deregulated the aircraft industry routes, flights, and fares in stages from 1978 to 1981. Before hubbing became the rage, "there was Delta in Atlanta, United in Chicago,

Continental in Denver, but not much more," observed George James, president of Airline Economics, Inc., Washington, D.C.[1]

Airlines quickly responded to the new geography of un-regulation. An airline could offer five hundred different route patterns, using only thirty planes, and changing passengers at one central HUB. By 1987, forty new HUBS had been created. For the next decade routes expanded via new HUBS, and passenger fares per mile dropped in constant dollars. It turned out that long-distance travelers especially benefited: the longer the distance, the greater the choice of HUBS. Brookings Institution calculated that by 1986 travelers were paying less, but service was uneven. "The big airlines moved out, the weed-eaters moved in," complained a Congressman from Missouri. "Now, if you want to go [from D.C.] to West Virginia to attend a dinner, you've got to give two days to it," lamented West Virginia Senator Robert C. Byrd.[2] Many small towns were left unserved, passengers complaining that to get There from Here you have to go to a distant Someplace Else first.

But HUB airports boomed. Many airline owners built or demanded new terminals, runways—or entire airports, as in Denver. HUB cities—Cincinnati with Delta, Charlotte with Piedmont, Baltimore, Dayton—all bragged of their accessibility compared with non-HUB rivals.

Creating a HUB turned upon a mixture of corporate strategy worked out in secret, reinforced by a city's willingness to expand its airport. Often a city woke up one morning facing an airline wanting to hub, thus filling the local airport's unused capacity. "A hub can mean that one carrier virtually 'owns' a city," observed *Condé Nast's Traveler.*[3]

Further, HUBS exaggerated peak-loading. Planes arrived in platoons, passengers disgorged by the hundreds, making necessary split-second connections, and bar-coded baggage-handling. The new HUB terminals had wider walkways for hurrying passengers toting their own non-checked bags. Stand-up snack counters proliferated.

Some competing airlines used HUBS as a competitive club; at O'Hare-Chicago they scheduled thirty-six planes to arrive at the popular time of 9:15 A.M., a patent impossibility. And when a major HUB became fog-bound, everything stalled; inbound traffic backed up across the continent, even overseas. Thousands of travelers "camped out" in the Atlanta and other Southeast airports when a two-day fog stalled traffic in the 1986 post-Christmas rush. In spite of widely adopted bar-codes to mark baggage, lost baggage increased.

As the term HUB moved more easily through the language, Augusta, Ga., was described (1978) as "the declining hub of a growing metro area of 375,000 residents."[4] The South Bronx (N.Y.) Hub was described locally as "a dynamic shopping area."[5]

Following deregulation, more than two hundred new airlines started up. Most were gone by the 1990s—quit or merged. Bankruptcies, takeovers, strikes intervened in the flow. As the first decade of deregulation ended, the air lanes were more and more dominated by a few huge carriers. Ambitious mayors, praising their own new runway, claimed it would usher in the fourth great transportation era. First there were ships, then railroads, autos, and now planes. HUB airports expanded, generating more noise and ancillary businesses; one multi-line repair center could create four thousand new jobs, the key to a new GROWTH AREA.

See also EARSHOT, in Chapter 4: Ephemera.

LIGHTING DISTRICT

THE LIGHT, it has been long assumed, will dispel THE DARK, the blues, depression, sorrow, and criminals. Shine enough light into dark corners of the mind or the city, and ignorance, along with criminality, will disappear.

Decriminalization by light is an old civic practice. In late eighteenth-century Paris, "By day, 1,500 uniformed police were on the streets. By night, 3,500 lanterns achieved the same result," records historian Wolfgang Schivelbusch.[1] "District lighting" was practiced from its U.S. beginnings in the 1880s when firms ran electric wires or conduits from hydro-generating plants to nearby businesses, a block or so at a time. (Gas companies fought this "invasion" of their turf.) If you were close, you got the light. And artificial lighting has expanded its ability to segregate neighbors, streets, corners, and whole districts by variations of light—its presence, intensity, color, and design. Outdoor lighting design has become a professional specialty.

"All lit up" is the phrase around Christmas holidays to describe a DOWNTOWN festooned with extra lights to brighten up the city. Such lights come in strands, strings, wrappings, spots, overhead, and underfoot in all colors. They flood, they spot, they distinguish, and segregate. They arrive usually after Thanksgiving, coincident with heavy advertising, and local promotion of what is called The Christmas [Shopping] Season. They arouse strong emotions: strident voices protest the "commercialization of Christmas" with artificial lights. History buffs recall medieval routs and riots with heavy drinks all 'round. Christians resurrect the Nativity scene and the three wise men for floodlit front yards and doorways, while dimly lit horrors and other macabre tableaux from the Middle Ages have migrated to Hallowe'en, when they flourish as traipsing children feed on goodie handouts.

At considerable cost, many a downtown committee pressures its merchant members to pay for Christmas display as "good for business." Oglebay Park at Wheeling, W.Va., attracted 1.3 million visitors (1990) to its winter Festival of Light beginning in early November. It became a model (offering how-to pamphlets) for a dozen other cities. Richard Bosch, noted lighting designer, has traveled from his well-lit hometown of Eindhoven, Holland, to prescribe city and festival light-ups the world around.

"District lighting" is offered by utility companies as a painless but not cheap antidote to slumping downtown retail sales (and to offset crime in many neighborhoods). By the 1990s LIGHTING DISTRICT was beginning to appear on official city maps. Merchants and utility firms often share the expense. Many self-conscious neighborhoods and small cities set themselves apart from their surroundings by "lighting effects."

In the suburbs and older residential districts, many an unsuspecting suburban family finds itself having moved, willy-nilly, into a Christmas LIGHTING DISTRICT in which all homeowners are expected to drape their houses, outbuildings, fences, trees, and shrubs in festoons of approved and not-too-gaudy accessories. In such places, spick-and-span upkeep, as well as high luminosity, is expected by one's peers. A result of such conversion is to generate traffic jams, attract surprise visitors and photographers from local media and light purveyors, all in the "spirit of Christmas." As often as not this turns into a competitive display of expensive seasonal merchandise. Such displays are generally said to be "purely voluntary," although, the timely LIGHT may be accompanied by not-so-neighborly coercion.

MIXED/MULTI-USE COMPLEX

Variants: megastructure, SUPERBLOCK, megamall.

Deeply mysterious, formerly beyond the realm of logical familiarity, now released from the bounds of formal expectations, COMPLEX has become a geographic oddball. It is a generalized term referring to any collection, aggregation, or organized cluster of building structures. The implied adjective is: "big," followed by housing COMPLEX, shopping COMPLEX, or such. It has come to suggest size and complexity greater than that of a mere center or a project. Long since arisen from the psychoanalyst's couch, no longer exclusively descriptive of a mental state, and no longer simply adjectival, COMPLEX has become an upwardly and outwardly mobile rendezvous for large-scale enterprisers. Along the way, its vagueness expands.

Once COMPLEX had entered THE SCENE, it became a BATTLEGROUND. On the one hand, architects and urban designers saw a skyscraper as an individual expression of their work—A Statement. On the other were those who saw it as part of an intellectual system. The battle had begun in pre-COMPLEX times of the 1960s when SUPERBLOCK was all the rage among urban redevelopers and city halls, all of them anxious to get new, bigger structures to replace what officially could be labeled slums.

Hither and yon, cities had begun adopting new urban design plans, often using upper-level skyways and, in colder climates, underground concourses to link downtown blocks into what could be renamed SUPERBLOCKS or COMPLEXES. The word COMPLEX itself came to be used by ecologists and public health experts as well as psychiatrists. The former spoke of the Malaria Complex—that array of climate, geographic conditions, and insects necessary to bring about a malaria epidemic among humans; or the food-chain complex, combining seed suppliers, cropfields, animals, cultivators, distributors, cooks, consumers, and garbage handlers. The sculptor Jean Tinguely's *Study No. 2 for an End of the World* was described by Guy Brett in *Kinetic Art* as a "self destroying machine complex."[1]

As their scope broadened, SUPERBLOCK and megastructure became components of this newer, larger thing called MULTI-USE project or COMPLEX. It borrows from Utopian thought, from skyscraper ideologies of the 1920s, and from thirty years of federal-assisted urban renewal, coupled with aggressive international capital-formation.

COMPLEXES have become awesome if not confusing presences across many a SKYLINE. One of the key projects in the rebuilding of Lowertown District, St. Paul, Minn., was the $100 million Galtier Plaza, opened in 1985 and described officially as a mixed-use square-block project, probably one of the last ventures of this size to still be called a project. In 1988 Indianapolis began its Circle Center COMPLEX southwest of Memorial Circle. This is a 3.5-block aggregation of old buildings, cleared sites, and underground as well as air rights for upwards of one million square feet of new retail and office spaces. All are to connect by skyways, a street-level mall, and lower-level concourses and parking garages. But when the year of its scheduled completion arrived in 1993, it still consisted of vast holes in the ground.

Putting a COMPLEX together is a complex process. It fuses long-term land leases, titles, and air rights; packages mortgages; allocates future rents; sorts out credits, creditors, and tenants; suborns or entices politicians; and often incorporates existing historic structures. All this may involve shotgun marriages of offices and apartments where none had coexisted.[2] It also requires new forms of joint-venture enterprise, pioneered forty years earlier by famed William J. Zeckendorf, Sr., in assembling the United Nations' site in New York City. These schemes often depended on big-capital sources: bank partners, hotel chains, and international consortia of speculators; and on convoluted sale-and-lease-back deals. Fluid capital played fast and loose with local sites and markets. A SUPERBLOCK, hurriedly thrown onto the crest of a building boom, is a notorious candidate for bankruptcy. A COMPLEX, if shrewdly planned for conversion and flexible multiple-uses, might survive and become a key element in the future city.

But not without subsidies. The typical large downtown COMPLEX would require city, county, state, and often federal subsidies to entice private capital into a risky venue. This was an ironic twist—new subsidies needed to offset the urban decentralization away from DOWNTOWN that was a direct result of subsidized suburban mortgages, and of federal interstate highways built during the prior forty years.

No longer was this a game for small investors. The way had been paved by Rockefeller Center, and expanded in 1957 when owners of the forty-one-story Pierre Hotel built a twenty-two-story office building next door in Manhattan, linked by a common concourse "reportedly for the first time in New York." (But, as waves of American tourists were discovering, the great Gallerias of Milan, Naples, and other European cities had done precisely that a century earlier.)

By the 1990s such linkages were not only common in the United States, but were essential to the flow of customers and capital between blocks and SUPERBLOCKS. And they often depended on friendships, complex contingencies, and risky futures. Thus a business journal urged that, "all elements of the downtown hospitality complex must work together" for success.[3]

In California, the new town of Irvine's Business COMPLEX, aimed at fifteen million square feet of new office and commercial space, was under way (1986) in various forms of predictability, its owners intending it to become the central business district of Orange County. In Los Angeles, a $120 million office and service center (i.e.

MIXED/MULTI-USE COMPLEX

When decayed parts of St. Paul, Minn., on the Mississippi River were rebuilt starting in the 1960s, the term **MULTI-USE COMPLEX** had hardly come into vogue. But by 1986 Lowertown had a name, a corporate sponsor, and a new **MULTI-USE COMPLEX** called Galtier Plaza, upper left, with housing, retail ("95 percent occupied" in 1992), offices, cinema, entertainment, a business college, a YMCA. By 1993, Lowertown was playing multiple roles as a 180-acre "urban village," having added some three thousand dwelling units and about five thousand jobs, and having rebuilt Mears Park (upper left center). Sketch courtesy of Lowertown Redevelopment Corp., St. Paul, Minn.

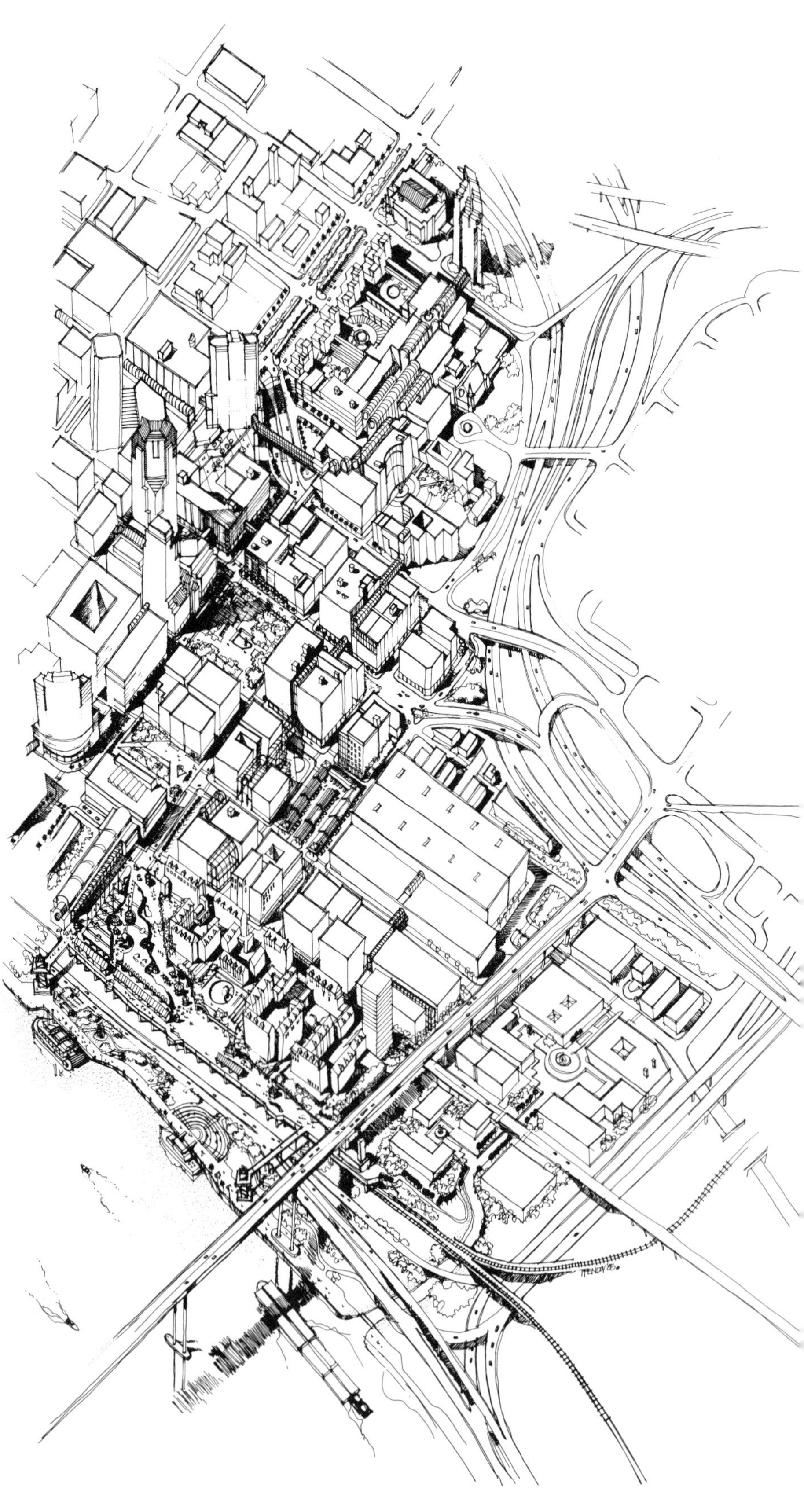

COMPLEX) was being built in 1988. Its third phase, "nicknamed 'Superblock' by its developers," occupied eighteen of forty-five acres formerly used by heavy industry.

By the 1980s the COMPLEX was expanding its clientele. In Atlanta, Ga., it was announced that businessmen in suburban Clayton County intended to capitalize on the fiftieth anniversary of the film *Gone with The Wind* and its mythical cotton plantation, Tara (which had actually been built on a Hollywood movie set). And how would they do so? By creating a "Gone with the Wind museum, park, and civic complex" near the fictional site of make-believe Tara Plantation south of Atlanta.[4] This was sneeringly put down by a local critic as "Six Flags over Tara"—a play on the phenomenon of a widely-imitated Texas theme park with overhead flags. That same week, a lithe and lissome young athletic director announced to a TV audience that she intended to go into business for herself with a venture called Exercise COMPLEX.

In a tizzy of capitalized complexities, the *Miami Herald* announced that, "The Art in Public Places Trust has set out to commission a major work of art as part of the new Government Center complex." Its three-block park would be the centerpiece of "this developing complex of city, state, and county buildings.... Government Center will be the main passenger transfer point for Metrorail and the Metromover"—all the foregoing further labeled as "a hub of governmental, commercial and recreational activity, and as a crossroads for people downtown."[5]

Seeking to capitalize on the Scandinavian generic term for DOWNTOWN, a project called Centrum opened in 1985 in Orange County, Cal.—not even close to DOWNTOWN, but at the HUB of four freeways.

Many COMPLEXES raised their new towers amid old DOWNTOWNS. Often they exploited AIR RIGHTS over expressways, using state bond issues, and local subsidies: e.g., Cobo Hall, Detroit; Copley Place and Hotel complex, Back Bay Center Boston, et al. Terms like "multi-million-dollar high-tech commercial and academic research complex" began to roll off the tongues and onto press releases of promoters in the 1980s. In Santa Monica, Cal., Southmark Pacific Corp. was busy topping out the first towers of its "$550 million multi-use complex" in the city's special office district. Los Angeles County in 1986 solicited bids for an expanded Civic Center COMPLEX DOWNTOWN.

Others got up and moved their COMPLEXES away from DOWNTOWN, into suburban and far-out locations offering big, empty sites. Even in depressed Dallas, Tex., in 1986, the rich owners of Southland Corporation (of "7-Eleven" drive-in fame) began construction of Cityplace, "a $1 billion complex of office and commercial buildings on a 130-acre North Dallas site."[6] Moving off-site to greener pasture, a Texas investment group bought eighty-five suburban acres in Jefferson County, Ky., in 1985 for a "$48 million retail, housing and office complex" near major interchanges.[7] By 1990 it was bankrupt. The Princeton, N.J., Forrestal Center (research, etc.) formed a MULTI-USE COMPLEX expected to have work force of 25,000 by the year 2000.

News writers have seized on all this terminology in an effort to dignify if not to upscale the local ventures they report. Thus William E. Schmidt, in the *New York Times*, described the Gwinnett County seat,

Lawrenceville, thirty-two miles from Atlanta, Ga., as a town where 40 percent of the residents "work in Gwinnett's new manufacturing plants and office complexes."[8]

COMPLEX moved into the lingo of public-housing managers in the 1980s, partly in response to public scorn of permissive management, and high-crime projects which came to be viewed as seedbeds of delinquency and drug-dealing. Within a decade, housing officials were shunning the word "projects" in favor of "developments" and later "housing communities." *A Glossary of Housing Terms* issued in 1937 at the height of New Deal housing ventures, included a long definition of project but no mention of COMPLEX.[9] But by 1990 the term housing COMPLEX was well ensconced in official lingo.

Along the way, the term COMPLEX found other uses: it was down-sized by private developers to fit smaller projects. These included a Holiday Inn of a mere 264 rooms plus five-story office building and atrium for a Chicago suburb; a scatter of Indiana state parks, forests, and caves called the Harrison-Crawford-Wyandotte COMPLEX; and a speculative Arabian Horse-Park-COMPLEX assembled and later bankrupted in Kentucky. For comparatively peanuts, a "Japan Firm Buys Aurora Complex for $13 Million," headlined the Denver *Rocky Mountain News*.[10] In the San Fernando Valley, Cal., a seven-acre venture was described as "the largest courtyard office complex" in the Valley. The term "courtyard" was used here, and by the Marriott Hotel chain in the 1980s as a homey touch to distinguish their new low-rise clusters from high-rising others.

While many a new COMPLEX boasts of its apartments, none quite matches the dream of Russian author Nikolai Chernyshevsky, whose 1863 novel, *What Is to Be Done?* raised a vision of expanding London's Crystal Palace. Each of its envisioned megastructures would contain apartments, industrial workshops, common eating, and recreational facilities including ballrooms for thousands of people.[11]

Nor is the COMPLEX end in sight. Huge public-private ventures, such as the Pick-Sloan Missouri River Project of the 1960s, could now be called, in the new fashion, "a flood control, river navigation, irrigation and electricity-generating complex" along the Upper Missouri River.[12] In the end, the widespreading use of COMPLEX surely derives in part from its applicability to a myriad of Opportunity Sites—but also from its neat fit into a one-column news headline.

Finally, the essence of a COMPLEX at The Center is control. Anything can go wrong in such a large-scale venture. Split-second timing is essential. Fire alarms and smoke sensors must work, emergency doors spring open, elevators not bobble or jounce. Full control of a COMPLEX depends on contracts, deals, connections, pressure, manipulations—on thousands of repeated, continuous human habits, contacts, and customs. It is an emerging microcosm of The Center itself.

See also SUPERBLOCK, in this chapter, below; and DEPRESSED AREA/REGION, in Chapter 7: Power Vacuum.

NO SMOKING AREA

It probably began after most Americans had acquired fairly predictable freedom from early violent death. By the prosperous 1960s most people could count on living into their seventies. Yet within the national euphoria, these were also stressful years. The nicotine in smokes seemed to have unique properties that made it "perfect" for coping with stress. Smoking was calming, and appeared to produce feelings of reward and well-being.

But suddenly other results of smoking began to penetrate the nirvana of those years of prosperity. Smoking was found to be a predictable, provable cause of cancer, heart disease, and early death. It became public knowledge that smokers' heartbeats increased by 10,000 to 20,000 over those of nonsmokers. The NO SMOKING AREA came on-scene fast when it became known by 1987 that smoking caused some 350,000 deaths per year. NO SMOKING AREA became a GROWTH AREA.

Such a prohibitory place as NO SMOKING AREA had deep roots. Early New England's blue laws from the 1600s penalized public smoking. Prohibition (of alcoholic drinks) in the United States had a side effect: between 1920 and 1930 several states prohibited the sale of tobacco as well. But tobacco moved into more pockets when cigarettes were packed into military K-Rations during World War II. Per capita consumption increased. American troops became tax-paid salesmen for tobacco in strange lands.

Smoking's decline of the late twentieth century began after the U.S. Surgeon General, following detailed surveys, published on January 11, 1964, the first general warning that "smoking is hazardous to your health." Later federal law required this to be appended to all ads aimed at tobacco smokers or prospects. Smoking in the United States began its decline in 1982, after a year in which U.S. smokers consumed 640 billion cigarettes. But not without many a struggle. As a headline writer noted, "Where there's smoke, there's ire."[1]

By the 1980s, under pressure from U.S. Surgeon General C. Everett Koop, who advocated "a smoke-free society by the year 2000," NO SMOKING AREAS had moved decisively into the public arena. Koop was joined by major medical and science associations in approving controls. More than two-thirds of Americans told pollsters they agreed. Airlines were for it, since they would not have to segregate smoking from nonsmoking passengers. The Congress at first banned smoking on long flights, then cut the permissible times, and finally (October 16, 1987) prohibited smoking on all flights within the forty-eight contiguous states. The Canadian government prohibited smoking on all flights as of July 1, 1990.

NO SMOKING AREAS proliferated, penetrating elevators and other confined places; shortly, waiting rooms, buses, and other conveyances were divided for smokers and nonsmokers. The U.S. Army, heavy with smokers, in 1986 published its own no smoking rules.

Smoking was fast becoming "deviant behavior" that was linked to death from cancer. But it also aroused others to reinvent deviance. Some restaurant owners confronted a 1987 no-smoking ban in Hollywood, Cal., by redefining their restaurants as "bars," to which the ordinance did not apply. A few "spent thousands of dollars on expensive liquor licenses to accomplish that."[2] Private office managers began segregating smokers, many of whom turned to the solace of their grandparents, snuff and chewing tobacco, thus boosting the sale of spittoons as a part of office decor and sanitation. Capitol Hill correspondents split over the issue in 1988; their Standing Committee voted to use smokeless ashtrays in the Press Galleries. (This amounted to "a smoke-at-your-desk-but-don't-walk-and-puff policy.")[3] But the ritzy Ritz-Carlton hotel chain introduced well-ventilated smokers' lounges in 1984, intending them for all its hotels thereafter.

Far from NO SMOKING AREAS, gloom descended upon the owners of tobacco patch and of cigarette factory, whose products faced further restrictions and competition for consumption territory. The six tobacco-growing states maintained the weakest no-smoking laws, according to a surgeon general's report in 1989. Their congressmen fought off attempts to cut out federal price supports for raising tobacco —a longtime subsidy. There was fuming and there was fussing in tobacco regions: Kentucky's legislators from tobacco-raising counties threatened in 1988 to cut funds for the University of Louisville for a no-smoking rule for common areas (stairwells, hallways, and reception areas) on its campus. Tobacco growers at Owensboro, Ky., said they might shift their tractor-buying habits after Ford Motor Company restricted smoking in its workplaces.[4]

The debate extended inevitably to the periphery, or the "sidestream" of smoking, where "environmental smoke" or ambient smoke was found to cause lung cancer among nonsmokers. Critics of smoking extended their attack to the smokers' environment. They claimed that a smoker's effluvium (exhaled smoke as well as "sidestream" smoke curling up from a cigarette, cigar, or pipe between puffs) could contaminate an entire room, plane, or airconditioning system.

On a world scale, smoking in 1986 was increasing 2.1 percent faster than world population. In eight of the more-developed countries, including the United States, cigarette consumption began dropping. But U.S. manufacturers stepped up exports to less-developed nations, notably China, that were acquiring Western goods, tensions, and habits.

In China, cigarette use grew 9 percent annually after 1983. U.S. cigarettes' share of Japan's smoke increased from 2 to 10 percent prior to 1988. World Watch, a U.S.-based environmental group, published warnings that "the worldwide smoking epidemic has taken a sudden turn for the worse."[5]

Along the way, the debate became more political. Private rights were threatened by the bans. A Smokers' Rights Alliance was formed at Mesa, Ariz., including 20 percent nonsmokers among the group's 750 smokers. Most newspaper editorials supported some form of smoking controls, but resisted bans on tobacco advertising. Social observers recorded the myriad social uses of smoking—as a conversation-opener and proximity device; for the display of physical moves and aroused sexual interest; as well as mere politesse and national or ethnic customs. What, they asked, is a social replacement for smoking? Whoever could come up with a nonthreatening answer might become a billionaire.

See also DOWNWIND, in Chapter 9: The Limits.

PARTY STREET

No two places going by the same name could be so different. One PARTY STREET turns into a RIOT SCENE; another PARTY STREET is a quiet neighborly barbecue and get-together, where horseshoe games and clangs are the big noisemaker. One starts mean and turns ugly; the other may be preceded by invitations, accompanied by gentility, and end up on the appointed hour.

One has its roots in medieval street frolics and semi-barbarous routs; the other in a tea party. One has its strong currents fueled by liquor and hostility "opposing currents that drink, fight, cut, scream, and growl" while summoned police try to keep it under control. The other is self-consciously set up "for the kids," meaning children-under-twelve accompanied by parents also on good behavior, oozing good manners. One is still selling beer at 3 A.M., legal or not; the other shuts down by dark.

At one PARTY STREET turned ugly in 1990, a spectator wrote, "Thousands of people roamed the street. Some like us, were there out of curiosity, but most were drunk (many underage), fighting, urinating on the walls, sexually engaged or worse! . . . we heard every sexual proposition known to man."[1]

One visiting columnist wrote, "Beer is everywhere. It is in kegs being moved in shopping carts, in coolers stacked on Radio Flyers. More noticeable, though, are the noises. Shrill shrieks that come in bursts, then gain momentum and travel in waves through the crowd. Breaking glass. Broken glass being kicked. Beer cans being kicked. And the wet thud of half-full beer cans landing on pavement."[2]

How two such social occasions can turn in opposite directions is largely worked out ahead of time. The rumptuopus, boisterous, dangerous PARTY STREET is likely to be adjacent to a public racetrack, fairgrounds, or loosely managed EVENT/FESTIVAL SITE; located in a wet precinct adjacent to dry territory; and in a neighborhood or community noted, for historic reasons, as a rough show, for unruly crowds, for its volcanic passion for violence. Nobody asks permission to close the street. It just gets blocked, and trespassing motorists get rocked. Sometimes it turns rough in response to cruising toughs "from away," spoiling to fight.

Such tendencies get known through local grapevines; they seldom penetrate local media; and when they do so, are treated as isolated, unpatterned occurrences—unless they get wholly out of hand, as in South-Central Los Angeles in May 1992.

The "nice" version reflects middle-class or upperclass values: muted talk, discreet flirting and drinking, the street blocked and signposted ahead of time with an OK from the local traffic department. Sometimes it coincides with neighborhood yard sales. And sometimes it is an effort by middle-class families to reclaim the street for a day from drug dealers.

The official answer to street disorder is to close streets whereupon the disorderly—especially teen-age cruisers—argue that there's "no place else to go." Many turn, in desperation or anger, to shopping mall parking lots, whose owners tend to call police first, and negotiate later if at all.

Rising to the occasions, a third type of PARTY STREET came into widespread use beginning in the 1970s, under promptings from Olde Timers, or Save Downtown groups. Under the banner of celebrations, festivals, strassenfests, fiestas, commemorations, they closed off the old main drag, or historic routes: Bourbon Street, New Orleans; Oliveras Street, Los Angeles; Oregon Street, Dayton, Ohio; around Pioneer Square, Seattle; Main Street, Louisville. Highly organized and well-policed, these enjoyed silent or open backing of city halls, police, and downtown retail and property groups. Historic preservationists pitched in with costumed paraders, competing musicians, and assorted hoop-de-do. Games, dancing, music took over Main Drag. Some chose Christmas week, some New Year's Eve; Dayton's biggest is on Halloween.

Another variation erupts when middle-to-upperclass families throw a money-raising bash for a Good Cause during a big-event weekend: Mardi Gras, Dogwood/Azalea Festival, Race Day, or the day of the Big Game. All goes well until the partying spills out of house and yard to block traffic, and lasts till 2 A.M. At that point teetotaling neighbors call police to control the public safety hazard.

Catching public attention in the 1990s was one of the nation's more noted PARTY STREETS: M Street in the Georgetown neighborhood of Washington, D.C. Observed *Washington Post* staff writer Gabriel Escobar: "This is one of the city's prime business districts, one of its main tourist attractions and a nationally known place to party." It is also "a place where street bullies look for fights along the business strip." Police said the local population of some 15,000 persons and 1,000 businesses "is dwarfed on weekend nights by 20,000 to 30,000 visitors." As the *Post*'s subheadline noted: "Night Rowdiness Stirs Fight over Georgetown's Future."[3]

RIOTSVILLE/RIOT CITY U.S.A./RIOT SCENE

These omnibus designations, free-floating in the journalism of the 1960s, came to designate any community or locale plagued or distinguished by frequent or large or spectacular or especially damaging riots, then arising from protests over the Vietnam War or from racial antagonisms, or both. It became a loose and freely used descriptor in conversation, but less likely to be found in local chamber-of-commerce publications, unless directed at a competing city. Often it was down-sized in local media reports to "troubled neighborhood" or similar evasions.

A make-believe RIOTSVILLE, U.S.A., had been set up at Fort Belvoir, Va., in the 1960s. According to the Associated Press, the U.S. Army "demonstrated its latest riot control tactics and equipment. The setting was Riotsville, U.S.A., a mockup of a city area swept by disorder. While about 3,000 persons observed from bleachers, a Riotsville mob made up of soldiers dressed as hippies set fire to buildings, overturned two cars, and looted stores. Then, with bayonets fixed, troops wearing black rubber gas masks arrived on the scene and controlled the 'mob' with tear gas."[1]

RIOT CITY has its less-violent counterparts—the suffix "city" works overtime in modern usage to expand, exaggerate, or dignify a local product (as in "artichoke city, quilt city, tobacco city") or a non-riotous event (as in "festival city"). These share time and space with CAPITALS of wide variety, such as the Spinach CAPITAL, etc. As thus used in many media, city and CAPITAL are interchangeable.

A decided shift in public descriptions of

rioting and of RIOT CITY came in 1992 following the televised police beating of Los Angeles black citizen Rodney King. It set off three days of widespread riot, arson, and looting in the south-central city. More than two thousand persons were injured, forty-four killed, and many city blocks burned out. Far more explicitly than in the past, national and local television concentrated on riot actions and specific locations. No longer did newspapers bury "unrest" news on back pages. Endlessly described in detail were RIOT SCENES here and elsewhere. It was unlikely that this generic locale would soon be wiped off the mental maps of millions of TV viewers. RIOT SCENE had exploded its way into the national conscience. Some would eventually make it onto local maps, attain historic designations, and become objects of pilgrimage and visitation. How soon the causes would be dissipated is another matter.

See also THE CAPITAL, in Chapter 1: Back There.

SAFE HOUSE

So undistinguished, unmemorable, and psychically invisible in and of itself is SAFE HOUSE, so perfectly camouflaged by its lack of any notable features whatsoever, that it is difficult to find, hard to remember, and lacking in telltale signs that stand out in the eyes and memories of passersby. It borrows no distinguishing or notable features from its surroundings. If anything, it repeats them to the edge of monotony. Into them it merges, it melts, it disappears. One uses terms like run-of-the-mill, a-dime-a-dozen, look-alikes, or invisible to describe SAFE HOUSE and its setting. It abides by the fixed rule among secret agents, refugees, and escapees: never call attention to yourself. Avoid eye contact!

The lessons of invisibility inherent in THE SETTING rub off on people who use a SAFE HOUSE: spies, intelligence couriers, criminals, refugees, et al. It is they who suffer if the cover of house or person is blown (unzipped, exposed). For a gripping visual encounter with a notorious SAFE HOUSE see the 1984 TV version of John Le Carré's spy novel, *Tinker, Tailor, Soldier, Spy*. It exhibits a London SAFE HOUSE used by The Circus, a counterintelligence group. That same discerning eye which can choose a SAFE HOUSE is equally adept at selecting a rendezvous for other subrosa goings-on. The woman charged with the armed holdup and $1.6 million robbery of a Brink's armored truck in 1981 had evaded arrest for 3 ½ years by using false identities and a variety of SAFE HOUSES, described as part of her support apparatus.[1]

Yet a SAFE HOUSE may become, despite its unobtrusive setting, notorious. Such was the SAFE HOUSE set up on Washington's main-drag Massachusetts Avenue by the Central Intelligence Agency—pointed out, with glee, by knowledgeable Washingtonians giving friends the insider's tour in the 1980s. Not to mention the typical white-frame semi-Colonial dime-a-dozen house in a Washington suburb, supposedly a "secret" SAFE HOUSE , but widely published, its cover blown during the Congressional hearings on the Oliver North case, 1988.

In the 1980s the term was extended to cover temporary, unobtrusive homes

SAFE HOUSE
Author Ronald Kessler uncovered this Central Intelligence Agency SAFE HOUSE "where the CIA kept high-ranking KGB officer Vitaly S. Yurchenko after he defected from the Soviet Union on August 1, 1985. The CIA later moved him from this townhouse near Washington to a more secluded SAFE HOUSE in the Coventry development near Fredericksburg, VA." Photo from Kessler's book, *Escape from the CIA: How the CIA Won and Lost the Most Important KGB Spy Ever to Defect to the U.S.* (Pocket Books, New York, 1991).

offered to women escaped from abusive husbands. Even with the Cold War displaced, no disappearance was anticipated for SAFE HOUSES, either in the United States or abroad. Intelligence agents, terrorists, refugees, and illegals would continue working their global nets, needing SAFE HOUSES in all corners of the turf-tightening world.

THE SCENE

Here is a shallow, one-dimensional purlieu of limited duration and shorter notoriety in which hustling, table-hopping, publicity-seeking singles, couples, and assorted gaggles of other-directed folk seek to outdo each other in the public gaze for gainful status employment.

THE SCENE has been converted into a widely used promotional device. "A stage for promoting fashion and new urban lifestyle," says Michael Jager in *Gentrification of the City.*[1] It requires as its basic raw material a pliant mass media, and a restless, ambitious status-striving population, anxious to know where "They" (their Significant Others) may be found, preferably after dark.

THE SCENE is ablaze with discontinuity. Its discos reverberate with today's Top Forty tunes. Yesterday's clothing fashions are to be raided for odd bits, leftovers from the last trend back, sold at shops called Second Time Around, Rear View Mirror, Yesteryear, or Nostalgia.

Here is also the haunt of "The With," that fluid coupling of two or more people. They may be linked by the occasion or the situation, but not necessarily by marriage, friendship, or intent. They are merely and solely "with" each other for the nonce, the moment—making THE SCENE. Their quixotic juxtaposition is the quicksilver on which gossip photographers flourish. "The With" is an ingenious coinage of sociologist Erving Goffman. He uses it to describe the limited-focus, mutual presence in public of two or more persons "who are perceived to be together" in "sociological proximity." Whoever is in a With is "bracketed" or "framed" by others' presence.[2]

Adepts at scene-making work to master the contrivance of the backdrop. They array and display themselves against bizarre, photogenic backgrounds that are calculated to bestir an editor's itch to publish.

Often as not scene-makers are artists, would-be artists, groupies, publicity and other agents, including photographers, social climbers, and assorted hangers-on whose mission, job, or enterprise is—to "make THE SCENE." This jobbery consists not so much of organizing or creating a particular setting, but of being seen, quoted, photographed, and remembered as having been there in propitious moments in the company of others more famous. The most notorious SCENES to be made are in New York, Toronto, Tokyo, London, Paris, and a few other world cities that have strings of trend-monger publications, TV shows, and publicity agents and their necessary gossip outlets—all anxious to out-whisper or outshout each other for attention and influence.

Making THE SCENE is not to be confused with the socially unacceptable act of "making a scene" by creating a disturbance, ruckus, or noisy dispute. Rather, it consists, in Manhattan, of gaining access to what are called "downtown magazines." These go (1988) by such names as *Paper,* "a cutting-edge monthly"; *Details;* the over-twenty-one *Interview* started by Andy Warhol; *Scene,* "a glitzy monthly from Fairchild" (Publications); *L.A. Style* from California; *Fad* "a San Francisco newsprint bimonthly with a hallucinogenic tone"; and *Equator,* with "a geriatric beatnik aura"—all described by Philip Weiss in *Columbia Journalism Review.*[3]

Looming behind all such impromptu, ephemeral SCENES are the larger settings within which they operate. Such may include great shopping streets, nightclub districts, an artists' quarter (not yet bond-issued as a formal arts district), trendy or emerging neighborhoods, run-down neighborhoods in the process of emerging, best addresses, experimental theater districts, and others of their ilk. In lesser cities lack-

THE SCENE

Whether you "make" it, photograph it, or paint it (as New Orleans's Lionel Lofton did here), THE SCENE can be all things to all people: as a venue for competitive display, as the locale of a party or carnival, or (as here) Mardi Gras in New Orleans. As American society seeks the proto-reality of virtual reality, new versions/editions of THE SCENE get merchandised as "revelry" via TV.

ing THE SCENE, manufactured events—openings, premieres, charity balls, etc.—serve similar purposes.

See also GENTRIFYING NEIGHBORHOOD, in Chapter 4: Ephemera.

SUPERBLOCK

Never quite a part of mainstream language, but ready and waiting to serve architects and big-scale developers, SUPERBLOCK had as its direct ancestor the first skyscraper. Its grandfather was W. L. B. Jenney, the architect whose Home Insurance Company tower of Chicago was the first completed steel-frame skyscraper. Finished in 1885 it had ten stories, with two added later. By 1929, U.S. cities boasted of 377 skyscrapers, of which 188 were in New York City. Some were already being called "supertowers."

In the successive building booms of the twentieth century, ready and waiting for mass-production was the term SUPERBLOCK. It was the product of widely merchandised mortgages, of pre-formed identical components, plus the grandiose plans of Le Corbusier, Walter Gropius, and other modern architects, coupled with open-ended zoning in many U.S. cities.

The term was much used in London's postwar rebuilding after 1945, but oddly, neither SUPERBLOCK nor COMPLEX was recorded in 1981 in *The Illustrated Book of Development Definitions.*[1] On the ground, however, SUPERBLOCK had been embodied in hundreds of projects—New York's Grand Central Terminal and adjacent Helmsley Tower; and Rockefeller Center, begun in 1931, two of the more famous. By the 1980s others had come onstream: The United Nations, New York; Place Bonaventure, Montreal; Prudential Center, Boston; Penn Center, Philadelphia; city hall, Toronto; Midtown Plaza, Rochester, N.Y.; Mellon Square Park, Pittsburgh; Marina City, Chicago; Mile High Center, Denver; The Omni, Atlanta; The Belvedere, Louisville; The Golden Gateway, and Rincon Center, San Francisco; and Courthouse Center, Columbus, Ind., originally labeled SuperBlock. Both Grand Central Terminal and Rockefeller Center have been further linked with surrounding blocks since they were built.

Dreams of superblockers had long permeated the shaping of most American DOWNTOWNS. Newspaper and science-fiction readers in the 1920s and 1930s were treated with futuristic sketches of hundred-story towers joined by high skywalks, airship moorings, and airplane landings, all linked up with sky trains and other sky-hooked hoopla. Industrial designer Norman Bel Geddes entranced the multitudes with his Futurama exhibit of a superblocked city for General Motors Corporation at the 1939 New York World's Fair. The film *Metropolis* showed soaring viaducts, airships shuttling commuters between skyscrapers. Even though most skyscrapers were one-off, single-block affairs, the notion and techniques of connecting blocks was spreading.

Along the way, SUPERBLOCK enraged many critics, including the authors of *The Exploding Metropolis* in 1958. By the 1960s, some of their predictions were being fulfilled: SUPERBLOCKS spread their influence and left their footprints, monotonizing many a DOWNTOWN. By the 1980s, central Manhattan was shadowed by monotonous skyscraping SUPERBLOCKS filling every cubic inch of allowable AIR-SPACE, masking sunlight near and far—a process the city was slow to regulate and powerless to control.

By the 1990s SUPERBLOCK was no longer the new, big boy on the block. The term itself had became a handy smear-word for opponents of super-scaled projects. In his own Chicago, Bertrand Goldberg, "the visionary architect who designed Marina City, the first mixed-use project in the United States to contain housing," was assailed by critics "who think his newest city-within-a-city is insular, overplanned and unresponsive to the city's real needs."[2]

Among critics of huge, single-use SUPERBLOCKS already built, "multi-use" became a catchword. In the act of jumping streets, SUPERBLOCK was becoming not just mixed-use but multi-use, acquiring fancy names, various and sometimes anonymous owners, turning more than one cheek to the world, leaving a bigger set of footprints, and assuming the risk of losing its identity among a host of other claimants for public attention.

Along the way, the emerging sub-profession of urban designers took a hand in SUPERBLOCK design and placement. Their efforts to arrange SUPERBLOCKS according to principles of spacing, viewpoints, proportion, and scale sometimes paid off in popularity, prizes, and prestige: Edmund Bacon, Philadelphia's planner, reached *Time* magazine's cover for his center-city design for SUPERBLOCKS. By the 1990s, as capital converged in a dozen or so "world cities," SUPERBLOCK began to be transmogrified into its larger offspring, the MULTI-USE COMPLEX.

See also MIXED/MULTI-USE COMPLEX, in this chapter, above.

TOURIST INFORMATION CENTER

An outpost of self-aggrandizement, such CENTERS were often little more than a hit-and-miss affair along the highways of North America. But as tourism expanded in the 1970s, so did these "welcome stations"—some grand and expensively wrought. In Southern states, they come in white-columned old plantation mansion style. Westward they run to stucco-adobe rancho fashions.

TOURIST INFORMATION CENTER
Also known as "rest stops," these facilities, located between interchanges, lubricate the highway system. What they offer tourists causes a constant tug-of-war between state tourism promoters and commercial operators who don't want the state to compete by selling food, drink, etc. This one uses Old South architecture to proclaim to Northern tourists that they're "Down South"—some thirty-five miles south of the Mason Dixon Line, on I-65 in Kentucky.

TOURIST INFORMATION CENTER

Not by information alone do tourism centers flourish, but by sale of everything from soup to nuts to service to lodging to endless souvenirs. Here visitors at Cumberland Falls State Resort Park, Ky., are greeted by mass-produced place-name signs suitable for posting in party rooms and patios, hunting lodges or other boisterous venues.

Located close to the state line, they occur prominently in states lately entered in the race to divert tourists off major highways. But all have learned a lesson from billboard locators: they invariably occupy high-visibility sites, silhouetted against a backdrop of heavy forest or a SKYLINE. Often, access to the TOURIST CENTER is more important than visibility itself: at New Bedford, Mass., the tower of TOURIST INFORMATION CENTER arises dramatically to challenge travelers on I-195 to turn off at the next exit.

Tourists learn to expect free toilets, temporary shelter, clean drinking water, automated pop and snacks, and a cornucopia of free maps and brochures that extoll, promote, entice, and exhort travelers to stay and spend in this particular state. What they do not get is uniformity except within a state.

Red Light District
SKID ROW
EASY $TREET
MALFUNCTION JUNCTION
DISASTER WAY

CHAPTER 3

Perks

Once upon a time the "perks" (i.e., the perquisites of office) began with kings and their royal henchmen and headsmen over exclusive rights and positions, places, and jobs. Today's Perks are more subtly distributed—through tax-abatements, leaseholds, contracts, easements, and other preferential attainments in and around the city center. New actors sneak or stride onstage to take over the prospect: here come the feds with funds for rehabbing old landmarks; or a new governor itching to expand a state-office building: Everybody Stand Back! Next it's city hall's turn, anxious to finance a permanent EVENT SITE. A mayor's committees and other power brokers twist the city plan thisaway, tug it thataway to favor a new CULTURAL ARTS DISTRICT. Local or out-of-town corporations, eager to establish a PRESENCE, bid up a prime site for their high-rise additions to the SKYLINE. Through all this process, the power of city planners gets called on more often to save The Center from its old firetraps, its new traffic jams, and other excesses.

The resulting scene, for those who understand the shifting rules, offers site-by-site evidence of deals made and pending, of local power plays and players.

AIR RIGHTS AREA

It's all potential. At first the "it" is invisible to the uneducated eye. But then, once that void of empty space is reconsidered, re-perceived, and legally defined as a building site, at that moment the right to control that particular space—that AIR RIGHTS AREA—becomes valuable indeed. The proper legal title will let you build a skyscraper in what is now unoccupied AIRSPACE above ground level, or in the void above an existing structure.

At that moment of insight, generations ago, the sky became the limit for packing more buildings and crowding more people into U.S. cities, especially New York. Nobody really cared much about air rights until the New York Central Railroad created a fortune in real estate—for itself and other property-holders—by developing and/or leasing the AIR RIGHTS AREA above its terminus at Forty-Second Street, Manhattan, and its tracks extending north on the Park Avenue axis. This followed an architectural competition held in 1903 for the Grand Central Terminal design.[1] Park Avenue became a hot property as apartment-builders filled the new space alongside it with stilted skyscrapers above the old railroad properties.[2] Thus arose Park Avenue as a new street and GOOD

ADDRESS; thus followed the first great wall of Park Avenue apartments, later to be mostly replaced by office towers; thus was built the Prudential Building in downtown Chicago, erected on air rights over the Illinois Central Railroad. And thus arrived a new address, "100 North Riverside," a $170 million project, proposed in 1988 to cover 1.25 acres of air rights west of the Chicago Loop.[3] In Manhattan, N.Y., the congregation of St. Bartholomew Episcopal Church on Park Avenue endured a long controversy over whether to sell the air rights over its valuable corner for over $100 million (1988 price)—an acrimonious case of separating church and real estate.

Air rights' next wave of popularity emerged from the highway-building boom of the 1960s. The key phrase was "highway joint development and multiple use." Confronted by citizen protest over many a destructive proposal—i.e., to dig an expressway across the front lawn of the National Capital, or to take thirty acres off downtown's tax rolls for an interchange—Congress jiggered highway laws to allow "joint development." This often meant selling or leasing the public's air rights above highways for a mix of public and private uses, with some asserted long-term public goal to be served. Thus emerged hundreds of new downtown projects: Seattle's Plaza Park; Manhattan's high-rise apartments over the George Washington Bridge approach; The Belvedere above I-64, Louisville, Ky.; the thirty-acre deck over I-10 Papago Freeway, Phoenix, Ariz., et al. Within thirty years the potential had been overexploited in Manhattan by proposals for, or construction of, huge buildings over the Staten Island Ferry terminal, over Hudson River and other old docks; and over uncounted slivers, strips and odd lots of public lands elsewhere.

Especially under the influence of the eight-year Ronald Reagan administration, the old idea that public air rights should be restricted to public use came under widespread abuse. Meanwhile, the marriage of the legal and building professions during the on-and-off-again building boom of the 1980s yielded a new wave of joint developments. These were MIXED/MULTI-USE COMPLEXES—curious hybrids that use air, surface, and underground rights, fee titles, leaseholds, and whatever government subsidy might be obtainable. One variation was Atlantic Center, a $530 million mixed retail and office project scheduled in 1988 to be built on air rights above the Long Island Rail Road terminal in Brooklyn.[4]

Hybridizing continued in 1988 when San Francisco's Standard Brand Paint

AIR RIGHTS AREA

Quick to capitalize on air rights over highways were independent state turnpike agencies and/or hotel chains, using, in this case, buildable space above the Massachusetts Turnpike and the adjoining Amtrack right-of-way, west of Boston. This "Gateway Center," built in 1969–70 and subsequently renamed "Sheraton Tara Hotel," has frontage along Washington Street in Newton, Mass. Photo by Michael M. Bernard.

Company built a three-story addition of forty-nine small apartments for low-income families above its store. The firm donated those air rights to a community foundation, receiving in exchange a $600,000 tax write-off.[5] Another variation in Chicago was called "a twelve-story blessing"—ten floors for self-parking atop the two-story Old St. Mary's Catholic Church at Wabash and Van Buren Avenues. The parking venture was designed to pay for the entire structure, including the new church portion. The St. Mary's parish is the oldest (1833) in Chicago. It was at first suggested that each parking floor be named for a saint, but a spokesman said "We gave up on finding ten saints in Chicago."[6]

See also MIXED/MULTI-USE COMPLEX, in Chapter 2: Patches.

CULTURAL ARTS DISTRICT

Variants: cultural area, center, complex, park, arts' park or district (the last-named not to be confused with artists' quarter).

During the 1970s, American arts groups—especially those supporting art museums, private galleries, the local ballet, and orchestra(s)—went through an orgy of self-examination and expansion. "The building of isolated cultural complexes in the form, more or less, of Arcadian campuses or sugarcoated shopping centers surrounded by landscaping and parking was a popular concept across the country . . . as if they were cultural cathedrals of sorts."[1] In a few large cities, especially New York and Toronto, the old artists' quarter offered a critical mass of artists, customers, cruisers, and hangers-on sufficient to bring about a flourishing gallery ROW, and attract museums, studios, and suppliers. In smaller cities, the artists' quarter turned out to be one or more blocks of galleries, lofts, shops, and suppliers on or just off a main drag, or snuggled up in cheap quarters close to a prestigious old museum or mansion district. It was enough to set city fathers to salivating at the prospect of a future GROWTH AREA.

Searching for new gimmicks to bring life (and higher taxes) back into mid-cities, the fathers found friends-in-arms among art-lovers. Out of this embrace came the new ARTS DISTRICTS where museums, galleries, theaters, etc., are viewed as growth-instruments. "Arts Districts Can Paint Downtowns the Color of Money," headlined *Governing* magazine in July 1988.

These DISTRICTS are Official, usually expensive, often "creatively" financed by sales-and-leasebacks, tax-increment financing, and/or bond issues. Most are situated close to DOWNTOWN. Cities often add parking garages to sweeten the deal and help pay off the loans. It took five years of intense lobbying to get Tucson's ARTS DISTRICT its first-year funding in 1988.

By the 1980s there were several new DISTRICTS, formally designated by city hall or others. Some expanded with a Byzantine mix of private gifts and a public finance. In Dallas, Tex., a sixty-acre redevelopment near DOWNTOWN was officially "targeted for major cultural facilities expansion" close to the Dallas Museum of Art, the Trammel Crow LTV center, and rapid-transit and highways.[2] Such new DISTRICTS were usually and officially declared to be Arts Turf, into which would be expensively packaged a collection of locally certified arts groups and their new buildings. Normally appearing on official plans, such DISTRICTS are not to be confused with the self-generated, impromptu, scruffy, and low-rent artists' quarter or HANGOUT which the newcomers may supplant or co-opt.

As the fine arts became subjected to the Big Bang theory of urban redevelopment, Calvin Tomkins, writing for the *New Yorker* (1983), found in Dallas that "the feeling persists that the arts have been appropriated here primarily to sell a massive real estate development."[3] In Los Angeles, urban redevelopment funds paid for the New Museum of Contemporary Art "explicitly conceived as a means of enhancing commercial success for adjacent downtown residential, hotel, and office construction."[4] Albuquerque, N.M., anxious in 1990 over cuts in nearby weapons industries, plus shrinking oil prices, "put a new performing arts center on the last piece of vacant land around Civic Plaza" located DOWNTOWN.[5]

Under such pressures, artists in many cities began to amoebify: some, attracted to big money, found a niche in the new DISTRICTS. Others did their thing in offbeat locations, a new gallery, "pioneering" a dingy warehouse block here, or old storefronts along there. Some learned to co-opt the media and thus fetch a new public off the beaten path. In Manhattan they lobbied successfully in 1986 in order to continue living and working in converted lofts—a privilege denied to non-artists. In St. Paul, Minn., a local coalition created thirty artists' studio-homes in a five-story converted shoe factory in Lowertown, "the city's arts district."

Intense competition grew out of such geographic tugs-of-war. In Manhattan circa 1982, The Fun opened in East (Greenwich) Village where "its mean streets and dingy clubs [were] still providing a sense of adventure . . . a tough turf of drugs, punk rock, and prostitution." By 1985 this was said to compete directly with the older studio-and-gallery mecca of SoHo (South-of-Houston). "Once upon a time, an artists' milieu was simply a quarter whose inhabitants could live on the cheap and carry on as they pleased. Nowadays, the milieu itself makes a movement of the art produced there."[6]

As matters proceeded, some ARTS DISTRICTS became barely distinguishable from the new MIXED/MULTI-USE COMPLEXES. Occasionally the entrepreneurial tail wagged the arty dog. In the ARTS DIS-

TRICT of downtown Phoenix, Ariz., Greyhound Dial Corporation began 1990 construction of two office towers with one million square feet, plus an adjoining one-story Phoenix Little Theater leased for $1 a year.

New centers could also serve territorial ambitions. At the edge of Fairfax City, Va., the George Mason University Center for the Arts, "of which the concert hall is the capstone, is part of [University President George] Johnson's strategy to put George Mason on the map."[7]

Meanwhile, in the generation between the Vietnam and Persian Gulf Wars, federal dollars had poured into fine arts via the National Endowment for the Arts. Artists working and showing in ARTS DISTRICTS often seemed to get the upper hand in hustling grants. The next generation in 1990 encountered rising pressures to control the content of arts "paid for by tax money."

None of the foregoing ventures are to be confused with the new cultural villages, a term used to designate a re-created assemblage "outdoor living museum" of folk buildings, usually from a countryside populated by distinct European ethnic groups. Such are the Shakertowns, Salem Villages and, more recently, the Ukrainian Cultural Heritage Village fifty kilometers east of Edmonton, Alberta. Only by fits and starts have newer ethnic groups penetrated the WASP-dominated boards that run established arts enterprises. Ethnic Chinese made belated headway in keeping or restoring their Chinatowns, or the Angel Island immigration station in San Francisco Bay (1990). There, 175,000 Chinese immigrants made their U.S. entry before the Chinese Exclusion Act passed in 1882.[8] So arts and culture continue to be redefined, and to find new uses as political or social stepping stones, and as boosters for real estate. The emerging and often politicized DISTRICTS serve as a medium of exchange. "The Arts Community" continues to be identified and solicited. Museums and galleries—strapped for funds—have branched out into party- and convention-hosting, or have added boutiques, sales galleries, public restaurants, and gala balls. The Museum of Modern Art, New York City, sold air rights for condominium apartments. Thus do The Arts mix and sometimes merge into multi-use. Amidst all such hoopla, art could be sidelined, buried—or end up in the hands of arts entrepreneurs running the new show.

DISTRICT

Unspoken but understood, the word "control" normally precedes the word DISTRICT. Other words—"improvement" "development" "management" or "taxing" may intervene, but the essence is control. DISTRICTS are also called "secret governments" and "the fastest growth sector of government."[1]

DISTRICT is widely used as an administrative convenience. The term generally describes an area ("the DISTRICT") subject to official regulation by an agency (also "the DISTRICT"), with legally defined geographic limits (the Metropolitan Sewer DISTRICT, the Cheyenne Irrigation DISTRICT). It levies variant taxes in the form of fees and charges. Often known only to insiders or specialists, it is sometimes the biggest game in town or countryside. In Medieval Latin *districtus* denoted the territory over which a feudal lord lorded.

Such "conveniences"—if school DISTRICTS are included—number around forty-five thousand nationwide. They grew in numbers 14 percent from 1977 to 1987[2] and 39 per cent since 1972—rates far exceeding those of municipalities. Their current purposes are as open-ended as is human inventiveness. There are DISTRICTS for political, school, administrative, or myriad special purposes. There are DISTRICTS military, magisterial, and judicial; DISTRICTS metropolitan, suburban, and agricultural. They have been organized to fight fires, to control soil erosion, to occupy foreign territory, and to protect historic areas. They are set up to supply electric power; to manage, and sometimes build, waste and sewage plants, parks, roads, forests, BEACHES, drainage, irrigation, and other public works; and for mosquito, predator, or epidemic control. Of 29,532 non-school DISTRICTS in 1987, 53 percent handled fires, sewers, and/or water. Many state laws provide for catchalls called improvement or development DISTRICTS, noted below. A map of local DISTRICTS can cover a county, obscuring all else—and jump county lines. Unknowing, travelers cross hundreds of local DISTRICTS, their borders mostly unmarked, except for the ubiquitous green roadside signs representing 2,994 soil conservation DISTRICTS.

As a handy (usually called "self-governing") device, a DISTRICT is often empowered by state and/or federal laws, its local officers generally appointed—although sometimes elected—as trustees or directors. Its infinite variations are subject to endless influences, often invisible to the taxpaying public.

At the one extreme are tiny DISTRICT associations in the arid Southwest called "acequias." They follow Spanish customs whereby local farmers allocate irrigation waters and clean out ditches, under a locally elected "mayordomo."[3] Things get very personal.

A newer twist was added in the 1950s when a rising young Georgia planner, Philip Hammer, conceived the idea of a "local planning and development district" as a bootstrap device for poor counties to pool meager funds and scattered powers. Georgia's legislature passed an enabling law in 1960, whereupon eleven counties in the Coosa River (Appalachian) Valley set up "the nation's first development district."[4]

This became a regional success story. Hundreds more such DISTRICTS soon were off and running under local power, becoming adept at "milking the feds" for funding. Cities far and wide voted to set up or expand tax-subsidized DISTRICTS for high-tech research, for minority businesses, for conventioneers, tourists, culture-mongers, racing fans, or downtown office- and store-owners. Some cities formed DISTRICTS to clean up downtown streets.

In turn, these provided the pattern for hundreds of local "enterprise zones" formed under state law and federal subsidy in the 1980s. These geographic gimmicks covered an incredible variety of urban places hit by depression, out-migration, disaster, or abandonment. As first proposed, they would attract new business to run-down city slums. From 1982 to 1989 some fifteen hundred zones were set up. Most of the eight-hundred-odd persisting in 1993 were formed by local governments as bait (low federal taxes, free utilities, and such) for footloose firms to colonize low-rated neighborhoods. Few solved the locational decrepitude they faced. Among these zones there was less self-governance than is found in most DISTRICTS.

Some pushed their limits far beyond the city. And in common with other arrivistes, they set up national association headquarters in the District of Columbia, where federal control is exhibited by denying residents full self-government.

At another extreme from the grass-rooted originals lies the huge Reedy Creek Improvement District which dominates Orlando and Orange County, Fla. It was set up quietly by the state legislature in 1967 after Disney Productions acquired some 27,400 acres near Orlando. "The Florida State Legislature passed three Acts. . . . [which] made the entire Disney property an improvement district"[5] and gave Disney major controls, usually held by governments, over utilities, fire and police protection, bond issues, etc. The DISTRICT became an umbrella for constructing Disney World and EPCOT (Experimental Prototype Community of Tomorrow). As the major landowner and political power base in the region, it controlled myriad tourist DESTINATIONS. "In effect it became the 68th county in the state."[6]

"'It was, in truth, a feifdom,' commented former Governor Claude Kirk, who helped speed the 1967 law creating the district."[7] Reedy Creek was first set up as a drainage DISTRICT, but its present status gives it overriding powers of regional management and control.

"Reedy Creek is only one of more than 700 special tax districts required to complete a comprehensive plan because of the impact Disney World has had on Central Florida."[8]

Unlike mayors or legislators, DISTRICT officers are politically invisible; "Board members can raise fees with little threat of being voted out of office," according to Amy Lamphere, in an article headlined "Districts Weather Recession Storms" in *City & State* journal, March 22, 1992. She cites an extreme case: Massachusetts's Water Resource Authority, its rates up 429 percent from 1985 to 1992.

Of the "top fifty" DISTRICTS noted by that journal in 1992, nine took in more than one billion (estimated) dollars apiece in revenue in 1992. The top earner was New York City Metropolitan Transit Authority ($4.9 billion). The top water/power DISTRICT was Los Angeles Department of Water and Power, $2.7 billion.

DISTRICT

What's so special about "this point?" It happens to be the city limits of Charlotte, N.C., where state law prohibits the use of "prison labor"—formerly called "chain-gang labor"—within city limits. Although these are not called "Chain-gang-control DISTRICTS," that is the practical effect.

Some DISTRICTS get expanded power and scope to become "authorities," usually anchored to a functional place. The famous Port of New York Authority under director Robert Moses acquired extraordinary bond-issuing and building powers in the 1940s–60s, beyond those of most DISTRICTS.[9] It was modeled on the huge Port of London Authority, became "America's first multipurpose authority," and may well have been an inspiration for Reedy Creek in Florida.

Loosely and widely applied, the term "DISTRICT" also serves to mark off special-activity sections of a city, as in the case of a central business, market, warehouse, financial, residential, or silk-stocking DISTRICT; as an outsider's put-down such as the now-disused "nigger DISTRICT"; or as a descriptor for neighborhoods of last resort such as tenement DISTRICT, red-light or whorehouse DISTRICT. Such latter places generally exist with tacit, illicit complicity of local authorities. As historian Patricia Limerick noted, "vice districts were among the more rewarding [nineteenth century] Western investment opportunities."[10]

It is also useful for designating a specific, locally recognized area with known limits, as in "I'm going down to the business DISTRICT for lunch." For someone at a headquarters, to say "I'm going out in the DISTRICT" is equivalent to "I'll be out in the field," implying "Don't call me, I'll call you." But this was before mobile phones ended the field's isolation.

Kevin Lynch tapped a rich vein of local awareness when he disassembled Boston, that prototypical American city, into its visual and functional components: paths, edges, nodes, DISTRICTS, and landmarks.[11] He saw an urban DISTRICT as a distinctly generic sort of place having a "common character" perceived in toto: civic center, medical center, skid ROW, transportation ROW, theater DISTRICT, West End, Back Bay. These places share commonalities such as a profile or SKYLINE, color, social status high or low, functions, age, or visual stigmata (Chinatown's or Little Tokyo's signs). Little did Lynch—with his sharp admiring eye for the indigenous, freewheeling, and locally unique—relish the fact that the next generation of image-conscious city planners would use his book as a guide to promote festival marketplaces and other up-market attractions that exert stringent look-alike control over color, styles, signs, and architectural features—sometimes using DISTRICT powers.

Nowhere has the U.S. Constitution reached into local affairs more drastically than in "redistricting"—the redrawing of Congressional voting DISTRICTS to accord with the "one person one vote" mandate of the U.S. Supreme Court in 1963. It happens every ten years, following population shifts disclosed by each U.S. Census. Following the Voting Rights Act of 1965, courts were being asked to redistrict, not just by head counts alone, but according to racial-ethnic makeup as shown by census. The resulting DISTRICTS—snaked out or stretched into bizarre shapes—began to meet Supreme Court resistance.

Thus the DISTRICT continues to expand its original power as a versatile political device, a place for governing self as well as others, and a legal enclosure for managing territory as well as nature.

EVENT/FESTIVAL SITE

Short of racial/ethnic riots, or political revolutions, few events transformed their SITES as did North American folk-rock festivals of the 1960s. Ten thousand assorted festivals were held in 1969.[1] For the next generation, all EVENT/FESTIVAL SITES would carry the extra emotional baggage of youthful tribal gatherings—at Altamont and Monterey, Cal., at Woodstock, N.Y., and their imitators.

The Woodstock, it was claimed, was "the largest spontaneous gathering of humanity in the history of the Western World," with some 350,000 assembled for three days.[2] Not only did the emotional aura that surrounded these places linger on, but the techniques, the know-how, and the huge profits to be made at such SITES and with such events, became embedded in national culture.

Crowd-handling and site management would be transformed. Crowd-assembly on this scale, it was believed, would pump new life into old DOWNTOWNS. Clinging to this belief, cities far and wide would mortgage their futures to build new public plazas, convention centers, exhibition grounds, and vast staging areas, many along old decaying waterfronts.

But here all similarity with Altamont-Monterey-Woodstock ended. Rock FESTIVAL SITES in the sixties and beyond had often been remote, undeveloped, and unprepared. In the summer of 1969, traffic jammed for thirty miles around Max Yasgur's isolated farm near Woodstock, N.Y. The air thickened with marijuana smoke, drugs were sold openly, the few toilets overflowed, the ground turned to mud—yet the crowds here (but not at riotous Altamont, the noted exception) mostly acted out the tribal message of the hippie generation: Make love, not war!

The audacious braggadocio, the assertion of superior insight, the certitude of defiance, the accepted conviction that the larger, older, (Vietnam-) warmongering society was corrupt, and that love was free, the living easy, and youth everlasting—all this fused into a passion for folk-rock music and the emotions it let loose at FESTIVAL SITES. Such was the fuel for festivals and the new wave of public events in the decade after 1969.

It was barely in time. According to Edith Matilda Thomas in 1970, "The circus is near death. . . . Audiences at outdoor sporting events are getting smaller. Political and religious gatherings . . . too militant, or boring, or both."[3] To the contrary, here were these new exhibitions of identity-formation on a giant scale. Such places identified what came to be called "The Woodstock Generation." Behind its formation were "major forces of money, unemployment and mobility, none of which can be shut off by refusing permits for rock festivals, by closing public parks overnight, or by other acts of repression against youth." It appeared that "identity-making [had] become a major preoccupation among millions of Americans."[4]

While there was no way to refuel Woodstock in the 1990s, the ripple effects lasted. Rock music and its successors became the mainstay of public-event calendars. No big-scale national convention or state fair could flourish without the music. Film, TV, and recording fortunes grew out of its fervor. And hundreds of local FESTIVAL SITES were built to capture the market.

In the process, many decrepit downtown waterfronts were converted to EVENT/FESTIVAL SITES. Some were barely able to accommodate the giant SETUPS and heavy equipment that grew out of post-Woodstock. Clearly mobility had expanded, the publicity was organized to keep the crowds coming. Towns and cities competed vigorously to keep their new SITES filled.

Meanwhile, city maps got marked by these new LANDING places. New jobs appeared at city hall as festival-managers became moving parts of the local growth machine. Typical of the new venues was the central plaza at Baltimore's waterfront HarborPlace. It was designated by the mayor's office after 1983 for public events—sometimes known locally as "The Crotch Show"—to promote the neighborhood: marathons, exhibits, charades, parades, fire-equipment testing, speech-making, fashion shows, etc.[5] Downtown Baltimore became a huge tourist attraction.

As civic boosters assumed leadership jobs in many U.S. cities during the recessions following 1973, they earmarked or commandeered assorted public places for fundraisers and other civic events. Many an old Main Street found new life. Fifth Market Place, a newly enclosed public concourse in the Westin Hotel, Cincinnati, was partly closed to the public in 1985 for local fundraising by volunteers manning telephones, tables, and public address systems, etc. In scores of mid-sized towns and cities, travelers suddenly found the main drag closed by city hall promoters as an EVENT SITE. Whether the hoopla could offset the continued, undramatic swarming of population out to SUBURBIA would be an open question.

See also PARTY STREET, in Chapter 2: Patches.

PARADE ROUTE

Variants: ceremonial parade, coronation, inaugural, holiday, protest, demonstration, memorial, victory.

It arrived impromptu, 'way back when: a two-bit horse race among friends down an unpaved nineteenth-century Main Street; or a five-minute pass-in-review for local veterans back from an 1812 battle. Or that emotional first Armistice Day Parade down New York's Fifth, and other Avenues, welcoming back Our Boys from Europe in 1918.

You'd never recognize it now. Once installed in local mores and budgets, PARADE ROUTE has been heavy-laden with burdens: history, hype, civic ambition, real estate promotions, and political push-and-shove. It acquires an overburden of group interests: Italians promote Columbus Day, the Irish St. Patrick's, military veterans their catchall Memorial (onetime Armistice) Day. Boosters of the "Indy 500" throng to race-day parade around Memorial Circle in Indianapolis. Boulder, Colo., advertises its own as the "shortest parade route in the world."

PARADE ROUTE is linear in space, extended in time, scheduled and planned, predicted and prescribed, sequenced, managed, predetermined, organized, policed, enforced, regulated, and widely published. It is also collaborative: that is, a product of volunteers, advertising/promotion, civic coercion, and arm-twisting. It is coordinated with other public activities

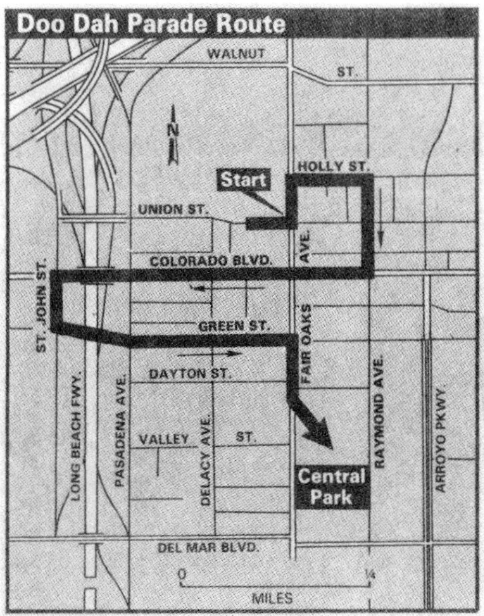

PARADE ROUTE

Fed up with all the civic hoopla associated with parades, the Doo Dah Parade in Pasadena, first organized in 1978, spoofs the whole idea with a "customary cast of zanies," including the Synchronized Briefcase Drill Team. During Mardi Gras, New Orleans and nearby towns offer so many parades (65) that the *Times Picayune* publishes a special section to show them all. Map copyright 1985, *Los Angeles Times.* Reprinted with permission.

and part of a larger event—a festival, anniversary, centennial, or other civic celebration.

Any definition of today's PARADE ROUTE must account for wide variety: the impromptu quality yet increasing structure of Carnival in Rio de Janeiro, its milder, more drunken counterparts in New Orleans and nearby Louisiana towns; and myriad ragtag local improvisations here today, gone tomorrow.

To parade is to have a plan. Those unplanned, impromptu, folksy, happenstantial affairs called parades of yesteryear have been taken in hand by an expanding cadre of professional organizers and events managers. City officials, usually police, issue permits; local police patrol barricades. There are endless and invariably capitalized Committees: Executive, Liaison, Planning, Policy, Floats, Band, Other Units, Manpower, Tickets, Legal, Safety, Lineup, Starting Line, Marshalling, Communications, and Dispersal. Cooperation with local police is essential to a good parade.[1]

Thanks to computerization, parade-pushers can phase their moves into urban traffic flows, can choose least-competitive routes, and divert counterflows by jiggering stoplight phases, blacking out, or selecting sites for stop signs and barricades.

Each route has its designated assembly area, official starting and reviewing location, sometimes an elaborate judges' pavilion, blocks of streetside bleachers, plus dispersal areas. A big city parade needs a half-mile extra length for the "dismantling zone." Designated streets are blocked off, crossing traffic is tightly controlled. So is the marching order: bands, clowns, military units, floats (each with its own float escort), historic horse-drawn vehicles, as well as Mickey Mouse, Kermit the Frog, Superman, and other folk heroes—each in a prescribed sequence, with a cleanup gang at the tail end. Dispersing without crowding can be a headache: Macy's in New York City provides costumes for its paraders—plus change-rooms at its store.[2]

Over time, the PARADE ROUTE has become a component of the larger, all-encompassing EVENT SITE, its route an element in a week-long program of local events, its sponsors hustling for national TV coverage. Minneapolis's Aquatennial Festival features events on Lake Calhoun and the locally modest Mississippi River, Kentucky's week-long Derby Festival splashes over into an Ohio River sternwheeler race. In larger cities little is left to happenstance: if you're not pre-registered, you don't parade: there's no room in the in-group.

Along most routes, street pavements assume a new, folkish identity—green stripe down Fifth Avenue and many another Main Drag for St. Patrick's Day. In staging areas, the pavement is elaborately diagrammed to show clubs, military units, and others where to stand and wait. Such routes acquire logos, symbols, souvenirs. Graffiti suddenly appear underfoot, plus confetti, banners, and planes towing streamers overhead. Hot dog, soft drink, and snack sellers foregather with their pushcarted and trailer-borne SETUPS. Of all such locales, Rio De Janeiro has gone the farthest toward making a permanent thing of PARADE ROUTE. That city spent millions in 1983–84 to build giant concrete grandstands along the path of its midwinter Carnival competition among parading samba bands. The late architect Lou Kahn, who insisted that "the street wants to be a building," would have seen this as the inevitable carrying out of his wishful prophesy.

Some parades pick up or follow well-known historical routes such as the Patriot's Path through Cambridge, Mass., or Paul Revere's Ride, in the Boston area. But few can match the scale and intensity of Queen Elizabeth of Tudor's grand parade, or "progress" through the historical districts of London on January 14, 1559. On that day, before her coronation, at successive old villages, she encountered a tableau, an arch, a host of dignitaries, children with poetry, and oratory. The progresses continued for a decade—through Coventry, Oxford, Warwick, Sandwich, Norwich, Sussex, Hampshire, and other locales.

Some parades acquire an aura well ahead of time. "In the beginning there will be dignity," predicted the *New York Times* before President Reagan lighted the refurbished Statue of Liberty at the two-hundredth celebration of Independence Day, July 4th.[3] "There will be trumpets and presidents, Gershwin and Sinatra, Kirk Douglas and Gregory Peck. Chief Justice Warren will give the oath of citizenship." Mikhail Baryshnikov then danced his first ballet as an American citizen. All this as part of a land-based parade-cum-fireworks on an extravagant scale programmed for TV, and allied to a giant parade of ships on the Hudson River.

Parades proliferated after 1890 when fraternal orders took up marching: Elks, Shriners, Odd Fellows, Granges, and police. Each new war supplied its battalions of marching veterans: Spanish-American, World Wars I–II, Korea, Vietnam. There were Labor Days, Charter Days, Mule Days (Clarksville, Tenn.), Apple Tuesdays, Sauerkraut Days, and Indian Days. Holland, Mich., holds its Tulip Time Festival in which Dutch-costumed folks scrub the PARADE ROUTE with brooms. The Sons of

PARADE ROUTE
Long-established PARADE ROUTES, such as Pasadena's Tournament of Roses parade, became semi-permanent parts of the local geography, reinforced by interminable coverage on TV. Meanwhile, the rise of organized protests in the 1960s (against racial discrimination, the death of Martin Luther King, Jr., the Vietnam war, etc.) escalated the planning and tactics used to "handle" parades. Copyright 1990, *Los Angeles Times.* Reprinted with permission.

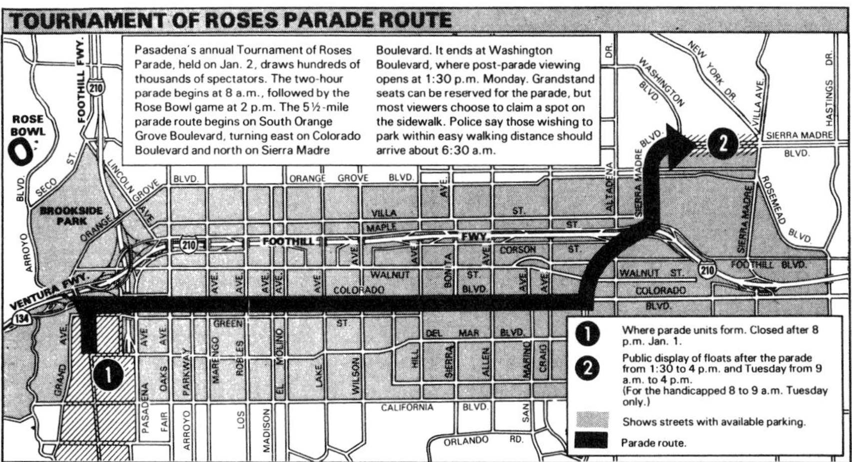

Norway hold their Constitution Day parade on Syttende Mai—May 17th. Irish from many states descend upon Savannah, Ga., for its St. Patrick's Day parade down famed Bull Street. For generations Macy's in New York held its Thanksgiving Day parade down Fifth Avenue, with help from up to 2,500 volunteers. There is the Pendleton, Ore., Roundup; Anchorage, Alaska, Fur Rendezvous; Memphis's Cotton Festival; and the greatest of these, Mardi Gras in New Orleans and lesser versions in smaller Cajun Country towns. In Mobile, Ala., "this afternoon's Joe Cain Procession begins the final push toward the Mardi Gras finale."[4]

Football fans join the Rose Bowl (Tournament of Roses) parade in Pasadena, Cal., which takes two hours to cover its five and a half miles, and had its one-hundredth anniversary in 1988. Now it has a take-off spoof event, the Doo Dah Parade begun in 1978, billed as a parody (occurring also in Pasadena, a locale many Easterners think of as itself a parody of real life). It would attract imitation nuns, witches, popes, such marching-malingering-and-swaggering groups as the World Premiere Mobile Volleyball Team with mobile nets and palm trees, the Texas Chainsaw Massacre Drool Team, and the Synchronized Briefcase Drill Team.

Main Drags have been the favored venue for PARADE ROUTES since early military parades. Independence Day, observed in church services since 1777 "moved to the street, where early practice favored morning burlesque parades of 'Horribles'" followed by military displays.[5] By the 1850s circus parades moved in, followed, after the Civil War, by the first Memorial Day parades (1869) and by the Grand Army of the Republic's reunions and parades of Union veterans.

If it's not a festival, it's a demonstration. Demonstrations and protest have added new features to the PARADE ROUTE, chiefly the presence of supporters, baiters and hecklers, police, FBI, platoons of media people and their apparatus, mixed at the site and along the route. Regulations for Atlanta's peaceful Dogwood Festival Parade fill four typewritten pages.

Street manners went down the drain at many marches in the 1960s when racial and anti-war antagonisms ended in riot. By the 1970s, most city police and National Guard outfits had been equipped, via federal law-enforcement subsidy, with expensive backup systems of walkie-talkie, remote TV, helicopter surveillance, plus extra arms and body armor where necessary. It's come a long way since New York's Finest—its mounted police—saddled up to keep order in the street. However, public disorder was not invented in the 1960s.

Pennsylvania Avenue in Washington, D.C., the PARADE ROUTE for all presidential inaugurals, became a favored target for protesters (anti–Vietnam War protesters, civil rightists, and mourners of Dr. Martin Luther King). Pennsylvania Avenue, in 1961 still a collection of old federal buildings intermixed with seedy run-down leftovers, was upgraded under pressure from President John Kennedy and Mrs. Lady Bird Johnson. A presidential commission under architect Nathaniel Owings proceeded to give America's Main Street, "'the nation's ceremonial way' something of 'a special character.'" Heavily re-treed, the route was to become "the image of our country in the tradition of America," predicted Owings.[6]

Since the Kennedy assassination in Dallas, the inaugural PARADE ROUTE in Washington is one of the most heavily guarded in America: Secret Service on the rooftops, access to open windows restricted, helicopters hovering at the ready. Any parade which attracts a president (all presidential visits stimulate paraders) is thereby upgraded to accommodate bigger crowds and the arrival of the Secret Service—afoot, in flight, or on wheels.

See also PARTY STREET, in Chapter 2: Patches; PRESIDENTIAL SITE, following.

PRESIDENTIAL SITE

The American PRESIDENTIAL SITE, long a rare and sacred icon, is in transition downward into mere celebrity. It is visited routinely, sometimes reverently, sometimes even cynically—a certified tourist attraction. For generations, whatever place touched, or was touched by, the Office or person of the President accumulated a holy glow. In its presence, tourists became pilgrims who convert ordinary trips into pilgrimages. Many still do.

Trappings of power and those who flourish them come and go. But sacrality endures long after limelights dim and places erode; and even after miscreants besmirch the holy offices.

"We now have hundreds of historical parks, museums, birthplaces, famous residences, tombs, BATTLEFIELDS, and other buildings and plots of ground that can be called nationalistic shrines," observes geographer Wilbur Zelinsky.[1] In a half-century, over eight hundred million visits to such nationalistic arenas were recorded in National Park Service annual reports (1930–81). Whether a decline since 1960 in such visits has long-term significance remains to be seen. By 1990 the White House had been replaced by the Vietnam Veterans Memorial as the top visitor attraction in Washington, D.C.

When presidents were cast in the role of the godhead of civil religion, PRESIDENTIAL SITES became tops on the sacred list. Once a scattering of homes in various conditions of upkeep with modest markers out front, PRESIDENTIAL SITES, mansions, etc., have not always been held in the sacred regard they commanded during the past half-century. Jefferson's Monticello, built in 1768, almost fell into ruin from neglect by the late 1800s and was not bought as a PRESIDENTIAL SITE until 1923 and opened a year later. Similar stories exist about James Buchanan's Wheatland, a museum in Lancaster, Pa., since 1936; George Washington's Mount Vernon, rescued by a ladies' association; and James Madison's Montpelier, opened in 1987.

Beginning with the so-called Hundred Days of Franklin D. Roosevelt (serving from 1932 to 1945), PRESIDENTIAL SITE vastly expanded into a place-ful collectivity, offering evidence of presidential mobility. Few places, aside from the federal PRESENCE, have had such explosive growth.

Franklin Roosevelt's Little White House at Warm Springs, Ga., was the first of many diminutives; his family mansion and summer home at Hyde Park, N.Y., joined a growing list of PRESIDENTIAL SITES beyond the White House and now spread across the nationalistic map.

Such PRESIDENTIAL SITES now include birthplaces, boyhood homes, camps, campaign trails, centers, command posts, compounds, cortege routes, "country," family homesteads, estates or ranch, family burying grounds and cemeteries; guest houses, hideaways, inaugural routes, libraries, mansions, memorials, parkways, residences, signed-here sites, slept-here sites, spoke-here sites, summer and winter White Houses, vacation homes, and yacht anchorages. Some are as large as small colleges.

And then there are assassination or attempt sites, such as Ford's Theater in Washington, where Lincoln was shot; or Blair House across Pennsylvania Avenue from the White House where Puerto Rican gunmen had a thwarted go at Truman; or the Washington sidewalk where Reagan absorbed a bullet in his lung in 1981 and lived. Even trees join the list by being presidentially planted (e.g., downtown Indianapolis, 1990, the site being renamed Presidential Place), or by offering shade at historical moments (e.g., the Washington Elm at Garden and Mason Streets, Cambridge, Mass., where the Revolutionary general took command of the Continental Army July 3, 1775).

Washington—in common with most early presidents—would be dumbfounded by the expansion of power of office since he held it. "By the early 1970s the American President had become on issues of war and peace the most absolute monarch (with the possible exception of Mao Tse-tung of China) among the great powers of the world," observes historian Arthur M. Schlesinger, Jr.[2]

Until the American Presidency began acquiring the trappings of royalty ("Leader of the Free World, Head of the Most Powerful Nation," etc.), a typical PRESIDENTIAL SITE would be exhibited on local landmark tours, but had hardly acquired world-class or sacred status. John Adams, the second president, would be astounded at the change—he who could walk out of the early White House to swim in Potomac River tidewaters just out back.

But that was long before presidents dared not, or chose not, to make a move without an entourage of hundreds or even thousands: Secret Service, local police, the media, and hangers-on of multiple hue and motivation. In 1901, after not only Lincoln, but also Presidents Garfield and McKinley had been assassinated, the Secret Service was charged with protecting the president. The president-elect was added to the list in 1913, and the president's immediate family in 1917. Congress wrapped the cloak of Secret Service protection around past presidents and their wives in 1965. The presidential aura, as well as its trappings and Secret Service protection, were extended to official candidates in the summer of 1968, after Robert Kennedy was killed while campaigning in Los Angeles.[3] Assassination attempts—together with twentieth-century wars and National Emergencies (declared and undeclared) that lasted for years—had transformed the presidency and expanded its sites. In response to a host of provocations, the presidency itself has moved into the forefront of everyday life and thought to a degree once unimaginable.

From 1939 to 1990 the Office of the President grew beyond the wildest dreams of its nineteenth-century occupants. In 1939 the White House Staff numbered only six; by the end of World War II it had reached forty-five.[4] By the Reagan Era—in spite of so-called Reaganomics suggesting a cutback in federal offices—the White House staff numbered 3,366 full-time employees.[5] The cost to maintain various PRESIDENTIAL SITES for George Bush in 1990 included an estimated $28,639,000 for "necessary expenses for the White House," and $6,773,000 for "the care, maintenance, repair and alteration . . . of the Executive Residence at the White House."[6] (There are no mortgage payments shown in the federal budget, though several presidents could have qualified for veterans' loans.)

And to protect the current PRESIDENTIAL SITE—wherever the president happened to be at any moment—employed a small army of advance men, route-scouts, security specialists with an occasional posse of full-time military personnel. By 1990 the U.S. Secret Service cost the taxpayer over $1 million per day (its annual budget was $367 million).[7] Some PRESIDENTIAL SITES had expanded to include GREENBELTS, security belts and zones, and surveillance areas, with sentries, guardhouses, radar, and mobile military units on call. In the riotous 1960s the security-girted White House itself occasionally became unapproachable. (The very idea that the holy of holies—the White House—could harbor criminal activity was unthinkable until "the imperial presidency" of Richard M. Nixon and the Watergate scandal.)

Local private donors had joined history buffs and preservationists to save or expand many a presidential birthplace, boyhood home, etc. By 1989 there were an estimated twenty-two presidential birthplaces labeled, protected, and mostly open to the public.

Even lost wars do not exempt PRESIDENTIAL SITES, such as the White House of the Confederacy, restored (again) in 1988 in Richmond, Va. Confederate President Jefferson Davis and his family fled the house in 1865 at the end of the Civil War. Later the victorious President Abraham Lincoln stopped by and, out of curiosity, tried out Davis's chair. He was rather tall (6 ft. 4 in.) for it. The house was restored with $3 million in public and private funds, and is listed as one of many Civil and Revolutionary War tourist attractions in Virginia.

Not all presidential moves are as expensively venerated as the inaugural PARADE ROUTE down Pennsylvania Avenue in Washington, D.C. More than $100 million in federal funds were spent during Kennedy and successor terms along the thirteen blocks between White House and National Capitol grounds—a costly redo of old buildings (Willard Hotel, Sears House) and grounds (Pershing Park). Huge new buildings—National Place (Marriott Hotel et al.), the Canadian Embassy—and SUPERBLOCKS now look down on this presidential pathway to power.

Newest among sacred PRESIDENTIAL SITES are libraries (or sometimes, in the current itch for centrality, called centers)—what Senator Lawton Chiles (Democrat, Fla.) saw becoming "monuments to the Pharaohs." Another description: "farewell equivalent of inaugural balls." The first of these (FDR) was dedicated in 1940. By 1990 there were nine: Herbert Hoover rates two: a library at his early hometown of West Branch, Iowa; and the better-known Hoover Institution on War, Revolution, and Peace, a research center at Stanford University, Cal. The following presidents have one each: Roosevelt, Hyde

Park, N.Y.; Truman, Independence, Mo.; Eisenhower, Abilene, Kans.; Kennedy, Boston; Johnson, Austin, Tex.; Ford, Ann Arbor, Mich.; Carter, Atlanta; and there was one under construction in 1990 for Reagan, in Ventura County, Cal.

Midway between modest and splendiferous is the Eisenhower Center at Abilene, Kans.—not his birthplace but boyhood home and preferred site. It includes a five-building complex dominating mid-town. Started in 1951, it has a monumental Dwight D. Eisenhower Library, Eisenhower Museum, chapel, visitors' center, and boyhood home on the original site. The Eisenhowers chose the chapel as their burial site. In the fall of 1987, the decision was made to build a Richard Nixon Library and Birthplace on a nine-acre tract in Yorba Linda, Cal. By 1987 all presidential libraries cost about $16 million a year to operate.[8] The Lyndon B. Johnson Library together with the adjacent Lyndon Baines Johnson School of Public Affairs at Austin, Tex., cost approximately $18 million to build on a contested site along Waller Creek on the University of Texas campus. (One co-ed, nude to the waist, was chained to a creek-side willow to vainly protest messing up the students' beloved creek.) The (Jimmy) Carter Center and museum in Atlanta, Ga., is a $25 million affair built on a thirty-acre wooded crest of Copen Hill, the knoll from which Gen. William T. Sherman had overseen the Civil War battle for Atlanta.

Today's centers come at today's monumental scale. They overshadow the relatively modest FDR library at Hyde Park (30,000 square feet costing a nominal $369,000 around 1940), and the larger Hoover Institute at Stanford University which raised little local opposition. Today's presidents keep more paper—25 million pages is said to be a norm these days—"more heft than history." Bigger sites plus heavy traffic combined to stir up local waves when sites were sought in the 1980s. Six of the last presidents wanted their centers on university campuses—each a limited and fought-over venue. The Kennedy Center, first proposed adjacent to Harvard on a contested site, ended up at Columbia Point miles away, costing some $21 million. More than $100 million was spent on the first eight centers (Hoover to Carter). The Reagan library/center with related endowments was expected (in 1987) to cost around $100 million.

The Hoover Institute at Stanford—an "independent institution within the frame" of the university—became a seedbed for conservative economic researchers. Its political connections to Republican policymakers luxuriated after Reagan's election. As a result, Stanford University faculty and students objected to being identified with the proposed on-campus Reagan Presidential Library, while Palo Alto and other neighbors objected to its size and traffic. In 1987 Stanford's faculty senate (26 to 4) and students (3 to 1) voted to reject. The Reagan sponsors then found in 1988 a 253-acre welcome by real estate developers, the Lusk Company, with a mixed-use project and an ocean-viewing bluff-top site in Ventura County, Cal. There the project got under way in 1990, the only recently built presidential library lacking tacit academic approval. The Carter Center (of Emory University, with Carter as University Distinguished Professor) was thrown into controversy when supporters tried to expand

Atlanta's nearby Ponce de Leon Avenue into a presidential parkway.

Once a locale becomes a PRESIDENTIAL SITE, everyday economics flies out the window. Land values may soar, tourism expand. The state of Kentucky calculated that tourists brought $806 million in 1989 to "Lincoln country," a fifteen-county region around little Abe's birthplace. Most Lincoln country never saw, hosted, bedded, or harbored that particular Lincoln. Well-off neighborhoods that face heavy impact from PRESIDENTIAL SITES tend to organize against them. When Kennebunkport, Maine—a town normally holding some 4,500 citizens—became the summer residence of George Bush, its population jumped, although some residents claimed the joint was already jumpin' before the Bush arrival. Such ballooning, especially by tourists and sightseers, accompanied the Jimmy Carter vacations at Plains, Ga., or migrations of the John Kennedy clan to its family compound at Hyannis Port, Mass., and the use of the Bush summer White House compound.

The reach of presidential aura is an old one; of the nation's more than 3,000 counties, 167 are named for presidents. There is a Washington County in 31 states, a Jefferson in 26, a Lincoln in 16, including several former Confederate states. Towns and cities, streets, and memorial highways named for presidents abound.

But memories are short. Fame flees as power ebbs or is misused; crowds wither when presidents depart. The drawing power of the Peanut Museum at Plains subsided when the peanut-growing Carter's presidency ended. The town, observed a *Time* reporter stretching for similes, "sunk back into the kudzu like Brigadoon."[9] For the most part, however, once a PRESIDENTIAL SITE always a PRESIDENTIAL SITE. Some sites keep their value and may even ascend to sacred status only A.P.—After Presidency—or postmortem. The Carter boyhood home is to become (in the 1990s) part of the Jimmy Carter National Historic Site and Preservation District. The longtime home of the Carters will join the District after they die—one of many postmortem sacred places.

By 1990 occasional countertrends were showing up in attendance. Zelinsky found the peak year for visitors at "nationalistic shrines" was not coincidentally the Bicentennial year, 1976. Lincoln's Tomb at Springfield, Ill., peaked in 1968, dropping by more than half in 1982. Wondered Zelinsky: was it "post-Bicentennial fatigue?"[10] For at least a half-century, the president had become a media personality, his goings and comings indispensable to media coverage of The News from Washington—or from whatever PRESIDENTIAL SITE was currently occupied. The president appeared regularly on celebrity listings—Best (or Worst) Dressed, etc. How far the de-consecration becoming visible in the 1990s would extend was unclear. How much disaffection would be laid to public fatigue, to short memories, or at the feet of unpopular or resigned or imperial presidents is yet to be measured.

SECURITY

Variants: security zone, gate, office, holdover.

Within a generation, SECURITY became a growth sector of the American environment. It survived the end of the Cold War, occupied prominent positions at airports, became a feature of thousands of neighborhoods, and carved a new slice out of family incomes and corporate budgets.

Prior to urban riots of the 1960s, SECURITY had been a benign presence. The word itself dates from around 1432 and appeared in Shakespeare's *Macbeth* from the witches' mouth: "Security is mortals' chiefest enemy." But that was long before the protection of material wealth became the overriding concern of modern western societies.

That was also long before lessons from the Vietnam War were pressed into stateside civilian duty to suppress civilian riots, with huge block-grant support from the then-new Federal Law Enforcement Assistance Administration. In the process, this "rich and complex word . . . moved from the spiritual domain into the material and finally into the political."[1]

Quickly, many police departments were equipped with rapid-fire military-derived weapons, helicopters, and some search-and-destroy tactics from Vietnam. Military phrases such as "firepower," "quick-response," and "securing the perimeter" appeared in police publications. When police got more weapons, in response to criminals' gunnery, it was described as an "arms race."

Fear of Russian attack "was the political idea that unified national life" for two generations after World War II. Thus international and domestic politics were joined by an obsessive wish to be secured against enemies within and without. The American Society for Industrial Security became the big daddy among specialized security associations that sprang up.

Following a rash of armed hijackings and bombings of airports and international planes in the 1970s, SECURITY became an essential part of air travel. Going through SECURITY became a familiar and often tedious ritual at airports. Its presence was marked by clearly elaborated gauntlets, zones, or gateways with metal-detection and X-ray apparatus, operated by uniformed guards, often armed. Before embarking from U.S. airports, and on international flights overseas, all passengers and their carry-on items had to be processed through SECURITY.

But SECURITY finds additional jobs at large. It may be an out-of-the-way room containing stored or lost valuables; as well as explosives, poisons, and dangerous creatures in transit. Also here repose prisoners, or suspects held pending the arrival of specialists—bomb-disposal experts, bailiffs, or, in the case of lost-and-found valuables, of anyone bearing proof of ownership. During a riot, a handy enclosure may be designated as temporary SECURITY.

At the height of the Cold War, many civil rights had been swept aside under the umbrella of "National Security." (In such propaganda usage, the term, often capitalized, appeared as a condition, not a place.) Such appeals struck linguist George Steiner as "ideological confrontations from which there is no return."[2] Historian Walter Laqueur predicted that: "The extraordinary measures taken against terrorism will become a permanent part of the landscape."[3] SECURITY also expanded geographically by the process called "remoting"—moving key government military targets out of Washington, D.C., and underground. Only after the Cold War

SECURITY

One of the fastest-growing generic places in America during the Cold War, SECURITY found its venue in countless lobbies, entries, buildings, neighborhoods, reservations, and regions (see AIRSPACE)—the most familiar being part of airport check-in. Some are more strict than others: You Have Been Warned.

ended was it publicly reported that the government had secretly built a huge underground bunker designed to accommodate Congress in the event of a nuclear attack by the USSR. Location: The Greenbrier, a posh West Virginia resort.

In a related geographic shift, the U.S. Navy's home-ports were "remoted" around the entire national coastline in the 1980s–90s. The stated purpose: to further spread out the expanded six-hundred-plus-ship Navy in the event of nuclear bombing.

Gradually, the term SECURITY became a handy marketing device. The phrase "twenty-four-hour security" began to crop up in late-twentieth-century ads for luxury apartments, condominiums, and resorts. It became an essential sales gimmick as an ageing population paid extra to remove itself from the world of perceived crime. "Concierge," once a distinctly Gallic term, migrated westward to settle in under the canopies of upper-income dwelling places, at first on the East Coast, but imitatively elsewhere. The term also expanded to accommodate a nationwide growth of private police forces, their equipment and vehicles (usually ready in a compound). Within a decade after the hijacking and bombings of the 1970s, "working in security" became a major occupation and SECURITY itself an important workplace. By the 1990s, SECURITY appeared to many observers as a national obsession—even after the Cold War ended, and the USSR as a powerful enemy faded into disarray.

For two generations, Americans who could afford it had been moving outward from the inner city, distancing themselves from "problem areas." In SUBURBIA they engaged in widespread beefing-up of "turf"—using distance, walls, fences, guard dogs, floodlights, padlocks, and electronic surveillance. Thousands of neighbors concerted to put up "crime stoppers" or "watch neighborhood" signs, and linked their home phones to instant-response police.

Much of this was a response to urban riots (Detroit, 1943, 1967; Watts–Los Angeles, 1965; Washington, D.C., 1968, 1992; Cleveland, 1968; and South Central Los Angeles 1992)—and to the growing disparity between well-off suburbanites and poor minorities in inner cities.

When the author devoted twenty-four pages in an earlier book to "Turf"—the rise of fenced, gated, guarded residential enclaves—it was considered a rather new phenomenon at the time, 1973.[4] *Defensible Space* by Oscar Newman became a best-seller in 1972 among academic books, instructing architects on "crime prevention through urban design." Selling and installing home-security devices became a growth industry in the 1970s. By 1993 the trend has gone national to a disturbing degree. "Hardening"—the encasement of places behind fences and walls, the sealing of windows—is everywhere. Vast realms of the American scene are blocked off.

All this struck foreign observers as a peculiarly American response to domestic violence—an effort to maintain the expensive luxury of low-density living patterns. Most European city cores had been shaped by centuries of wars, riots, and revolution. Many evolved as tightly walled, convoluted mazes of narrow streets, easily (and often, in the Middle Ages) defended. Many of these—aside from the war-torn Balkans and other high-risk areas—are jammed

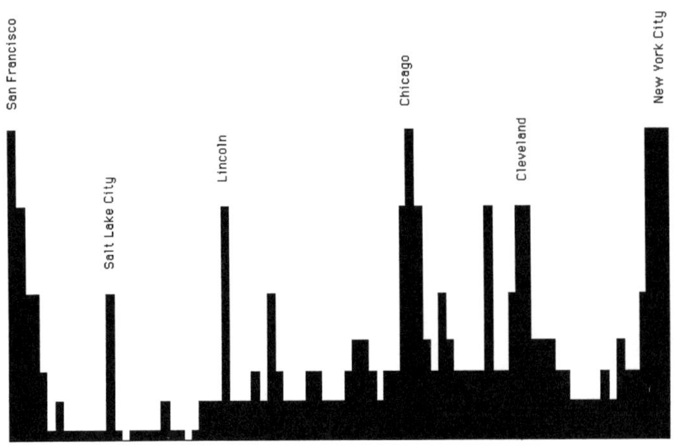

Interstate 80 Northern Transcontinental

Interstate 5 Pacific Coast

with summer tourists. Here Americans on vacation can enjoy the unaccustomed proximity of strangers who would be rigorously kept distant from their own protected stateside doorsteps.

Back home, where they are blessed if not obsessed with open space, Americans face another major test of their ability to keep an open society while shutting off house, home, and neighborhoods from the still-teeming life around them.

SKYLINE

Somewhere near the top of every civic promoter's want-list, somewhere up there in the stratospheric demands of bond-issue salesmen, there occurs the magic word SKYLINE. A city without SKYLINE is asserted to be no city at all. A DOWNTOWN with no SKYLINE is a contradiction in terms. SKYLINE suggests the sky's no limit to urban expansion; it connotes downtown derring-doers with an unlimited line of credit; the city as playground for the godly architect and the ambitious corporation. There's got to be plenty of executive suites-with-a-view from on high up the corporate ladder.

Photographed from a distance, SKYLINE is designed to impress, to beckon, to cast a long shadow indeed. Even the most modest semi-high-rise must be photographed dramatically (no other photographers need apply) to suggest the onward-and-upwardness of it all. By contrast, Tulsa, Okla., was once described as being "so flat that the net at the municipal tennis court forms a skyline."[1] The concentrated cores of skyscrapers in Dallas, Houston, Oklahoma City, Phoenix, and Chicago also benefit from their flatland situations.

In everyday usage, SKYLINE indicates the extent of the eye's reach to that point where a city's visible roofline meets the sky. This visible locale, which changes with every structure added or subtracted, is of surpassing importance in establishing a city's visual image. Civic boosters would become tongue-tied if unable to boast of "this latest, greatest, addition to Our Skyline." This possessive attitude was expanded by historian Lewis Mumford in his influential "Sky Line" columns in the *New Yorker* magazine (1931–63). He and his successor Brendan Gill (1989–) used the term in its widest generic sense, which we

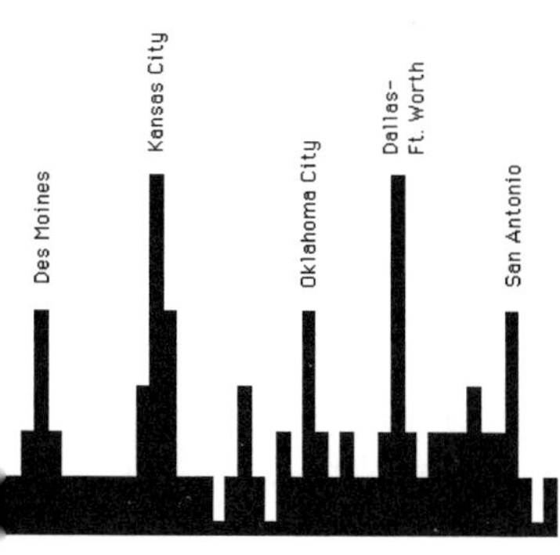

Interstate 35 Interior Plains

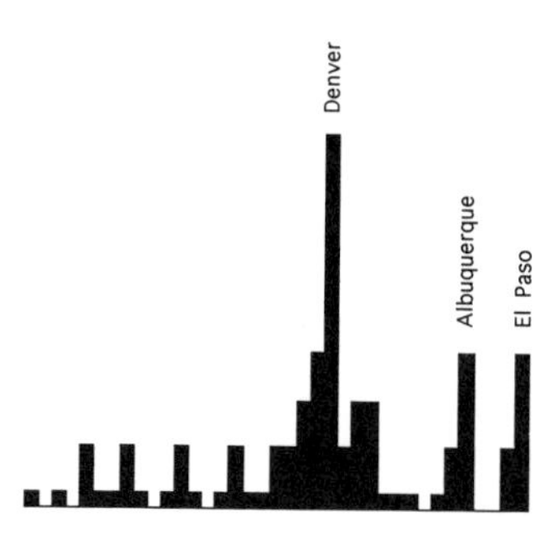

Interstates 15 and 25
Rocky Mountain-Great Plains

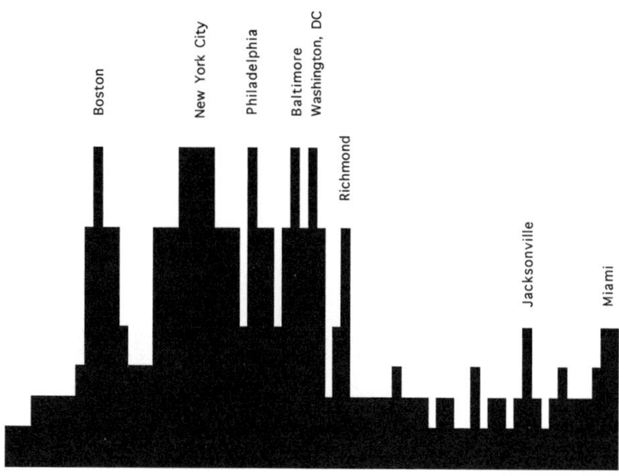

Interstate 95 East Coast

have adopted here. Sometimes interchangeable with horizon, SKYLINE, as used here, is distinctly structured, urban, manmade and in the nineteenth century came to symbolize the traditional Central Business District. The term now is often used to embrace a much wider area.

By 1989, many SKYLINES were hard to tell apart, reporter Robin Garr noting that his native Louisville shared "generic skylines and suburbs" with Buffalo and Sacramento.[2] "Record" heights vary with the place and size. Irvine, Cal., a new town built on a giant family ranch beginning in the 1960s, had achieved by 1986 its first high-rise of sixteen stories. A local headline noted, "Irvine Tower Reflects Rising-Skyline Trend."[3]

The present looks of SKYLINE did not come easily. The first skyscrapers had arrived on the heels of technology: new steel-frame construction, powered elevators, and electricity for lighting all spaces. All these became fully available during the 1890s, along with huge populations flocking off the land to crowded cities. In 1894 the Architectural League in New York City "proposed a law against the erection of skyscrapers which made the streets a hotbed of malaria and were themselves eyesores."[4]

Such distrust preceded by almost a century the growing antagonism toward skyscrapers reemerging in the 1990s, especially in New York City. A magnetic target for loose capital from all the fearful world, Manhattan had become notorious for its shadowed canyons, its showy "signature" towers with their one doorway per block, its bleak sidewalks and pallid pedestrians scurrying toward distant sunlit spots. (Their maneuvers are skilfully recorded in *The Social Life of Public Spaces,* a fine film by William H. Whyte, Jr.) The effect was as though street-level Manhattan had been lifted up and moved 15 degrees closer to the Arctic Circle, away from the sun.

In older cities such as Philadelphia and Pittsburgh, new towers on the SKYLINE were still taken locally as proof of progress. Local history buffs in Philadelphia had been outraged when the new sixty-story One Liberty Place broke the twenty-one-year-old, local but nonlegal, forty-story rule in 1986. The rule was a gentlemen's agreement struck by Edmund Bacon, local planning director in the 1960s. In Bacon's

SKYLINE

Even the continental United States exhibits its own cranky, individualistic SKYLINE when "altitudes" are attached to places where people pile up. SKYLINES vary, depending on where you traverse the land. These population profiles were computer-mapped by census geographer Donald Dahmann, Washington, D.C., who observes that there's "plenty of frontier left" (i.e., land with fewer than two persons per acre).

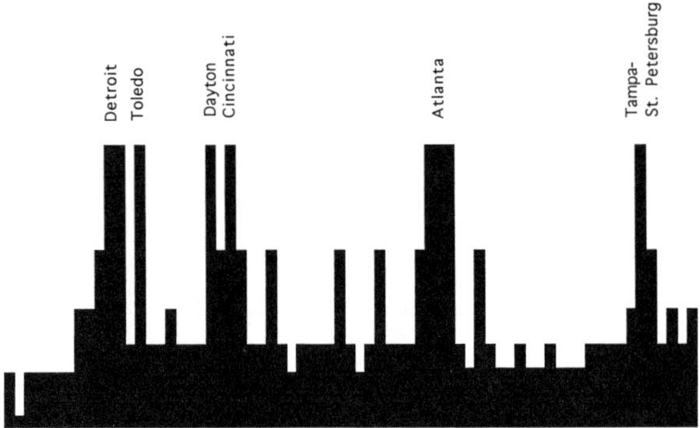

Interstate 75 Great Lakes-Southeast

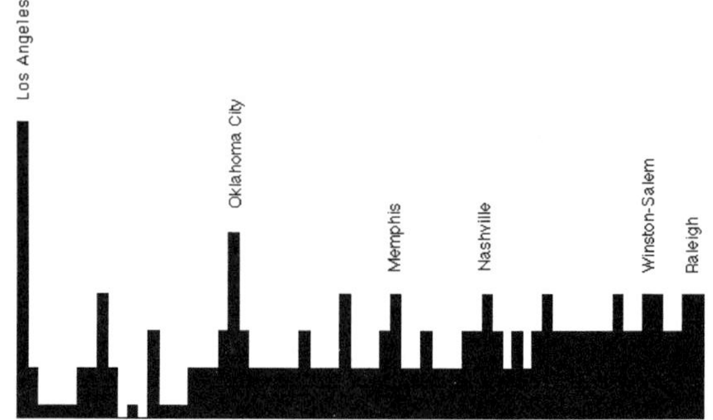

Interstate 40 Southern Transcontinental

Philadelphia, as it was called, nothing could go higher than William Penn's hat on his statue atop city hall, 491 feet above Broad Street. But local anxiety to rise above a moribund economy and a soggy national image broke that rule. As on many another lofty BATTLEFIELD, the new tower's promoters wrapped their project with the highfalutin rhetoric of "A New Day Dawning" above 491 feet elevation.

River cities manage to present SKYLINES as entrance symbols, simply because their approaching bridges and viaducts offer superb viewing sites from which visitors see the SKYLINE: as in New York; St. Louis; Columbus, Ohio; Mobile and Montgomery, Ala.; Nashville; Louisville, Ky.; Little Rock, Ark.; Minneapolis. And many a new interstate expressway opens up original vistas for first-time arrivals from the local airport: Atlanta and San Francisco notably among others.

On Western plains and prairies, SKYLINE may consist of grain elevators, or such rare free-standing towers as Price Brothers' oil headquarters in Bartlesville, Okla., or the State Capitol tower in Lincoln, Nebr. They soar suddenly in splendid isolation to tell us where the power lies.

In addition to power, whatever lies beneath a SKYLINE is high in emotional content. The SKYLINE has long been a testing ground for individual as well as collective ego. In New York notably, as well as in other world cities, competing corporations sought to declare their presence at the very tiptop of their new possessions: the dirigible mast atop the Empire State Building of the 1930s was the first of many, capped a half-century later by architect Philip Johnson's Chippendale highboy top for AT&T's headquarters in Manhattan.

After the great London Fire of 1666, Sir Christopher Wren's rebuilt St. Paul's Cathedral dome and his many lovely, slender church spires punctuated the medieval SKYLINE of London. For this insensitivity to the past he was attacked by some contemporaries, though venerated by later historians. Yet in 1990 England's Prince (of Wales) Charles launched his own architectural crusade against modern metal-and-glass towers which by now had achieved an elevation that dominated the old Wren understory. In Honolulu, Cancun, or San Juan, vacationers arriving for the first time—unless well-briefed by other than travel agents—are usually shocked at the highly-risen SKYLINE looming over their fancied tropical isle.

On the urban scene, does SKYLINE present dozens of spider-legged rooftop water tanks as in New England mill towns, or in textile and tobacco towns of the South, all with their local memories of fearful fires? Or rooftop sawdust hoppers as on the furniture factories of High Point, N.C.? Or chemical tanks and ominous vent-stacks like those peering above the flat roofs of Silicon Valley, Cal.? SKYLINES such as these, the way they cluster or wander freestanding off to the horizon, tell tall stories, often of waste-handling or fire-fighting. But the truth of their origins—how they came to be—often fades from memories, obscured by promotional rewrites of local history.

Each SKYLINE carries telltales of bygones. Distant barns and silos mark the 1840s section-line roads given new visibility from Interstate 70 across Missouri and Kansas. In Chicago's Loop, the prevalence of rooftop water tanks, still visible from far out on Lake Michigan into the 1980s, reveals the effect of a former fire-prevention code that required such rooftop tanks to supplement city water pressure for fire-fighting. So long as this rule persisted, builders shrewdly stopped adding floors at this financial breakpoint.

Los Angeles suffered at the hands of insufferable New York and European critics for generations as "having no skyline"—chiefly because its SKYLINE consisted of nontraditional scatters and spines of high-rises. But then, at long last, an old decayed Victorian mansion district, Bunker Hill, was razed. Cut into broad platforms, Bunker Hill became the competition ground for dozens of tall skyscraping hotels, offices, headquarters: First Interstate, Arco, Union Bank Square in the 1970s; Wells Fargo, Crocker Center, California Plaza in the 1980s. From a distance this SKYLINE, too, began to look like New York's, Chicago's, and London's. In their early days Los Angeles's new forty-story

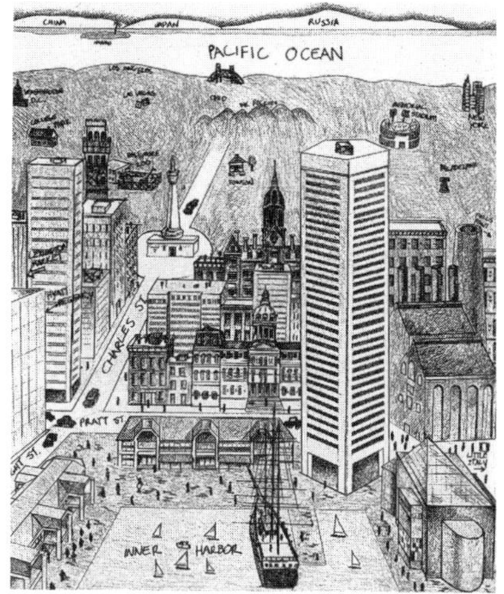

SKYLINE

Most cities' bag-of-tricks to entice tourists and footloose investors includes a SKYLINE. It can be made to turn tricks as a symbol, if not proof, of city-hood or metro status. These cartoon postcards are among scores produced by Harvey Hutter & Co., of Ossining, N.Y., for gimmie-something-different tourist souvenir-hunters. (Clockwise from top left: Baltimore, Md., Denver. Colo., Boston, Mass., and Charlotte, N.C.)

towers were popular pylons for passing helicopters. To many Californians the new DOWNTOWN was "distinctive," which in California had no connotation of quality. Just different. *Los Angeles Times* critic Sam Hall Kaplan saw it as "fractured" into a series of isolated clusters, badly in need of ground-level linkages.

Tax codes continued into the 1990s to affect both the height, location, and duration of skyscrapers. Professor Leland Roth, University of Oregon, observed: "The ethic of building today seems to be, 'Don't build anything that lasts beyond the depreciation period allowed by the Internal Revenue Service.'"[5]

Seldom does a modern SKYLINE please all. In suburbs resisting annexation, among fearful parents of callow teenagers, or among uneasy newcomers, the SKYLINE symbolizes big-city risk, strangers, power exerted beyond local control. Especially in the new world cities (not to be confused with New World cities), much of the SKYLINE may be controlled by foreign syndicates, fueled by laundered cash, quickly thrown onto the market or into bankruptcy in response to distant purposes.

But there are other stories to be seen. In combat areas the invariable rule for survivors is "stay off the SKYLINE"—a distinctive term when shouted as a warning. It can be translated by politicians into "in sensitive positions, maintain a low profile." In smaller towns, churches are often "the most significant architectural buildings . . . in a skyline."[6] In open country, bird-watchers look for the favored perch of hawks and other raptors—bare treetops and telephone poles silhouetted against the sky along a SKYLINE formed by hedgerows and utility rights-of-way along the edge of fields.

The final word on SKYLINES in the 1990s would be electronic. It had become possible, and cheap, to construct or deconstruct electronic images showing how any proposed new structure, or the absence of a familiar landmark, would affect the looks of SKYLINE. In his critiques of high-rising threats to his onetime haunts in Philadelphia, planner Ed Bacon used computer-generated images with telling effect. But soon the web grew more tangled. Developers, anxious to downplay the effects of their proposals, could tweak and twist their images in an apparently "real" color photograph to appear less noxious. Promotion-minded others could "enhance" their proposed projects, making them appear less intrusive, or more prominent, as occasion demands. Soon enough it would no longer be possible to introduce a simple photograph as evidence in a zoning hearing or legal dispute. Soon enough, no photo would be accepted without proof that the original exposure had not been artificially and perhaps illegally "enhanced." SKYLINE was becoming the latest electronic BATTLEGROUND, the latest repository for mistrust of the printed page, and of the electronic image.

SECTION TWO

The Front

TH

By the 1990s it was clear that—in one especially "American" sort of place—our lifestyles and settlement patterns had reached the end of their joint honeymoon. Particularly in this place, our future has been put in question, our lifestyles challenged, our settlement patterns reexamined in the harsh light of a new world order.

This is the place where the energies of the city and the country mix, merge, and compete; where the number of cross-grain commuters exceeds the number going into and out of the old Center; where struggles for control and political wars between places are yet to be resolved. For the first time in our history, here is where "anti-city" population has accumulated the voting power to dominate metropolitan life. Here is the major zone of geopolitical tension for life within the United States.

This complex of places carries many labels, none of them precise: SUBURBIA, exurbia, the outer city, metro, the slurbs, the growth ring, metroplex, edge city, interurbia, commutershed, and zone of urban penetration. It incorporates GROWTH AREAS, ANNEXATION AREAS, and competing market areas. Somewhere lies that disputed terrain called THE EDGE OF TOWN. One word-slinging U.S. mayor called it mouse country—where country mice entertain their city cousins.

▪RUGGLE FOR CONTROL

All these specialized places comprise the larger place, one where flows of energy from all directions compete, mix, and merge: the urban Front. Some of the most dynamic Fronts occur between two competing cities such as Minneapolis and St. Paul, Minn. Over the years their Fronts have begun to merge in a zone of busy competition and increasing interaction: the Twin Cities.

To grasp the reality of the urban Front it helps to visualize modern cities as geographic devices for distributing surplus energy. Some do it well, some badly. But all of them must have—or they die—a zone of energy transfer, competition, and often traumatic change. This zone resembles nothing more than a battlefront or weather front, where powerful forces collide, mix, and merge. They produce spectacular and/or awful results—often unpredictably. Every Front has its winners and losers. Fronts appear under many guises. From all directions they collect and distribute energy by means of orders, message systems, services, roads, expressways, pipelines, and wires. The atmosphere itself spreads urban energy widely via microwave, television, pollution, noise, and light.

Here on the urban Front is where the struggle between developers and sustainers will be played out. Not exclusively here, of course, for the politics of environmental protection, of income distribution, human densities, and the U.S. role in world affairs—these are finding local venues all across the grain of life.

For generations, Americans have enjoyed lifestyles on The Front that aroused the envy of other nations. But now as never before, the future of those spread-out lifestyles has been put in question, and their costs showcased—excess fuel consumption, air and water pollution, and destructive land uses. America's waste of resources became, for the first time in international affairs, a bargaining chip at the 1992 Earth Summit in Rio de Janeiro. Said the Third World: Before you and other rich nations tell us to stop cutting our forests, you Americans—with only 5 percent of the world's population—stop using up 25 percent of the world's annual energy flow! . . . You and your gas-guzzling cars, fancy suburbs, and wasteful habits!

To give up, kicking and screaming, some of these luxuries—and the waste of energy involved—Americans will be testing their ability to practice, rather than to preach, sustainable development and the planning that it requires. Much of that proving ground will lie outside The Center, out in that middle distance, the focus of much new development: The Front. It is on these Fronts that GROWTH AREAS compete for OPEN SITES and drastically change

THE VIEW. Here the debates focus on GROWTH CONTROL DISTRICTS, on ANNEXATION AREAS, and around FLOODWAYS and SPECULATIVE SITES. Who shall noisily intrude into my EARSHOT; where do we draw the line?

No Edge at the Edge

Within these frontal zones of urban unrest, the expression "edge of the city" has lost contact with the new reality. Most so-called EDGES are invisible. If there's no sign, you can't tell. With few exceptions, such as Toronto, Vancouver, or Columbia, Md., THE EDGE presents no sudden, sharply defined, well-managed, visible shift from urban to rural or vice versa. "Farmland protection" is usually a token joke. To travel these Fronts one crosses dozens of borders: of cities, towns, districts, zones, and commutersheds. Their flows come from all directions: over here is an army base with thousands of jobs funded by federal taxes. Will it be deactivated? Over there the U.S. Corps of Engineers maintains a flood control dam. Will they release enough water downstream for the whitewater rafters? At the interchange, the shopping center was built by Canadians now gone bankrupt. Who's next? Down the road, bulldozers expand the HUB airport, paid for by three counties, six cities, two states, and the feds. All contribute energy to—and demand it from—The Front.

In only two generations The Front has grown more complex and contentious. The so-called suburban population of the United States has expanded to dominate that of traditional cities. The Center has been drained of its old working classes and their factories, while many of its retail shops and offices moved to The Front. Within The Front old springs and watercourses have dried up—for U.S.-style urbanization has proved itself an efficient form of desiccation. In the drying-out process, irrigation and the search for scarce waters have crossed the western twenty-inch rainfall line that runs down through Kansas-Nebraska and spread into the East, where cities compete for precious watersheds, sometimes in adjacent states. The thirsty West, which has short-circuited waters flowing into Mexico, now looks across its northern border for future water from Canada.

The situation is nonetheless fluid. Adjacency and proximity still count for much on The Front where THE VIEW is both valuable and negotiable. Whatever moves into those chancey places called CHANGING NEIGHBORHOODS, GROWTH AREAS, or IMPACT AREAS is subject to intense scrutiny if not pressure.

The Omnidirectional Push and Shove

In this land of flux, many forces are in contention; their tugging and hauling runs in all directions. Not just in-and-out. The pressures and tensions exist between and among variegated suburbs, exurbs, strips, clusters, stringtowns, commuter villages, old county seats, onetime resorts, and no-longer-remote office and industrial parks.

Tensions build up out here among the frontal places we examine in this book: ACTIVE ZONES, ANNEXATION AREAS, GENTRIFYING NEIGHBORHOODS, OPEN SITES, SPECULATIVE SITES, and TEMPORARY HOUSING DEVELOPMENTS. Ethnic enclaves no longer cluster just around The Center. Political power is too often fragmented, and "local solutions" don't work. Soon enough, there's a demand for yet another outer loop, beltway, or BYPASS to connect all the above places around the older center, or some new DISTRICT to solve local problems.

The American public—mobile, disaffiliated, responsive to fad, fashion, and fluctuating incomes—has by no means given up its attachment to mobility or its search for new and better mixes of The Center and Out There. Out on The Front, such generic places as CONVENIENT LOCATIONS may be off-the-beaten-path tomorrow. Convenience is quixotic, evanescent, subject to reinterpretation by every decision to build or rearrange highways and route numbers. It shifts with every new detour, SHORTCUT, or one-way street. The current, or changing, convenience of every urban location is routinely analyzed by computers in the offices of United Parcel Service, Federal Express, and other pickup-and-deliverers. Their drivers are instructed by radio, guided by computer-driven maps showing how to reach their next stop most directly. Soon, private motorists can have their own inboard computermaps to guide them on or off "smart highways" and SHORTCUTS.

The OPEN SITES of yesteryear may have disappeared; yesterday's EDGE OF TOWN may today be swallowed up by new subdivisions, its SPECULATIVE SITES become passe. In some states ANNEXATION AREAS will be actively encouraged, while folks who live in proposed ANNEXATION AREAS elsewhere may be chronically litigious, resisting The Center while wrestling with, or even inventing, rogue forms of self-government. Here they are, trying out thousands of "security" devices—from manned gatehouses to exclusionary zoning.

Many parts of The Front, embedded in a state of flux, will become STUDY AREAS of long duration. Millions of land parcels will be earmarked for review, options, appraisal, and reappraisal; for remapping, up-or-down-zoning, and for endless rehearings. Everything in THE VIEW may be up for lease or up for grabs. Land values may be quietly sucked away—or built up—with little notice given to present occupants. If you thought things had settled down, look again.

The Front will continue to provide testing grounds for community and regional planning, an old tradition pushed aside in the land-speculation rush from the 1950s to the 1980s. By the 1990s its legal basis was shifting to environmental protection as part of "the general welfare," so that the right not-to-be swamped by floodwaters from an erosive project upstream, or by stinks from a DUMP upwind, gains political support.

Tax assessors are consistently slow to reflect these shifting tides; just as taxpayers resist admitting that tides can work against them. But taxes are dynamic in their own way—increasingly a lever of political change and geographic flow. And all these scenes will now be sattelite-photographed and computer-mapped. More power goes to those who analyze electronic information about places, and can access the expanding new geographic information systems.

The Death of Privacy

The penumbra of data surrounding every home and office address expands daily, especially in middle-to-upper income neighborhoods of The Front. These are the targets for researchers of every stripe. Those places called "home" and "office" are now coded, indexed, cross-indexed, mapped, analyzed, and interpreted by hundreds of salespeople, tax ferrets, government snoops, and by market specialists who generate junk mail. Information about front-runners is increasingly "public": These are the upwardly mobile people—trend-setters, oft-quoted sources, and the fast-pacers. They make The Front work for them. But The Front is also a target for data-mongers who plumb its rich yield of incomes, buying habits, and other demographic data.

More often than not, the power to manage and to massage this information about changing Fronts lies in already-powerful hands: those of planning and governing officials, large corporations, credit agencies, and national research centers. Only the most determined (and well-financed) networking "computer-cottagers" can afford equal access to place-ful data.

Homeowners, first-timers as well as empty-nesters, will calmly or frantically search the new computernets to find (or to avoid) GROWTH AREAS. People without choice (as well as some who like it) may well end up in TWILIGHT ZONES—or contribute to their making.

All The Front might be considered a CHANGING NEIGHBORHOOD. Pressures upon it never cease. It is forever vulnerable to tomorrow, and its more canny residents tend to become joiners and petition-signers at a moment's notice. They post "Watch Neighborhood" signs on utility poles. As block captains they scout their beats and survey their streets for carriers of strange viruses—political or other strands—from The Center, or from Out There.

All this hugger-mugger suggests that those who are adept at maneuvering in changing Fronts are learning secrets for surviving in the ever-more-complex urbanizing world of tomorrow. People of The Front need all the friends and all the local organizations they can muster or create, so as to match or counter the external forces—distant corporations or governments—that can quickly move into The Front and send them reeling.

Learning whom to trust is most essential along The Front. Here, as elsewhere, trust builds up mainly from repeated dealings with significant others, more often under stress, and essentially under rules of civil encounter. Body language works best at close quarters. Love at a distance is never enough. The placid, uneventful life of people immured within stable neighborhoods, living behind the spreading walls and gates of private "compounds," is poor training for the unstable futures that lie ahead. Outside these limited spheres, The Front offers a unique learning environment that encourages the mastery of survival traits for all who cross the grain of an American city.

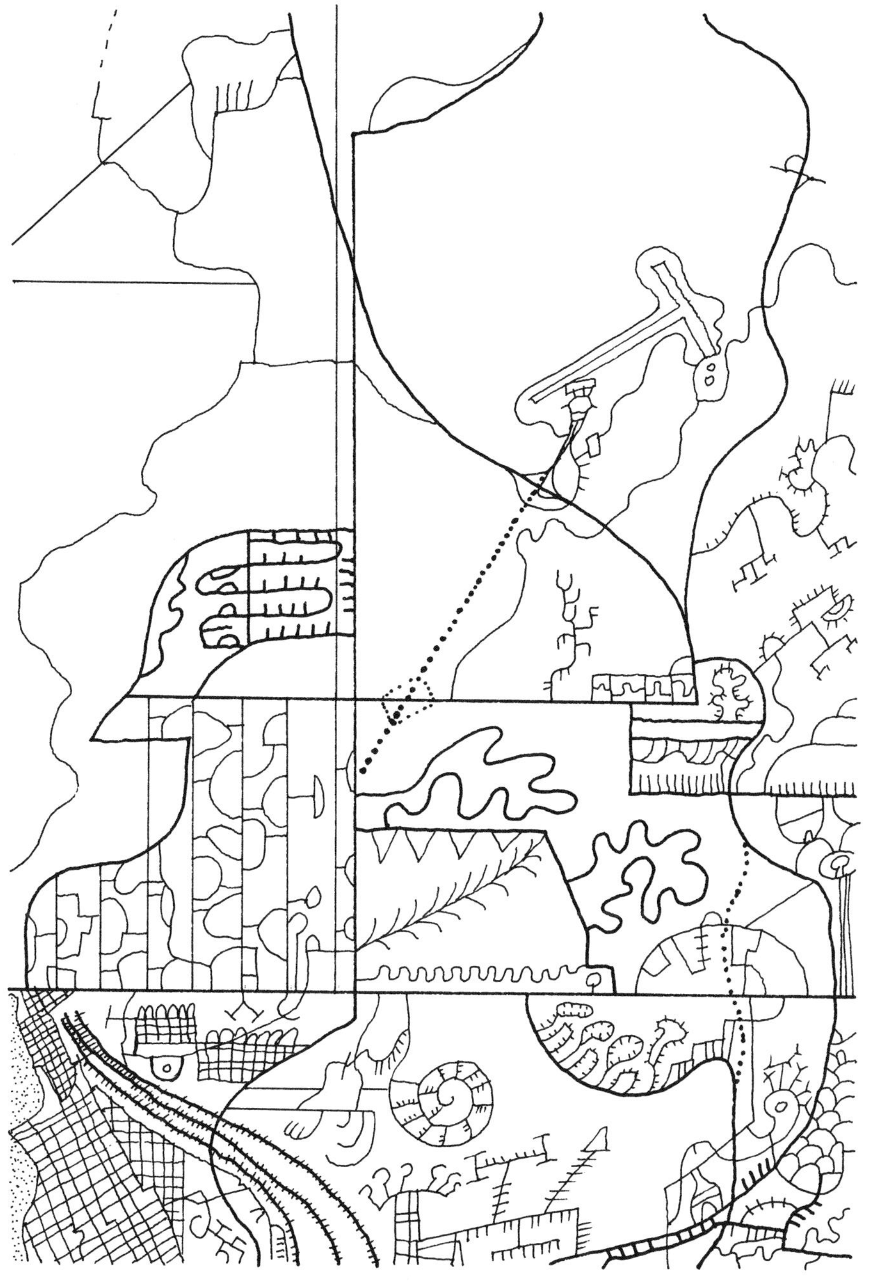

Out of date the moment it was first published, this "futures" map of THE (prototypical) EDGE OF TOWN portends imminent changes (shown in dotted lines) required to serve the unfinished new airport. Soon to come are connector roads, interchanges, speculative land purchases, and change-of-zoning applications, accompanied by local hysteria and lawsuits. Map by artist-cartographer John Himmelfarb.

CHAPTER 4

Ephemera

There are no Baedeker Guides to places that are here today, gone tomorrow. The study of ephemeral places is nobody's baby, an academic waif-and-stray with no learned society to offer sanctuary. Surrounded as we are by the fleeting images of TV, by roadside scenes that flit and flicker past, and grown accustomed to half glimpses, we shrug, and say "That's the way it is." But there's more to be said. Ephemeral places are the products of a restless, destructive, uprooting capitalist industrial system that forces all places to become, or to respond to, the market. The other unifying element that connects all man-made generic places is indeed time. This is the basic ordering element—the genus-at-large connecting us each to the universe.

But ephemeral places contain the makings of many futures—often leaving time-lines and traces behind on the landscape. Such places remind us of—as they demonstrate—the touch-and-go nature of all cities. Here are BOOMTOWNS leveled off, FLOODWAYS dried up, THE SETUP disassembled, TOADS finding new mating ground, WRECK SITES turned into statistics. All is in a state of flux—but some of it takes root to become another and more lasting kind of place. Learning to accustom one's eyes to futures aborning is a way of coming to grips with whatever comes next, and thus with time itself.

ARRIVAL ZONE

Here is a place everyone has experienced; but it has no name. It is a kind of place familiar to millions of travelers, but marked on few maps, and recognized by only a handful of specialists. It is the psychological, geographic but unofficial ARRIVAL ZONE of a town or city.

Ignominiously, most highway travelers approach North American cities with no ceremony, no welcome, few greetings beyond roadside billboards, a city limits sign, or gaggle of luncheon-club trademarks. The "Welcome Zone" has no lobby to promote it, few scholars to study it, but many cartoonists to lambaste its clutter. (The rare exception is Wilbur Zelinsky, geographer, of Pennsylvania State University, whose nationwide study of "welcoming signs" is an excellent source.)[1]

Consequently, a strange city lying ahead over the horizon, just out of sight beyond its expressway ramps, is a problem area to travelers. "Getting there" is a chore.

Only at the state line does "Welcome" get the environment it deserves—in the familiar state-sponsored TOURIST INFOR-

MATION CENTERS, travelers' rest areas, and such official pull-offs. However, these facilities are typically located far from cities, in open country, easily visible a mile down the road, but offering little insight into the next city ahead. Often this location is a political compromise between competing counties, and/or follows some official mileage rule. But seldom does it announce the next city: seldom do state and city politics join hands at this logical point to make travelers welcome.

Unlike highway rest stops—where travelers share toilets, but little beyond the road experience ("Wow, the cops are really out today!"), and a flood of resort/tourism handouts—a welcome station would dramatize the city's SKYLINE ahead, with maps, computer displays, how-to-get-there advice, and civic promotions.

Most travelers recall those few and memorable and unmarked ARRIVAL ZONES at elevated spots where, suddenly, the next town or city arises from the horizon.

Examples:

(1) At the start or along causeways or bridges approaching waterside cities: Cincinnati, Cleveland, Columbus, Ohio; Galveston, Tex.; Jacksonville, Fla.; Kansas City, Kans.; Louisville, Ky.; Memphis, Tenn.; Miami, Fla.; Mobile, Ala.; Portland, Maine; Pittsburgh, Pa.; Toronto, Ontario; Washington, D.C.

(2) At high spots along routes (often elevated expressways) between airports and DOWNTOWN: Albuquerque, N.M.; Billings, Mont.; Boston, Mass.; San Francisco, Cal.; Denver, Colo.

(3) Or from elevated roads and causeways along a river valley, as at Montgomery, Ala.; Hartford, Conn.; Vancouver, B.C.; Seattle, Wash.; Portland, Ore.

Some glimpses of DESTINATIONS are tantalizingly brief; the "Welcome" flashes quickly and is gone: Atlanta from the south, Cincinnati through a notch on I-75 in the Kentucky hills, Manhattan from the New Jersey Palisades before the highway drops into a tunnel.

Along most highways the moment of arrival is unannounced, unmarked, gone in a few moments. Yet each offers the next generation of highway improvers a place for new welcome stations where the DESTINATION is visible on the SKYLINE just ahead. Often they still appear at the last highway cut through the hills around a valley city.

But changes have shown up. A pioneering study led the way in 1964: *The View from the Road* by Donald Appleyard, Kevin Lynch, and John R. Myer.[2] Wayne Attoe followed in 1981 with *Skylines: Understand-*

ARRIVAL ZONE

Futuristic scene: a traveler's view of the next city-just-ahead on today's trip. This pre-viewing room is located in a TOURIST INFORMATION CENTER. It also offers computerized maps showing the city SKYLINE, with routes-on-call to various DESTINATIONS. Punch in your DESTINATION and get a print-out map showing how to get there. As of 1993, not yet available to the general public. Concept-design by the author; computer montage by Ronald D. Facktor, Video Perspectives, Inc., Louisville, Ky.

ing and Moulding Urban Silhouettes.[3] In the 1970s "The Athens of the West," Columbus, Ind., acquired large acreage around its arrival highway from the west as a greenbelted entry district, but without an accompanying welcome station.

By the 1990s there were more signs of change. The term "gateways" acquired a new valence. City officials and others grew concerned about first impressions. Tallahassee conducted a competition in 1991 to improve its "city entrances." The planning office of Rochester, N.Y., commissioned a Gateway Study, with the optimistic hope that "a city, through its gateways, may gain control over its image."[4] Many urbanists still thought of "making their cities observable" in terms of improved road signs. But full-fledged ARRIVAL ZONES were still far from reality.

BOOMTOWN

Monuments to human greed . . . hellholes of corruption . . . "Fair, O fair domain of man's adventure!" Back of every BOOMTOWN described in any such terms lies the blinding speed of capital movements, of fast bucks pellmell across town or around the world. When fast money begins to flow, mobile mobs assume it will never stop. Thus, both simple and complex are tales of BOOMTOWN. Some grow fast on quickly skimmed local assets, and in the process reach out to new and wider markets. Others flourish, then fade into obscurity or petulance. Over much of the original and later Western frontiers, anxious hordes of settlers/speculators swept across, skimmed the first easy wealth, then vanished, with latecomers picking up leftovers.

On early pages of local histories, many a Great Boom is dimly or vividly recorded. New York City captured some 68 percent of the more than three million mostly European immigrants arriving in the United States between 1840 and 1855.[1] Such was the rebuilding boom that a man born in New York in 1816 would find forty years later "nothing, absolutely nothing of the New York he knew," wrote the editors of *Harper's Monthly.*[2] Early New York appeared a GHOST TOWN to its survivors, whereas newcomers flocked to its recurring booms.

Much of the continent is today dotted with remnant GHOST TOWNS, some buried amid latecomers as in New York, some surviving only in histories and old maps.

The Twin Cities of Minneapolis–St. Paul started as a 155,520-acre cession from the Indians at a waterfalls site discovered in 1680 by Father Louis Hennepin. It emerged from military reserve and army post (Fort Snelling) to mill town (first lumber, then grain), then staging area for lumber rafts down the Mississippi River. Water power made possible the growing cities, the value of their products quadrupling between 1870 and 1880. The number of waterpowered mills jumped from eight in 1869 to eighteen by 1876. Then came foreign competition: by 1930 the milling district had shrunk visibly, and the world's largest new mill was being built not in Minneapolis but in Hungary.

But the Twin Cities would soon rapidly coalesce into a modern bipolar metropolitan area. Together they combined the goings-on of a shipping port and a wholesale distribution center with a reach of 1,100 miles into the vast Upper Northwest. They even did the once unthinkable—joined amicably to build a single sports stadium, and formed a regional metropolitan council for governance in the 1970s.

Other BOOMTOWNS lose both boom and bloom, but the reputation lingers long after boom departs. Bayfield, Wis., was a backwater town that attempted in the 1980s to restructure its appeal to tourists. It was Indian land until a treaty in 1855 shunted the Indians off to reservations. Its grid streets cut out of the Wisconsin forest in 1856, Bayfield quickly became a booming sawmill town and shipping port. A single mill cut up to 4½ million board feet per season, so that the edge of Lake Superior became a giant sawdust field. Commercial fishing was bringing a quarter million dollars a year by 1881, and Lakes steamers were offloading crowds of summer tourists from cities farther south. Local red sandstone sold at good prices, and was shipped off to New York for fashionable blocks of brownstone row-houses that by the 1990s would fetch millions apiece in good Manhattan locations.

But by 1920 most of the forests were timbered to death, the lumber boom was tapering off, the market for brownstone busted. Farmers moved in on cutover land, but eventually quit the scraggly farms on thin soils. The last big clambake on nearby Madeline Island was held around 1890, the last Indian Festival in the 1920s. Fishing fell off, and was almost killed off by an invasion of trout-destroying sea lamprey in the early 1950s.

Across the landscape appear many variations on this basic theme of turnover: Oak Ridge, Tenn., built in secret to produce World War II's atomic bombs that pulverized Hiroshima and Nagasaki, was maintained into postwar years as a one-client research town for the Atomic Energy Commission. Albuquerque, N.M., was once a mecca for Eastern tuberculars; a few of their scattered tiny cabins survive today among thronging sun-seeking tourists. "Jobhunting fires" helped fuel many a Western town's temporary boom in the form of emergency government relief: "The sudden infusion of dollars from a major fire could transform a stagnant hamlet into a boomtown for a week or two."[3]

In Alaska, the North Slope Borough, about the size of Minnesota, flowed with wealth based on oil taxes in the 1980s. Its budget in 1985 allowed spending sixteen times as much per capita for its 7,500 residents as did non-oilport Anchorage for its 244,000 citizens. All this made the town of Barrow the highest-cost-area of all the fifty states.[4] But by the 1990s the oil boom had tapered, incomes fell off, many former BOOMTOWNS were strapped for cash.

Improved gold-mining techniques in the 1980s set up a series of new BOOMTOWNS across the Old West: Jamestown, Cal., Wenatchee, Wash., Deadwood, S.D. New methods of grinding and refining low-grade ore could recover about a quarter ounce of gold per ton of raw ore, or five ounces of concentrate from twenty tons of ore. Wenatchee's "state-of-the-art mill complex" was producing in 1985 at the rate of five hundred ounces per day, worth the then-current price of $320 per ounce, with recovery costs of $150 per ounce.[5] What was called "the biggest mineral rush in North American history" followed a 1991 diamond strike around Yellowknife, former gold-boom town in Canada's Northwest Territory.[6] "Having survived gold booms, logging booms, and railroad booms, Centennial [Wyo.] sits quietly at the foot of the mountains waiting for the next wave to roll through. . . . the town's only 'police car'—an immobile junker painted as a decoy."[7]

Other BOOMTOWNS reflect quick infusions and leakage of migrating capital: Houston, Tex., under the impetus of high oil prices in the 1970s ("at the top of the pile" of U.S. cities in 1974) and the Manned Spacecraft Center's millions in the 1980s grew notoriously as the skyscraper city of the Southwest.[8] But its economy crashed in the 1980s following oil prices and the savings-and-loan fiasco. The once-dazzling Shamrock Hotel, built by millionaire wildcatter Glenn McCarthy, was sold at a knockdown price in 1985 to Texas Medical Center. And, finally, holdout Houston accepted what it had always insisted it could do without: zoning.

Along Connecticut's Gold Coast, new skyscrapers broke the old sedate SKYLINE of Hartford, Conn., as in the 1970s Hartford became a mecca for firms moving out of high-rent, high-risk Manhattan coming down off its last boom. Hartford's boom peaked by 1989 when housing costs had gone up sixfold, pricing many office workers into one-hour commutes, which they had quit Manhattan jobs to avoid.

City boosters cling to the BOOMTOWN image long after it no longer fits reality. A city's fabric reflects layers of building types left over from passing booms—a mosaic of pioneer homesteads, Victorian mansions, Craftsman cottages, the last war's "temporaries," suburban condos by the thousands, aluminum-reclad offices of the 1960s, Post-Modern paste-ups of the 1980s, and other reminders of mass-produced fads and fashions now over the hill. Many leftovers become the lodestones to new historic districts, such as Lower Manhattan's cast-iron storefront district, and Miami Beach's Art Deco district, woebegone in the 1960s, jammed with new night spots in the 1990s.

"One consequence of the Great Crash of 1929 is that *boom* has come to connote an unrealistically rapid advance in prices, which will probably be followed by a sharp fall," wrote Geoffrey Hughes.[9]

Once the bizarre savings-and-loan scandal of the 1980s broke into public view, booms and BOOMTOWNS came under harsher media scrutiny than in the Good Old Days. Newly injected into American BOOMTOWN lingo in the 1990s was a radicalized media language, sometimes called post-Marxian: a sharp-edged transition from fawning press coverage in the booming 1920s and 1950s. Here and abroad, BOOMTOWNS based on cheap extraction of public resources—minerals, waters, access—would encounter more explicit language, regulations, and competitors.

See also DECLINING AREA, in Chapter 1: Back There; GROWTH AREA, in Chapter 5: Testing Grounds; and GHOST TOWN, in Chapter 7: Power Vacuum.

CHANGING NEIGHBORHOOD
A handy indicator of a neighborhood's downward mobility is the demolition of old stables, garages, or other alley outbuildings, and cars (often old models or junkers) tucked up tightly to the house for protection. If, however, new two-car garages and elaborate flower beds appear, they signal another transitional process: gentrification.

CHANGING NEIGHBORHOOD

It all depends on how one says it, the load of innuendo one adds, the expression on one's face. "Changing" alone means little: all neighborhoods, even those known as "stable," go through changes. The term is useful especially to sociologists for denoting a neighborhood in process of shifting from a higher to lower socioeconomic status. Also useful in real estate circles as a code word to conceal from the public an ongoing loss of real estate values.

Yet it has other uses: to indicate a border zone between high and low density or intensity in land uses; to indicate farmland being bought up by city speculators; or a neighborhood in the throes of any socioeconomic change, such as an invasion by new people intent on "upgrading" the neighborhood. During widespread gentrification in England and the United States in 1960s, this term lost something of its pejorative status and shifted into neutral to indicate any neighborhood in transition from one occupancy or land-use to another. But even when used neutrally, it was taken by suspicious middle-class observers to mean "in transition"—and not for the better.

Seen from the viewpoint of newcomers, these places may be Opportunity Sites for investment, for fixing-up, and for enticing like-minded others to move in. By those of lesser status already there, the process is viewed as an "invasion" that instigates "forced resettlement" of natives. In New York City, it was reported that artists "tend, to an enormous degree, to define the resurgent neighborhoods of the city."[1] Big-city artists, with their entourage of social climbers, publicity and real estate agents, can publicize and upscale a district beyond the pocketbook of those already there.

See also GENTRIFYING NEIGHBORHOOD, in this chapter, below.

COMMUNITY BONFIRE

Possibly the largest collective COMMUNITY BONFIRE comes alight on Christmas Eve along the Mississippi River levee near Gramercy, La., overseen by the Gramercy Volunteer Fire Department. Groups assemble weeks ahead to build pyres, towers, and effigies, such as this stick-built ship and replica of 1927 White fire truck—all to be burnt when the local major pyro-domo gives the order to light-'em-up. In peak years, traffic from New Orleans and Baton Rouge backed up eighteen miles. Photo, Gramercy Volunteer Fire Department.

COMMUNITY BONFIRE

Long before large BONFIRES became administered and highly regulated affairs, Stanford University celebrated football events with group enterprises such as this 1910 structure. The bonfires began accidentally in 1898, then became an annual event. Photo courtesy of Stanford University Archives.

COMMUNITY BONFIRE

To celebrate a victory, jubilee, or important birth date, BONFIRES have burned under various names and impulses since heathen times. "Lighting bonfires on Christmas Eve in St. James Parish [Louisiana, on the Mississippi River levee] has been a holiday ritual since the 1880s. . . . The most popular explanation is that the bonfires were lighted to light the way for Papa Noel, the Cajun version of Santa Claus."[1] Others say they lighted the Mississippi River steamboat route. After World War II, informal BONFIRES expanded into larger competitive constructions, mostly around twenty-five feet high. Some are tightly tucked pyramids, others take the shape of buildings, boats, railroad train cars, etc.

Much earlier BONFIRES were more utilitarian: for burning bones, corpses, heretics, Bibles, or proscribed books such as those of Martin Luther. In the original annual Scottish "banefire" or "bonfire" in the burgh of Hawick "old bones were regularly collected and stored up, down to c. 1800." As early as 1473 it was spelled "banefire," and the spelling "bonefire" was common down to circa 1760.[2]

Today's COMMUNITY BONFIRE is sumptuous, exciting, magnetic, competitive. It occupies a special place in local memories, and occurs in a startling variety of locales: football fields, pastures, municipal parks, and farm sites donated for the occasion. Many colleges and universities have permitted or sponsored Big Event BONFIRES—to celebrate a football victory, or let off steam after exams. Sometimes whole houses—usually isolated derelicts or designated eyesores in redevelopment areas—are converted to BONFIRES to train local firemen.

Stanford (Cal.) University's giant annual BONFIRE began by accident in 1898, and by 1910 had expanded into a seventy-foot construction, its flames visible for miles.[3] This was discontinued in the 1970s under pressure from environmentalists—and growing sensitivity to fires in California's hillside suburbs. By the 1990s outdoor burning was outlawed in most U.S. urban areas, but some BONFIRES persist during suitable weathers with—and sometimes without—special permits.

Fire still retains its magical powers—not least to disrupt schedules—as exhibited whenever one occurs during rush hour, or produces highly visible plumes and flames. "Policing" often means keeping the public away from hoses, equipment, and firemen, not to mention flames and falling structures.

DESTINATION

One revolution is now behind us—the multiplication of the amount of food one farmer could produce by one day's labor. Another is taking place: the multiplication of places one person can occupy or modify in one day's travel. The former was the multiplication of sustenance; the latter is the multiplication of presence. We can "be" in hundreds of places within one day.

Every move we make, each departure and arrival at a DESTINATION is a form of multiplication. To "make one's move," to choose one's DESTINATION, is to choose life over sloth, there over here, prospect over retrospect.

We are where we go. DESTINATIONS

determine destiny. A move-making society such as ours is immersed in transition, free movement, cheap travel, open access, right-of-entry, rights-of-way, the end-of-the-road—with "Getting There." We are more excited by futures than by history; by the end of the road than its beginning.

As a consequence, life has become "a trip" in more than the psychedelic sense of the word. It has sequences, progressions, successions, trajectories, and sometimes predictability. Making a move has been one of our most popular forms of problem-solving—by moving the cause, moving the solution, or moving ourselves. Ever since man took over the horse, and then horsepower in machines, we have made environments work for us as behavior settings for regular, periodic, recurring movements. This has added much to the risk, for risk expands with every move, and especially with speed. We expect 65 mph on highways, 500-plus in the air, supersonic in space, and instantaneous in communication.

Thus DESTINATIONS loom larger in our budgets, more prominently in our planning. In poverty or in wealth, our working geography is dominated by DESTINATIONS. Most of our occasions, events, trips, or ventures will fall under their shadow. They are the focus of our schedules, the goal of our movements, the end-product of our spent energy. To get from here to there, to arrive at somewhere else, one must go past or pass through many interim spots and finally end up someplace. Even people who run around a track, walk around the block, or just "mess around"—using up gas, taking up time—are all going someplace. To ask someone "Where are you from?" may intrude into yesterdays better left forgotten. None of your business, buster! But to ask "Where are you going?" —that's more like it: exploratory, anticipatory, forward-looking. It opens up prospects and possibilities; it invites answers that range from fact to Fantasia.

We put special names on these moves, calling them runs ("over to the grocery"), trips, swings, tours, calls, errands, visits, beats, forays, commutes, or just travel; and in doing so we follow our rounds, circuits, orbits, treks, and routes. Along the way, we may drop in, stop by, detour, take a SHORTCUT, make a delivery or a pickup, or change route.

Choosing to move rather than to stay has long history. Greek myths, those gripping early versions of the romantic novel, involved pursuit, capture, search, abduction, escape, and reunion—all overlaid with a heavy dose of coincidences: ". . . and suddenly." Those wandering warriors, those separated lovers confronted endless and shifting DESTINATIONS which beckoned over years of travail and through epic poems lasting far into the night.

Movement gave physical form to the linear east-west or "long-lot" shape of the first American Colonies. This settlement form we still see today in exaggerated aspect in Canadian farms along the St. Lawrence River and in the Cajun (French Acadian) Country of Louisiana. This east-west linear shape of colonies, extending west from the Atlantic Coast, encouraged movement from coast to interior and back, especially by river. Then in the 1830s came the National Road (later U.S. 40) from Maryland to St. Louis to promote settlements beyond Appalachia and outward via Western rivers.

Corridor, not cluster; strip, not containment, became our favorite settlement pattern, especially since cheap fuels made it easy. Early New England and Virginia colonists tried to compress their citizens into compact towns. But our "independistas"—from Roger Williams leading his flock out of Massachusetts to found the Hartford Colony, down to today's suburbanites out along Strip City—have favored stringtowns, riverbank towns, and other attenuated growth corridors. Advocates of "compact settlement" have been outvoted, time and again, by millions of others who could use cheap transport for escape into low-density patterns Out There where land was low-cost and movement easy.

Thanks to cheap oil and mobility, all places in the United States have become, one way or another, DESTINATIONS. In this era of constant travel, movement, uprooting, resettling, each life is filled with arrivals and departures, with travel planning, and leave-takings. The average American car covered 10,729 miles in 1989, the typical long-haul trucker 78,900 miles and, in 1990, 435 million passengers flew 432.4 billion miles on domestic airlines to uncounted thousands of travel DESTINATIONS.[1]

In this competitive, time-bound, goal-oriented society, finish line is more important than THE START. In the Horatio Alger tradition, where you end up is more important than where you began. (In those stories, it is assumed that you started off poor, Out There someplace.) Even in the most banal TV sitcom, the plot thickens with travel plans, big moves, and a wild chase, while in the last minutes we arrive at the Big House—prison for some, mansion-on-the-hill for others.

The prevalence of DESTINATIONS is a measure of the tightening up of urbanized life in America. Most of our moves over public roads and highways are regulated by "origin-destination" studies which clock and map our travel patterns. These dictate the way old highways are improved or abandoned, new routes plotted, and stoplights coordinated.

The more intensely our lives and moves are organized, the more value we place on DESTINATIONS, on timely arrival, and on what we do once we're there. "Tourist Attractions" offer prime examples: Disney World is a famous DESTINATION dependent upon highway interchanges—which Disney controls—and on airline schedules which supply millions of Disney customers. All DESTINATIONS are at the end of somebody's trip, some traveler's trajectory.

"Looking ahead" has become an essential part of this mobile everyday life. Millions of families now possess or call upon electronic calendars and reminders, computerized schedules, agenda, and checklists, endless variations on the AAA Triptik. There is no way, short of deliberate seclusion, to avoid the scheduled end-of-the-day, and whatever DESTINATION to which it delivers us.

The fact that cheap petroleum is disappearing throws all future movements, along with myriad DESTINATIONS, into question. And the invention of strategies to reduce the need for coming-and-going has barely begun. It shows signs of becoming a worldwide preoccupation for years to come.

DROP ZONE

Distinctive places are created where people discard stuff: stolen goods, throwaways, identities. Here, on an expressway exit ramp, still caught up by the exhilaration of expressway speed, impatient motorists, halted at the first off-ramp stoplight, fret and fume, open their doors, and empty out their ashtrays and trashbags.

DROP ZONE

"DROP ZONE" occurs not merely in air-to-ground combat talk among paratroopers, who wait there for supplies to be dropped from planes. This occurs also in civilian talk about a place not to be found on one's handy-dandy city map. DROP ZONE is a geographic term, a coinage for this book, adapted from the term "zone of discard" recommended by city planner William Knack of Chicago, and originating with the geographer Raymond E. Murphy in his book *The American City*. We use it to describe a zone of transition, where land values and productive human activities drop off or diminish. It is often called a DECLINING AREA, perhaps THE BOONDOCKS, or even no-man's-land. Murphy's "zone of discard" is a bit upscale from DROP ZONE. Here, he says, "one finds pawnshops, family clothing stores, bars, low-grade restaurants, bus stations, and cheap movies. . . ."[1]

Several DROP ZONES may exist in the form of non-growth or derelict rings or sectors around a city center. Such places give off signals to would-be dumpers, who feel free to drop their loads of trash. Often the location and surroundings are conducive to surreptitious goings-on: DROP ZONES tend to appear in out-of-the-way places, vacant back lots, odd corners, dingles, dead ends, at topographic quirks—old quarries, swamps, or declivities—rips, nooks and crannies in the city fabric. Often they reveal old boundaries where two towns, precincts, or subdivisions abutted each other in an odd lash-up of surveys, titles, loyalties, and boundaries now overgrown and trashed.

At a DROP ZONE, car thieves feel more secure in their break-ins; fly-by-night repairmen get away with shoddy work when city inspectors close their eyes to goings-on in the DROP ZONE. Closer scrutiny of a DROP ZONE shows it often doubles as a refuge for poor minorities, the homeless, and other defenseless urbanites. But some DROP ZONES attract migrants from opposite ends of the socioeconomic scale: ruffians and refugees on the one hand, as well as young architects, artists, or hippies of the 1960s and their successors, on the other. To the former, it's a hideaway, out of sight of process-servers. To the latter, a place to declare one's individuality.

One need not be a sociologist to spot a

DROP ZONE

The "dropping-off-place" for wheelchair and other infirm patrons to this huge State Fairground (Louisville, Ky.) is well-if-not-excessively removed from the actual entrance in the background.

miniature DROP ZONE. Just stop your car at the end of the off-ramp the next time you exit a local expressway. Roll down your left-hand window. Look down at the gutter. It's full of cigarette butts, and sometimes bottles. It's where careless slobs, impatiently waiting for the stoplight to change, open their left-hand doors, dump their empties, and empty their dashboard ash trays into the gutter. That is your neighborhood DROP ZONE.

Quite another sort of DROP ZONE is a low spot in a roadway close to where this is being written: a spot where rainwaters from a nearby slope accumulate and drop their small deposits of silt. When it dries out, this DROP ZONE yields a dozen or so wheelbarrow loads of silt per year for the author's nearby garden.

Geographer William Warntz identified another specimen DROP ZONE along the Ohio River, calling it an "Income Front."[2] Here occurs the sharpest drop in incomes and capital formation across the borders of the mainland States—a line still marking off the borders of the old Confederacy from the victorious and wealthier Union states.

A final DROP ZONE: the north-south interstates through middle Georgia. By the time over-the-road truck-trailers from the frozen North reach this point, they begin shedding heavy chunks of winter ice from underneath fenders and undercarriage. This particular zone of discard may be a hundred miles long, depending on weather.

DROP ZONES are akin to a variant called "spill zone"—that place on the outside of a curve where the contents of fast-turning trucks are likely to spill off onto the roadway: at Eugene, Ore., where log trucks covered with snow come off the highway into town, spilling snow (and in an occasional spectacular accident, an entire load of logs) onto the street. Or in Lexington and other Kentucky towns where tobacco-loaded trucks swing into town for fall auctions, spilling loose leaves along every curve (a diminishing sight as tight tobacco-baling takes over). Or places in coal-mining districts where enough loose coal drops onto roadways to replenish the buckets of poor people out scavenging. Or roads leading to a cotton-ginning town such as Fitzgerald, Ga., their edges white with fresh-picked cotton blown off of giant wire cages being hauled to the gin.

In spy fiction as well as real life, a simple unadorned drop is a designated spot where Secret Agent A secretes a highly valuable document, the spot known only to Secret Agent B who is trained not to be observed in the stealthy act of retrieval.

See also DUMP, in Chapter 7: Power Vacuum.

EARSHOT

Variants: noise impact area/zone, echo chamber.

Noise, as distinct from sound, is something other people make and we dislike. The innocuous term echo chamber, since it originated indoors in experimental acoustic laboratories, has moved outdoors to include noise areas, noise impact zones, bounce-back zones, various forms of DOWNWIND, and other imprecisely named places of high noise levels, the most annoying being those subject to echo effect.

"Since noise was recognised as a problem in Britain in the early 1960s, it has been treated as a pollutant. Noise abatement zones are broadly modelled on smokeless zones. But noise is not like other pollutants. We can predict that a certain amount of effluent in our rivers will cause a certain amount of damage. We can't always do that with noise."[1]

Among the more subjective influences coming at us, noise and its perception are highly personal: A British committee in 1960 defined it as, "a sound which is undesired by the recipient." Memorably, in a P. G. Wodehouse novel, a golfer complained of "the roar of butterflies." Britain's countryside code was revised around 1981 to suggest: "Make as little noise as you can." The author's 1983 "Olmsted Code" of conduct for users of U.S. public parks designed by the famed Frederick Law Olmsted included the admonition: "No artificial noisemakers."

While not pinpointed as the main factor, echo chamber and its effects are significant in landing and approach zones to airports; along expressways with impacts up to a mile away; and near firing ranges, heavy industry districts, and schoolyards. A big city itself, combining millions of motors, engines, pumps, and movements, is a major noise generator.

Aircraft noise became the most invasive racket in urban areas by the 1970s, especially following deregulation of the air transport industry. In dozens of cities, noise impact zones were plotted and debated. Anti-noise protests brought on quieter aircraft engines, plus local controls over noisy take-off procedures. Many airport authorities bought out neighboring objectors. A handful of the enlightened planted GREENBELTS to absorb noise from airports and industrial zones.

The typical U.S. interstate (four-lane, divided highway) in 1990 produced an average sound level of approximately 55 decibels, with 75 decibels expected when widened to six lanes in urbanized areas. By the 1970s, excessive build-up of noise along urban expressways brought selective introduction of noise barriers—prefabricated concrete, masonry, or metal. In response to local political pressure, these were installed notably alongside higher-priced residential areas where, residents claimed, excess noise would lower property values.

However, people grow accustomed to living in an echo chamber. Old-timers around airports notice noise less than newcomers. Noise is also a sign of life: in Brazil, to say that a town is *movimentado* (bustling and thus noisy) is to give it praise. The idea that noise is bad and quiet is good runs counter to the notion that activity (life) is good and inactivity (death) is bad. Whether noisy music from rock and roll and punk rock discos is positively dangerous to human hearing continues to be vigorously debated. Adult antagonism toward "that goddam noise"—as well as to many other sounds, human or otherwise—usually expresses social disapproval of the sources.

See also DOWNWIND, in Chapter 9: The Limits.

FLOODWAY

Variants: flood/floodable area, flood zone, flood site(s), flowage easement, bottoms, impoundment, inundation area, flood retention area.

Rainstorms crossing North America are as certain as death and taxes. New records are set every year—somewhere. Meanwhile, the works of man—cities, roads, roofs, airports, sidewalks, and ditches—reduce the greened landscape's original capacity to absorb rainfall. They speed the storm's run off the land, confine it to narrowed valleys, and heighten its peak flows. Yet the public continues to be surprised by the inevitable results of all this, called floods.

Thus, the so-called hundred-year flood (or "flood-of-the-century") appears to come more often these days. The old definition ran, "A hundred-year-frequency flood has a 1 per cent chance of being equalled or exceeded in any one year" and "a 30 per cent chance of flooding over the life of a 30-year mortgage."[1] In 1993 there occurred in the United States the sixth flood to be labeled a "200-year flood." All who live in FLOODWAYS would do well to learn the new lingo. Rather quickly, their real estate can be converted to "positional non-assets."

While few people brag about living in the FLOODWAY, millions do so: a national inventory, 1959–62, estimated floodplains in the mainland United States totaled 134,156,000 acres, or 5.8 percent of the nation's land area.[2]

Historically, the Rivers and Harbors Act of 1899 has been so amended that the U.S. Engineers' purview extends far beyond its historical reach into "navigable waterways," and extends upstream into millions of suburban and urbanizing tracts. Meanwhile, FLOODWAYS' inhabitants—the determined, the foolhardy, the ignorant, and the helpless—are reinforced by newcomers who refuse to read warnings and heed advice. These are often quick to complain that "They"—distant and despised governments—are to blame for the latest flood and its damage.

The oddity of FLOODWAY politics persists. "Flood" is a man-made term for a nat-

FLOODWAY

Airview of urbanizing portions of the Ohio River Valley during recurrent high waters. After it was published in a regional newspaper in 1964, the author of the accompanying news account (and of this book) was threatened with physical assault for "hurting the business" of home-builders working this dynamic FLOODWAY between Fort Knox and Louisville, Ky. Louisville *Courier-Journal* photo by Billy Davis.

ural condition. There is no "flooding problem" until people move into the FLOODWAY, and, come the next high waters, seek public reparation for the costly results of their own actions. This is the "problem" to which politicians are trained to respond—rather than to get at its human causes: i.e., the choice of risky locations.

Meanwhile, weather predictions have become an obsession with TV watchers. Predicting hurricanes, tornadoes, storms, and ensuing floods since the 1950s has become a growth industry. It attracted toothsome, winsome, and occasionally well-informed "weathercasters" to hundreds of TV/radio stations. In the ensuing glut of "disaster" reporting, local flooding is now scrutinized competitively and in great detail. But between floods, any media critique of the continued settlement of people into flood-prone areas is criticized as "sensational" and hurtful to local growth.

For more than two hundred years, much flood protection in North America consisted of speeding downstream as much of the flood as possible. Local and other governments have dealt with natural flooding by streambed manipulation, by floodwater-dam building, and by "local protection," chiefly single-purpose levee and floodwall structures, pumps, and diversions. Some are spectacular: the Atchafalaya escape route may someday permanently change the course of the Mississippi across Louisiana. Most flood-prone areas have long since been surveyed if not prescribed-for by the U.S. Corps of Engineers. As expected in floodtimes, the Engineers, National Guards, Red Cross, and local emergency workers are "called out" for flood rescue, whereupon the whole cycle begins again.

Among FLOODWAY'S components, visible and invisible, is the flowage easement. This is a bundle of rights as well as a geographic place, located where an adjacent river or reservoir is expected to rise and expand during floodtimes. It is usually defined, and sometimes purchased, by public agencies to forestall building or other intrusions upon slopes or bottoms adjacent to the public water body. Sometimes, after the latest flood, local governments try to install flowage easement across already urbanized tracts. All hell breaks loose when residents—some living illegally in a local FLOODWAY—insist "nobody told us" about the easement, and politick against it.

By the 1990s the movement of population into FLOODWAYS, flowage easements, coastal beaches, and other such risky places guaranteed that the annual cost of flood damage would continue to go up. Bloody predictions by environmentalists about "the flood next time" were borne out in wet-year 1993 when the Missouri-Mississippi river system broke out its largest Midwest flood in history. Its waters created a thousand-mile FLOODWAY as large as Lake Michigan, inundated more than a hundred towns, and topped, breached, or damaged eight hundred man-made levees. Congress provided $110 million "to help people in the flood-ravaged Midwest move to higher ground" by buying out their flooded properties and setting them aside for parks or wetlands.[3] Oft-flooded parts of two hundred communities would be removed.[4] Specialists compared this and future disasters to Bangladesh. No longer could America's great rivers diffuse themselves harmlessly across open landscape, nor could such floods any longer be dismissed as "acts of God." FLOODWAY had come to stay as an uncomfortable, vengeful neighbor.

GENTRIFYING NEIGHBORHOOD

This former slum, lower-class neighborhood, or once-fashionable but long-since-declined housing area is usually part of a mature but still-changing city. In an oft-recounted scenario, it is "reclaimed," "redeveloped," or invaded by ambitious couples a.k.a. "yuppies," young upwardly mobile professionals and/or intellectuals, eager to do-over older properties. By thus investing their presence, good taste, money, and furnishings, they think to "bring the neighborhood back," often to a glory it never knew—and profit in the process from higher prices for their property. The change has often been dramatic: "Summer chairs on the sidewalk, television out on the stoop, and children's street games are replaced with herringbone pavements, fake gas lamps, wrought iron window railings, and a deathly hush on the street."[1]

Gentrification, the practice and the term, migrated from England to its colonies, the modern use of the term being first charted in 1964, in England, by sociologist Ruth Glass. That was when younger generation country gentry moved "back to the city" into run-down, eighteenth- or nineteenth-century row-house sections of London, importing mod colors, cafe curtains, and blatant lifestyles glibly publicized by upper-class media. It was a process well supported by developers and well studied by sociologists. Glass observed that, in the Twenties, "One by one, many of the working class quarters of London have been invaded by the middle classes.... Once this process of 'gentrification' starts in a district it goes on rapidly...."[2]

Gentrification can be understood as a form of urban epidemic, spread by human contact, and supported openly or covertly by mortgage-lenders and officials—who sometimes collaborate in initiating the process. But, unlike lethal diseases which required injections, sneezing, coughing, or rats and lice to spread their bacillus or virus, gentrification spread via the book, the talk, and the illustration. Charting its course became a preoccupation with the big-city media. Thus the penetration by gentrification of London and major U.S. cities could be called a "mediated epidemic."

Shopping, as practiced among the newcoming gentry, was a social event; whereas among those they displaced, shopping had also been hard work. Armed with special tactics—overinvesting for the future, "seeing the potential" of a run-down block, organizing to extract extra services from officials, calculating for the long run—the newcomers brandished such weapons as were unavailable to those being dispossessed. To the new gentry, entry seemed easy.

Official redevelopment, like blitzkrieg, drew its impact from its speed. Gentrification took longer, but seemed sudden. The old residents, clinging to old ways, hardly knew what hit them until too late. Often the in-migrants proved to be the vanguard, conscious or not, for large investors and/or speculators waiting for the right moment to demolish blocks of long-held slums for new high-rent properties next to the newly gentrifying district. The process worked in New York, Philadelphia, Boston, San Francisco, and other U.S. cities as it had in London and Paris. City promoters seized on gentrification as a device to "save the city," or to "revive downtown." Banks and other moneylenders seized on "redlining" as a covert device for denying loans to old properties in neighborhoods "ripe for redevelopment." By the 1990s, U.S. gentrification had become widespread as a not-so-hidden agenda of many city officials anxious to replace the growing ranks of the vulnerable poor, deviant, and anonymous with outgoing, free-spending, image-making white-collar taxpayers and their upgraded houses and trend-setting shopping districts.

But after its Atlantic crossing, gentrification encountered resistance, especially from residents of older ethnic enclaves in North America, and from dispossessed blacks angered by earlier urban renewal; they called it "black removal." Gentrification came to be synonymous with "dispos-

GUNSHOT
"Whose Woods?" headlined the *New York Times* when it published this scene—a joggers' woods trail within GUNSHOT of hunters. The occasion was a deer-season spate of accidental-shooting deaths, including that of an exurban mother hanging up laundry outdoors within GUNSHOT of hunters. They mistook clean, fluttering diapers for a white-tailed deer. Sketch copyright 1992 *New York Times.* Reprinted with permission.

session," thus contributing to the rise in neighborhood protection groups, and laws to back them up. Some studies showed that gentrifiers moved not so much "back from the suburbs" or from outer Gentry Country, as from other locations within the city. Those displaced tended to be more varied than the in-movers. Target areas seemed to be "primarily white, lower-middle-class, socially heterogeneous neighborhoods,"[3] according to one study. To others, it was merely more evidence of the grip by which a white upper class and its bankers held power over urban real estate.

The newcomers solicited new neighbors "more like us." As certain city populations added newcoming refugees and ethnic mixes from Pacific and Caribbean hearthlands, displacement has expanded. Entire towns and regions can become targets, accompanied by old and new tensions and compromises. For many existing residents, lulled by inflation and by decades of political promises that they were "entitled" to ever-rising property values, any neighborhood change was a threat. For others, "mobilizing against the logic of supply and demand is tantamount to mobilizing against the rain."[4] Yet the processes of disinvestment and abandonment, the imposition of non-compensated costs, and conversion of housing stocks continue apace.

Some commentators see all this in Marxist-militaristic terms. Neil Smith views the 1980s history of Manhattan's Lower East Side as an invasion by real estate speculators who "enlisted the cavalry of city government" for "reclaiming the land and quelling the natives."[5]

The famous Black enclave of Harlem in Manhattan, N.Y., was beginning by 1980 to show signs of gentrification. "It is not the first struggle over the gentrification of Harlem and it certainly won't be the last," observed a *New York Times* reporter in 1987. Here and elsewhere, there is more to come.

GUNSHOT

According to the usual definition, gunshot is the sound you hear when a gun goes off. But that's not the whole story. Another usage, "within GUNSHOT," adds the geographic dimension, with its variants known as IMPACT AREA, danger zone, downrange, and target area. Such terms became all too familiar to media watchers and readers in the 1990s in the aftermath of the Los Angeles (1992) and other riots. This usage usefully diverts our attention away from the noise of the shot to a much larger venue—

the location of the gun, the gunner, and their potential targets, intended or not. It raises such questions as: how far are we from the gun-barrel? Are we within GUNSHOT? Minnesota has had laws making it illegal to discharge a firearm within five hundred feet of a farmstead—one definition of GUNSHOT.

Thus alerted, one is immediately in expansive territory. In 1992, there were approximately two hundred million guns in private hands in the United States. Gun sales were running around four to five million per year, which included about two million handguns. Gunshot—the activity itself—was claiming about a thousand lives each year. For people who are black, between the ages fifteen and nineteen, shooting became the leading cause of death in the United States. GUNSHOT had become a widespread danger zone.

In Los Angeles, after the summer riots, a newspaper poll found that "9 percent of adults . . . carried a firearm during the riots"[1] and "one in four homes has a gun."[2] People living in the line of fire learned to cope; some children learned to sleep on the floor to stay out of GUNSHOT. Thus GUNSHOT as a geographic place became part of the psychic redefinition of territory that takes place in all combat zones. This change of perception alters the aura, the reputation, and the language of violent places. It withers over time as memories of the last shootout diminish.

If we may coin a phrase, within hand-GUNSHOT is a killing ground with a radius of roughly two hundred feet, depending on the type of gun, and the aim of the gunner. But a handgun might carry further than two hundred feet, and a hunting rifle has a deliberate killing range of five hundred yards. In deer season, local residents feel uneasy just hearing guns go off a mile away. Conservatively calculated, GUNSHOT for a handgun covers about two and a half acres. This suggests that another five million acres of the United States comes within GUNSHOT of new weapons every year.

This also suggests that, at this rate, most metropolitan residents of the United States—except for Alaskans and other remote boondockers—will soon be within GUNSHOT of somebody with a gun. This has brought the American populace a long way since 1990, when most of us were thought to be within inter-ballistic-missile-type-GUNSHOT of the Russians. Now we're at peace with Russia, while GUNSHOT has moved closer to home.

HANGOUT

It all begins when two or more friends or acquaintances hang around together in one place, or a set of particular places, which come to be known by the term HANGOUT. The usage, to hang out or to hang around (or round), has been traced to 1847, and H. L. Mencken was still hyphenating it in *The American Language,* 1937, as "the place where tramps and hoboes foregather is a jungle or hang-out."[1] This eventually came to be called hobo jungle.

Some HANGOUTS attract loners, and some involve consenting partners. There's a fine line between HANGOUT and hideaway. However, once past lovers' lanes, trysting places, and rendezvous, the intensity of action steps up. HANGOUT begins to go public. It exhibits a graduating scale of public involvement, exposure, and publicity. At each stage in its expansion, it attracts more strangers. It becomes an element in that publicity-driven place called THE SCENE. It comes in many forms, each more exposed, more complex than the last—from pickup place to watering hole, underground party spot/place, and on to stamping (stomping) ground, MEETING PLACE, gathering place, chic promenade (with variations called great shopping street or ladies' mile), and eventually to PARTY STREET and cruising street/zone, the latter frequented by competitive, self-conscious young male showoffs in exotic cars.

By this time and in these places, HANGOUT has changed almost beyond recognition. It attracts notice from other hangers-out, from motorists, merchants, neighbors, police, and possibly rival gangs. It may be mapped by local media, but almost never gets onto gas-station maps. HANGOUTS are distinctly and concretely tied to a particular sort of place; they are, in fact, a particular sort of place. It is hardly likely (but not impossible) that an enterprising TV show or newspaper will pin the label HANGOUT onto its Teen-Age Weekend feature, perhaps imitating the *Chicago Tribune,* which has a classified ad section called "Meeting Place."

Over time, an occasional HANGOUT will graduate into an upper class with designer labels, becoming known as a fashionable shopping street, exclusive resort, executive retreat, club district, millionaires' row, or the like. In every generation, real estate developers strive—and mostly fail—to start one of these from scratch.

As HANGOUT goes public, it forms a set of Saturday-night specials: places and events geared for Saturday night fever and Saturday night crowds. Saturday has deep roots in primitive worlds before Christ, in whose time one day was set aside for sacred matters, separate from the profane world of the work week. Its roots are reflected in the word Saturnalia, which names a time of heavy drinking if not debauchery, of "unrestrained often licentious celebration or spectacle."[2] It has continued as a time for rituals, for nostalgia, for "going out," for being themselves—and the least-favored time for long-distance telephoning. People who are alone at home feel most lonesome on Saturday night, a realm explored by Susan Orlean in her book *Saturday Night.*[3] Others go where the action is—often a known HANGOUT. His Appalachian town "has a history of 'shoot your neighbor, howl and hoot on Saturday night,'" but no

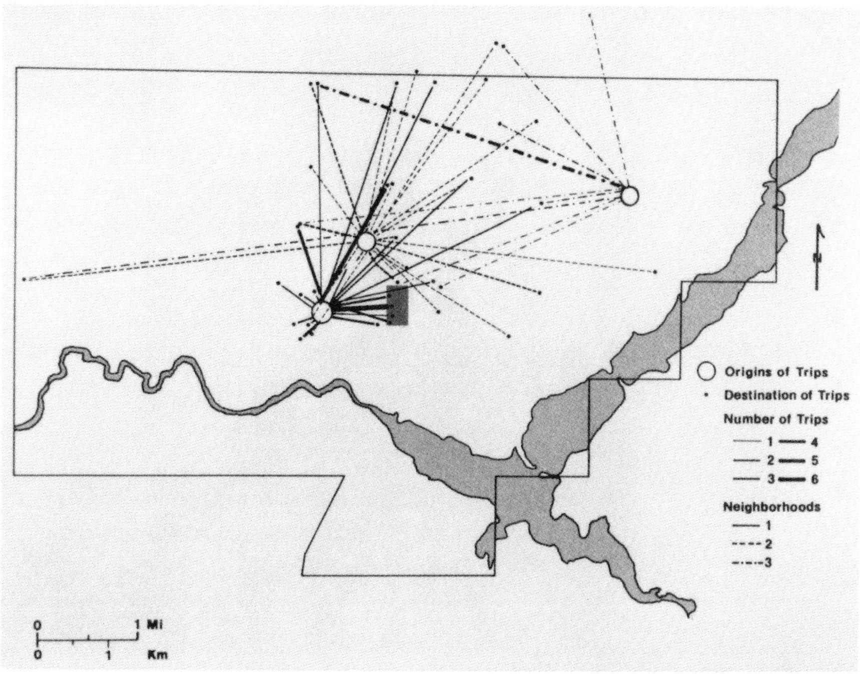

HANGOUT
Journey-to-drink patterns separate profitable from bankrupt watering holes, in this case located in Decatur, Ill. "Those residing in neighborhood 1 [bottom] are of lower social status and tend to patronize taverns in their own neighborhood or in the nearby downtown area (CBD). Those in neighborhood 2 [upper middle] virtually ignore the places patronized by their neighbors to the south." John F. Rooney, Jr., Wilbur Zelinsky, and Dean R. Louder, *This Remarkable Continent* (Texas A&M Univ. Press, 1982), p. 105.

longer, according to Mayor Lester Smith of Jackson, Ky.[4]

Saturday night-places are distinguished, in part, by their ability to attract frequenters from a distance. At the mundane level of eating-out on Saturday nights, thousands of diners hang out at The Hilltop, a twelve-acre restaurant complex on U.S. Highway 1, Saugus, Mass., coming from Maine, New Hampshire, Rhode Island, and all over Massachusetts, up to seventy miles away.[5]

If, however, you choose to hang out underground for diversions, you must learn to stay in that special zone known as one-step-ahead-of-the-law. Here are located those "Hidden Hot Spots," described by *Los Angeles Times* writer Marcida Dodson, known as underground party spots/sites. The parties occur, chiefly on Saturday night, at empty warehouses and other offbeat places taken over as "roving night clubs." They solicit party-goers via leaflets and word of mouth, with perhaps a $20 entry charge and the chance of being raided, before the night is out, by cops looking for illegal drugs, liquor, or trespass. The attraction here is "cutting-edge, high-energy, acid-house dance music" surrounded by funky, avant-garde decor in a furtive, out-of-the-way place.[6] These places and their goings-on seem to have originated in London, moving west via New York and Chicago.

But somewhere along in here, a shift takes place. The above-ground HANGOUT is no longer merely an element in THE SCENE; it expands beyond all recognition into the enterprise realm of official programming and events management. Here we enter the world of fundraisers and promoters, of the bash, the big event, of EVENT SITES and fairgrounds, and the convention center/complex. By the time it gets out here in the big-time big-scale, the HANGOUT has been co-opted by mainstream politics, bond issues, and civic enterprise. Here, one can barely hang-out in Mencken's sense of the word, and only in the presence of grandiose buildings, heavy policing, printed programs, and crowd control. It's a long way from lovers' lane, hobo jungle, and the underground.

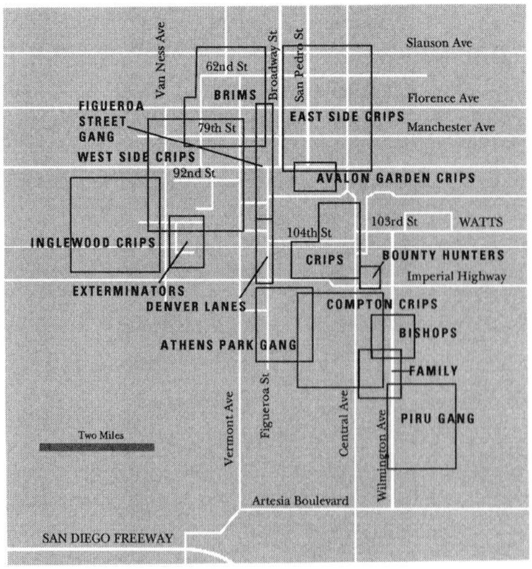

HANGOUT
By 1992, Los Angeles police had identified 230 Black and Latino gangs and 81 Asiatic gangs in the Los Angeles area. Twenty years earlier, local gangs had already staked out their territories shown on this map. Many of these territories overlapped areas of highest unemployment in the region. Mike Davis, *City of Quartz* (London: Verso, 1990; New York: Vintage Books, 1992, p. 301). Map redrawn by Joan Sommers.

See also MEETING PLACE, in Chapter 1: Back There; THE SCENE, in Chapter 2: Patches.

HURRICANE PATH

Deep down within the dark recesses of a local rainstorm somewhere in or around the Caribbean Sea, a tiny puff of a swirl of an eddy grapples in moist embrace with a waif of a zephyr of cooler air. From this nebulous union emerges a windy surge. It reaches out, gathers strength. It emerges as

HURRICANE PATH
Picked up and left by Hurricane Camille (1969) in a prominent waterfront highway location at Pass Christian, Miss., this fishing boat makes the most of its difficult-to-salvage location, far from its homeport, as advertising for a nearby business.

a dynamic nexus—a mix of warm and cooler winds to become a tropical hurricane headed north. A hurricane, after all, is nature's way of transferring surplus energy from the tropics to the north. Once widely held to be at the mercy of the gods, HURRICANE PATH is one of the most intensively scrutinized—and yet still mysterious—routes known to man.

Hurricanes generally make their moves from August to October. But no one can yet predict where the fickle next will go. The PATH is likely to extend ten miles outward from its "eye" or center. It is generally curved, but it can zigzag, or circle back toward the equator; it can stall, loop, or hit the same place twice. Hurricanes originate somewhere in the South Atlantic via the Caribbean Sea, but their endings may be thousands of miles apart, from the coast of Mexico to easternmost New England and Canada. Winds rotate counterclockwise within the north-moving path and thus are faster on the eastern edge than on the west. (Picture the hurricane moving north at 40 mph and the winds rotating at 160 mph. The cumulative speed will equal 200 mph on the eastern edge—exploding frame houses, throwing large boats ashore, etc.—and only 60 mph on the western edge.) One hurricane may suck up millions of tons of moisture and deposit it as driven rain—up to two inches an hour. Hurricane Carol in 1954 swamped one-third of Providence, R.I., under eight to ten feet of storm surging water—a replay of the unnamed hurricane of September 21, 1938.

Human adjustment to hurricanes has always been in part linguistic. In English, the name *hurricane* derives from the Spanish, whose explorers encountered them while crossing the Atlantic. It was *huracan,* or the Taino term *hurakan.* (In the western Pacific, it's a typhoon; to the Philippinos, a *baguios;* to Australians, a willy-willy.) In early stage, it is called merely a "disturbance." When it qualifies for hurricane status, with wind velocity of 73 mph, it gets named in alphabetical order by the National Hurricane Center. Female names were used from 1953 until 1978, and alternating female-male names attached ever since. Once named, it attracts intense and expensive scrutiny from a growing watch-and-predict industry. Thousands of eyes, radars, satellite cameras, and other airborne and surface observers chart its course, guess its damage potential, and issue warnings that dominate TV as the crisis builds. Most media "encourage us to talk of them as if they were angry harpies bend on wreaking havoc in our peaceful households," observe Wallace Kaufman and Orrin Pilkey.[1] An unforgettable example was E. B. White's essay "The Eye of Edna," recounting the trepidations of waiting through the breathless, hyped-up, blow-by-blow radio accounts of Edna's every move—only to be let down when The Eye finally arrived.[2]

A HURRICANE PATH changes almost by the hour. It may veer toward Charleston-Savannah where a twenty-foot surge in 1893 killed 2,000 people. Or Florida's Lake Okeechobee which a 1928 hurricane emptied, killing 1,836 people. Or Puerto Rico (as did heavily damaging Hugo in 1989). Or the Florida Keys, where the "Yankee Storm" of 1935 killed 408 workers on the Keys Highway. Or Galveston, Tex., as did the lethal, unnamed hurricane of 1900 that killed 6,000 people and dashed Galveston's hopes to become the New York of Texas. Or it may parallel the East Coast, then swing unannounced into New England (1938) to fell forests by the square mile. Hurricane Camille, 1969, after doing heavy damage to the Louisiana-Mississippi coast, swung inland to the upper watersheds of Virginia, causing flooding on the James and other rivers.

Or, it may strike with such well-advertised warnings that people are able to escape, as in the case of Hurricane Andrew (August 24–26, 1992), when only ten deaths were reported. Yet that proved to be the costliest hurricane in U.S. history—some 200,000 homeless, Homestead Airbase virtually destroyed, and "South Florida left looking like the end of the world."[3]

Tumultuous seas and up-to-forty-foot waves move along the domed back of a storm surge in HURRICANE PATH. They wreak as much localized damage as winds, especially to homeowners and developers foolish enough to build structures in such known-risk locations as beaches, low-lying margins, or exposed dunes. In the PATH, entire beaches may be downsized to become offshore sandbars. "View properties" disappear along with their views.

It took six great storms hitting the Atlantic coast in the disastrous season of 1954–55 to stimulate formation of the National Hurricane Research Project, now the National Hurricane Center in Coral Gables, Fla. From this Center and its outposts flows a huge output of films, tapes, radar images, predictions, and statistics. Every new or incipient HURRICANE PATH is plotted, charted, rated, ranked, freeze-framed—and scrutinized by fear-driven millions who watch TV.

Fear, however, was not in charge when thousands of cheap, shoddy houses were built in what turned out to be Andrew's South Florida HURRICANE PATH in 1992. "A group of engineers announced that thousands were made homeless unnecessarily: their houses had collapsed in moderate winds because of shoddy [building] practices. Another group . . . found that most roofs that suffered damage simply hadn't been nailed on properly. . . . The truly frightening thing [is] . . . that if you die in the storm's fury, it is quite likely your own fault, or someone else's."[4] In the aftermath, eleven insurance firms failed outright, and thirty-four withdrew or cut back their coverage. "Everyone's rates went up."[5]

To some jaundiced observers, Congressional hearings to liberalize Federal disaster insurance against hurricanes, floods, quakes, had an ominous ring. James Barth, professor of finance, Auburn (Ala.) University, observed, "In one way, it would be like the savings and loan situation. People may engage in socially undesirable behavior if they know the Federal Government will be there to bail them out."[6] Asks Barth, "Why should the rest of us bail them out time and again?"[7]

Yet in spite of all this coverage, the flow of U.S. population to its most risky margins continues. The migration includes newcoming millions who are ignorant of survival tactics in a HURRICANE PATH. Thus the cost of restoring the PATH to something like its pre-blow status keeps going up. "Coastal zone management" under a federal law of 1971 is spotty and in some places a local joke. If global warming continues, ocean levels will rise, along with hurricane damage. The chief mitigation tactic is getting-the-hell-out. Hundreds of coastal communities, especially on vulnerable barrier islands, now practice various degrees of exodus, some in good time, some in confusion. Later, after the PATH has been cleared, it tends to disappear from public view. Regulations to stop rebuilding in precarious locales get bent by developers' pressure. Few beach resorts permit historical markers that might discourage future visitors and builders from rebuilding—in next time's PATH.

See also EMERGENCY CENTER and DISASTER AREA, in Chapter 2: Patches; IMPACT AREA, following; THE BEACH, in Chapter 9: The Limits.

IMPACT AREA

A recent invention, IMPACT AREA has expanded far beyond the BATTLEFIELD where it once designated that comparatively small patch of earth disturbed by a bursting artillery shell or bomb. It now includes entire regions (e.g., "the Frost Belt, impacted by the departure of heavy manufacturing"), having come into widest use as an environmental IMPACT AREA officially designated under federal environmental protection act(s) of the 1970s. IMPACT AREA went public widely in the 1970s when federal agencies promoting airports, highways, dams, etc., were required to file an environmental impact statement and be subject to public hearings. This generated new sub-specialty consultants, as well as place-conscious local groups finding themselves designated to be in an IMPACT AREA.

Thus in a property-conscious society, burgeoning with "Watch Neighborhoods," keyed up to suspect all strangers, the impact of distant Big Government was, more than ever, seen as a threat.

Sometimes it was real, and other times

the product of local paranoia. As neighborhood self-consciousness was whipped up in the 1970s, IMPACT AREA could be variously defined and located—an oil-soaked riverbank three hundred miles downstream from a refinery's oil spill; a clutch of houses close to new early-morning takeoffs from the local airport; or Antarctica with its penguins carrying remnant chemicals from half a globe away.

See also EARSHOT, in this chapter, above.

THE SETTING

This forms the background—the plausibility structure—for a foreground of what is assumed to be an expensive, if not elite, pattern of human activities. The term and its usage suggest interior decorators madly smoothing pillows in the Master's Suite of the Great House before the photographers arrive from a ritzy architectural magazine. Out front, in the circular driveway, there's the photographer's flunky unloading specimen potted plants from the station wagon to enliven a dull corner. Up rolls a delivery van with expensive china table SETTINGS (another usage) for the dinner table—loaned in return for the trade-name dropped casually into the magazine captions.

In such usage, SETTING connotes great oaks enframing a white-columned mansion carefully stage-set behind a nattily-turned-out couple astride their thoroughbred horses or in his BMW convertible, waiting for The Hunt to begin. Or it is peopled with upper-class models sporting new fashions in old-family SETTINGS.

Usage: "No time, expense, materials or workmanship were spared in transforming this secluded mountain setting into a retreat of incomparable beauty."[1] The short-lived American edition of *Geo* magazine, under new owners anxious to convey an upscale aura for its wealthy world-traveling readers, changed an earlier, pedestrian headline about the reconstructed Italian hill town of Orvieto to read: "Jewel in a New Setting."[2]

SETTINGS cannot exist in isolation. They cry out to be accompanied by what they're set up for: negotiable objects, persons, or goings-on suggested by, and evocative of, the SETTING itself. Thus SETTING invites adjectives to take up positions just upstream from the noun. And so we find appropriate, photogenic, romantic, stylish, suitable, upscale, and wholesome SETTINGS. But when the mood and status change, SETTING down-shifts into more plebeian surroundings—beat-up, low-class, run-down, trashy.

THE SETUP

When the rock group struts onstage before a yowling mob, when the fifties' dance band (The New Kings of Swing) settles down amidst its loudspeakers, when the advance team comes to stake out the foundations of an $800,000, prefabricated, ready-to-be-assembled six-story stage-cum-backdrop, all are involved in THE SETUP. It can be simple or extravagant, but less and less likely is it to be an impromptu affair, a jury-rigged stage-set of sticks and canvas. THE SETUP has become the indispensable high-tech, knockdown setting for expensive-to-produce, high-ticketed affairs once called outdoor concerts.

THE SETUP

Eight stories high and longer than a football field, this techno-baroque SETUP for the Rolling Stones' "Steel Wheels" American tour (1989) was designed by London-based Fisher-Park, Ltd. "We take a barren jock-style stadium and transform it into a fantasy environment," said Mark Fisher. It looks like, observed Bonnie S. Schwartz in *Metropolis* magazine, March 1990, "an otherworldly spaceship . . . a futuristic oil rig seized by fashion-conscious pirates." Photo by Mark Fisher; courtesy of Fisher Park.

While The Event remains the main thing, it must be reinforced by the space-demanding SETUP, with its own fans, and its provisions for back-lit stage effects, extravaganzas of strobe lights and lightning flashes, and multiple TV cameras for the essential videotaped record. Selling videocassettes of tonight's and other taped performances will pay for all this technical prep and electronic hoopla.

Platoons if not regiments of grips, electricians, gofers, and structural-steel workers come into play—sometimes after extensive unionized wrangles over who-does-what. Crews drive in their huge over-the-road trailer-vans, days ahead of time, to line the streets where the mediated event takes place: rock concert, pop singer groups, big name presentations. Out come the electric cables, the special lights, electronic controls, the $10,000 amplifiers to awaken the ungrateful dead blocks away. Up in their stretch limos are chauffeured the stars, shepherded past the crowds by prepaid police and hardhatted grips. Over it all looms THE SETUP, its workable chunks now conjoined, where firecrackers safely burst, blank-shooting cannon roar, and dry-ice clouds spread their eerie up-lighted glow. Not for itself alone does THE SETUP need space, but for its emergency generators, special trailers, equipment trucks and vans, often fenced and floodlit for security. Not for tonight's crowd alone does THE SETUP do its multiple jobs. It must survive hail and gale, riot and ruckus, ever-ready for the next gig (in a locale known to the crew as just another gig-stop).

On their 1989 North American tour, the Rolling Stones SETUP was three hundred feet high, twice the size of most touring sets, made of wasteland junk and evoking "shuttle-launching gantries, North Sea oil rigs, derelict industrial buildings and unrestrained urban sprawl," according to Mark Fisher and Jonathan Park, London designers.[1]

As observed by Bonnie S. Schwartz in *Metropolis* magazine, "more of a mobile piece of architecture than anything else, the structure was a study in contrasts, with two asymmetrical wings flanking the band's performing area. And although it seemed permanent and firmly rooted in a way unlike conventional rigs, its solid, thick steel girders were configured to reveal the surprising hollowness of the structure . . . [and the] barren and alienating angularity of the performance area."[2]

Such upscaling is infectious. "Starlight Express," billed as an "enormously big musical—the most expensive road show ever," required eleven forty-eight-foot tractor-trailers to move its 1990 company from city to city.[3] (The typical road show needs five.) Although staged indoors, not out, it required fifty tons of sets, equipment, and costumes. Its major grid for lighting and mechanicals weighed twenty tons.

Taking its cue from THE SETUP, the promoters of Crystal Cathedral located their $100 million cathedral complex in the Garden District of Anaheim, Cal., jogging distance from Disneyland. Dominating the complex is a giant, lacy steel structure, a masterpiece of transparency, containing 11,400 windows, seating 3,000 persons, and having a span of 420 feet. A large cross sits atop a twelve-story tower. "The chancel is essentially a state-of-the-art staging area. Flexible. A giant screen provides everyone with a close-up of speakers and flashes the lines of hymns and announcements."[4]

But by no means does THE SETUP specialize in good works or good words. Anarchia intrudes. Jean Tinguely, the great French mobile sculptor, offered an early anarchic model with his self-destructing machine complex, *Study No. 2 for an End of the World* on the Nevada desert, 1962; and self-destructive others in New York and Paris. The Seattle Center for Constructive Art sponsored another SETUP, June 1990, at an outdoor "Carnival of Displaced Devotion" put on by Survival Research Labs. Among its mechanized goings-on: electronic robots locked in battle, a Big Bone lurching and dragging itself across the stage, a "pulse-jet" effusing a subsonic rumble, and an earth-digger Screw Machine. Such put-on structures that ramble, gargle, pose, and pout had propelled THE SETUP into exploding fourth dimensions from which it would not soon return. Spacewalkers, moon-rakers, inter-planetoidal adventurers would continue offering role models for THE SETUP.

THE SITE

Forever optimistic in its usage, THE SITE is implicated, willy-nilly, in the future. In everyday utterance, THE SITE is incomplete in itself; it lacks its following linked word "of . . ." One looks at open land and takes it at face value: Not much happening there. Crops growing, cows grazing. But the moment someone labels open land as a SITE, this becomes a statement of implication and intent. Something is to be added: It is now THE SITE of . . . a future interchange, shopping center, redevelopment, subdivision, or airport.

In earlier usage by building architects, THE SITE was strictly secondary to the structure; it was ancillary if not an appendage. As architects used the term, no SITES were complete without buildings—their buildings.

Beyond the architectural profession, site planning entered technical language by way of the planning profession in Canada in the 1960s, and found its way thence into the lingo of landscape architects. Many called themselves site planners as a linguistic tool for entering the booming subdivision business of the 1960s. This dismayed old-timers who thought landscape architecture needed no further embellishment. But the term had already come into general usage among builders and developers, since site planning was a nonlicensed specialty open to all comers. By the 1990s some state laws regulating land surveyors were expanded to cover anyone self-labeled a site planner.

Meanwhile, all SITES lost their innocence in the building booms after World War II. All SITES came under suspicion among the new breed of environmental watchdogs who flourished following passage of the U.S. national Environmental Policy Act of 1969. Radical environmentalists looked at nature in all its glory and saw it everywhere endangered by the process of siting. And in truth only small portions of the American landscape were permanently restricted against becoming SITES for development.

TOADS

Temporary, obsolete, abandoned, derelict sites (TOADS) proliferate across the American landscape, but had not become a col-

TOADS

Running east-west across northern Illinois, a bumpy ten-mile stretch of U.S. 20, east of Chicago, suffered drastic cuts in traffic when travelers deserted it for SHORTCUT interstates nearby. These empty, broken-out road signs are tokens of derelict stores, motels, and gas stations that remained, summer 1989.

lective noun (much less an acronym) until identified and coined by Michael R. Greenberg, Frank J. Popper, and Bernadette M. West in 1990. This trio—two planners and a public health man—identified the onset of coined TOADS as "A New American Urban Epidemic."[1] It is spread by the speed of abandonment, dereliction, disregard, dumping, and other land-uses-and-misuses. TOADS lie "at the bottom of a land use cycle" in a society specializing in fast turnover.[2]

Few if any cities recognize TOADS by that name. TOADS owe their on-ground existence to the use, abandonment, and reuse of real estate as exaggerated by a capitalistic system. As named objects of scorn, TOADS are coinages of convenience, constructed by scholars anxious to slow down the social production of TOADy sites. Production expanded after World War II with federal and local subsidies which promoted SUBURBIA, the exodus from cities—and the increase of urban TOADS.

TOADS lack identity until they become a problem; then they attract notice and naming. They occur in three varieties: (1) one-time valuable assets (steel mills) now abandoned; (2) former noxious neighbors (slaughterhouses) now empty but reeking of the past; and (3) sites that are merely vacant but overgrown, dead-ended, landlocked, overlooked, invisible, or inaccessible. The TOADS epidemic, it was predicted, would "only spread and intensify if the [urban] service cutbacks causing it continue."[3] Three out of four selected mid-size cities in New Jersey reported in 1990 that the epidemic spread from larger cities. Lacking strong federal programs to aid cities, the TOADS epidemic, it appeared, would remain a local, not "federal," problem.

TOADS are not to be confused with, but have affinities with, LULUS (locally unwanted land uses), as identified in the 1980s.[4]

See also DECLINING AREA, in Chapter 1: Back There; DROP ZONE, in this chapter, above; ABANDONED FARM/AREA/ TOWN, in Chapter 7: Power Vacuum; LULU, in Chapter 9: The Limits.

VACANT LOT

A result of overproduction, notably in the 1920s, VACANT LOTS proliferated when entire suburbs were left unpopulated by the 1929 Crash or depopulated by the resulting depression. VACANT LOTS and tracts, large and small, are byproducts of building boom-and-bust cycles in a capitalistic society. In the United States, the oversupply of vacant, subdivided land from 1929 lasted through the Great Depression in most cities until the postwar building boom of the late 1940s and early 1950s.

Suddenly, however, as any local housing boom expanded outward to, and beyond, old city limits, there were no more jungles, hideouts, wastelands, or nooks and crannies beloved by all children. It was that final disappearance of the last VACANT LOT in hundreds of neighborhoods that awakened U.S. families in the 1960s to the poverty or lack of local recreational facilities. Such awakenings led toward the eventual formation of neighborhood councils, of "tot-lot" associations, of parent delegations to planning-and-zoning meetings. It bore further fruit in the growth of "Neighborhood Watch" groups and their ubiquitous signs.

In 1963 it was estimated that VACANT LOTS still made up one-fifth the area of forty-eight American cities with over 100,000 population.[1] These were remnants, unbuildables, corporate or institutional reserves, or were being held for speculation.

But in the 1970s vacant lottery moved to the newer suburbs where the mass-production of homesites occupied thousands of acres. Here the old functions of VACANT LOTS were absorbed by large and as-yet undeveloped vacant tracts. These became the haunts not so much of neighborhood kids but of roving dumpers, bikers, hunters, poachers, and encroachers, as well as bands of bird watchers. Each considered such sites its personal opportunity site.

Vacant tracts of larger size played an important part in U.S. history. There were huge tracts acquired by British speculators and Eastern capitalists before and after the Revolution. There were Mexican lands to be squatted upon by U.S. settlers, and claimed after the War with Mexico. There were 183,386,240 acres to be had cheap in the Northwest Territory after the Oregon settlement in 1846, and especially after the Homestead Act of 1862.[2] In this tradition, any legally recorded VACANT LOTS or larger tracts that are not being put to constructive use are seen as fair game for trespass and speculation. If the title is cloudy, so much the better for specialists in that commodity.

VIEWSHED/THE VIEW

Once it was no more than an unnamed part of the visible universe, an element in what was generically called THE VIEW. But VIEWSHED (or seen area) has mushroomed into widespread technical presence, sometimes protected by law or regulation, especially in Western States.

Often VIEWSHED took its first form as the result of federal/local requirements for environmental-impact statements. What would be the visual impact—of this new highway, of that ski slope, or a clear-cut hillside—upon residents and tourists? What expectations do viewers bring to these scenes?

In response to such demands, especially from the giant landowning National Park Service and U.S. Forest Service, computer-aided mapping had a boom in the 1970s, vastly expanding the vision of the Scottish biologist Sir Patrick Geddes whose Outlook Tower in Edinburgh had become a famous vantage point for urban surveys in the early twentieth century. (Earlier concern for THE VIEW had come chiefly from military observers and artists.)

New satellite images flooded into the market: the first Symposium on Remote Sensing of the Environment, held in February, 1962, excited a new generation of landscape analysts. VIEWSHEDS took on regional importance, became political footballs, especially after passage of the 1969 national Environmental Policy Act with its required environmental impact statements.

VIEWSHED/THE VIEW
Instructive sign at the World's Fair, Knoxville, Tenn., telling camera-toting tourists just where to stand to get the properly framed "Perfect Shot." Such photos usually resemble the perspectivists' view, perfected and popularized by seventeenth-century Italian Renaissance painters.

VIEWSHED/THE VIEW

The Renaissance-inspired "VIEW-from-the-Road"—in this case, the prestigious, suburban West Paces Ferry Road area of Atlanta, Ga.—boosts the price of real estate, distances the house and its owners from the passing throng, and is apt to be picked up by elitist house-and/or-garden magazines. Photo by Lawson P. Calhoun, Jr.

Hundreds of such statements, with studies, before-and-after maps, and computer-generated views, poured out of universities and design/engineering offices. Burt Litton, University of California landscape architect, defined VIEWSHED as a visual corridor, a routed, physically bounded area of landscape visible to an observer.[1] Nancy Hardesty, Palo Alto, Cal., landscape architect, defined it in master-planning Portola Valley Ranch (a subdivision): "The natural scenic lands within a defined visual boundary as viewed from public corridors."[2] U.S. Forest Service specialists widened the definition to include views from single points such as overlooks, not necessarily from highways. Other VIEWSHEDS could be perceived only from the air. A skilled cartographer-planner such as Joseph Skvaril of Edmonton, Alberta, could dazzle Canadian audiences in the 1980s with his sketched VIEWSHEDS, using tricks of perspective to improve THE VIEW.

When Philadelphia chose in 1958 among competing designs for its Society Hill apartment towers, local planners counted in each proposal the number of windows having access to THE VIEW. The number of "view windows" in three apartment towers by architect I. M. Pei exceeded those in other proposals. Pei won, the towers are now Society Hill landmarks. Thus the term VIEWSHED moved from West to East, becoming a useful tool for urban designers as well as for wilderness protectors.

It became possible, given new computer programs, and the UC-Berkeley's Environmental Simulation Lab, to produce movie film versions of VIEWSHED, with THE VIEW changing as the observer moves from one viewpoint to the next. Once the computer map showed the extent of present and proposed VIEWSHEDS, scenic highways could be designed, and historical views defined into law for protection. Around San Gorgonio Pass, Cal., consultants in 1982 proposed fifty-five square miles of "critical visual areas." Much of the area was declared off-limits to wind-energy development: no wind turbines permitted.[3]

But protecting THE VIEW alone was often not enough. Each scene could be subdivided for aesthetic content into "character types" and subtypes; into foreground, middleground, background; into dominant elements and textures. Gradually scenery was transformed, becoming part of a Visual Resource Management System, and much of the federal West had been computer-mapped so as to be managed or protected.

Once hauled into court, such VIEWSHED documents eased the way for protesters to stop or modify proposed developments that would obtrude into their share of THE VIEW. VIEWSHED and THE VIEW are omnidirectional, as skyscraper developers found whenever their proposed shaft intruded into neighbors' VIEWSHED. Once sacrosanct and beyond the range of aesthetic criticism, skyscrapers in the United States and abroad have become targets for critics located in their VIEWSHED. "Local outrage over how an immense skyscraper would cast a shadow over New York's Central Park helped kill that design" in 1987.[4] Analysis of VIEWSHEDS moved overseas. Even in Kenya, a nation hungry for development, an environmental critic attacked a proposed sixty-story tower for the Kenya Times in Nairobi, "the tallest skyscraper in Africa." In thirty years, the number of Kenya's cities of a half-million has grown from three to twenty-eight. Outrage over a skyscraper is "a harbinger of what may happen as Africa becomes increasingly urbanized and its open land disappears."[5]

See also WINDFARM, in Chapter 8: Opportunity Sites.

WRECK SITE

There is always a question as to where the wreck actually occurred—legally, that is. Was it at the moment of first glancing impact when treetops sheared off the plane's wingtip, or did it include a quarter-mile path of broken trees and flayed aluminum bits of a disintegrated plane's fuselage . . . or did it extend off the highway, where the bodies were pried out of two cars locked in heavy metal embrace? Afterwards, WRECK SITE may become a mere statistic, perhaps the focus of an annual visitation by survivors, or the locale of simple white crosses put up on the roadside by survivors or who-knows-who to mark yet another automobile casualty. This latter practice appears to be dying out under the overwhelming impact of deaths on the road: fewer relatives find it compelling to keep up with the increasing occasions for white crosses—and with the reluctance of highway officials to allow such "intrusions" into the right of way.

But knowledgeable travelers come to recognize such violence-prone situations as accidents-about-to-happen: a stretch of highway a predictable distance past the last heavy-drinking after-hours bar, or a bad

WRECK SITE

Not all WRECK SITES can be marked with a simple "X." This map shows the search area of the Atlantic Ocean off Cape Canaveral in which searches were conducted in 1986 for the exploded fragments of the space shuttle *Challenger* and its crew. The area comprises about 450 square miles—possibly the most extensive WRECK SITE since remnants of the Spanish Armada foundered against the rocky, foggy coasts of Scotland and Ireland in 1588. National Aeronautics and Space Administration, Washington, D.C.

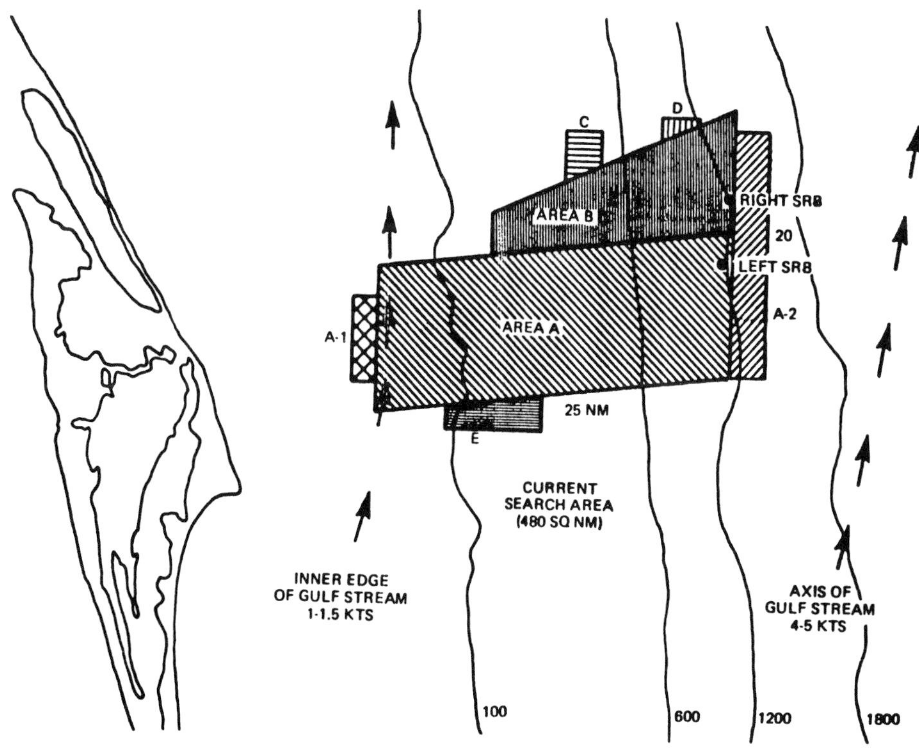

WRECK SITE

Familiar and often illegal, crosses put up along roadsides memorialize fatalities, either in groups or alone, as here in the median of I-65, southern Indiana, viewed to the north against a backdrop of older wreck jobs and junkers at upper left. Photo by William Strode.

curve with an outside camber, or a remote four-way intersection with a traffic light and a downhill run that encourages red-light-jumpers from at least two directions. They take notice of that small and much-repaired house at the outside of dead man's curve where the owner, beset by vehicles and drivers who fail to make the curve, has studded his yard with boulders and logs to restrict the approach zone.

Once the wreck has occurred, WRECK SITE begins to specialize in an orderly sequence of events: road shoulder becomes a parking lane, the closest house a first aid station, and wreckage-strewn landscape a part of the official records, and a parts yard for scavengers. Afterwards, a flurry of lawsuits will usher in the expanding breed of forensic specialists, accident buffs, and expert witnesses. Says Bernard Ross, a founder of Failure Analysis Associates, specializing in industrial accidents: "The world is infinitely more complicated every year, and the 'expert witness' of just a few years back is now nearly obsolete. It takes a team of experts to completely analyze an accident to ensure that something has not been overlooked. Then all the pieces have to be fit together—like the pieces of a puzzle."[1] That fit must be precise enough to satisfy a jury: more than ever before, damage suits arising from WRECK SITES end up in court, or at least in hearings before insurance adjusters.

To those involved in puzzling out such an apparently simple locale as a WRECK SITE, the place will never again look the same. And when all the parties come to court, their memories may have distorted it out of all recognition. Many lawsuit parties now pay to bring computerized reenactment of significant wrecks before a jury.

WRECK SITE took on unexpected dimensions when suddenly the spacecraft Challenger exploded off Cape Canaveral on January 28, 1986. This area was officially designated as a search area divided into debris fields by NASA (National Aeronautics and Space Administration). Total area: approximately 480 square miles. This was possibly the largest WRECK SITE since that fateful August of 1588 when the Spanish Armada, under storm, distributed itself in wreckage for hundreds of miles along the rocky northern coasts of England, Scotland, and Ireland before its few survivors straggled back to Spain.

CHAPTER 5

Testing Grounds

Here—but not exclusively here, of course—is where millions of Americans try on new lifestyles, new forms of shelter, movement, exploration, and speculation. ANNEXATION AREAS can bring on constant litigation, while GROWTH AREAS and oversupply of SPECULATIVE SITES may bring on GROWTH CONTROL DISTRICTS in various guises. Not everyone keeps score; millions of experiments fail or fold with no public records kept. But the itch to differ with the past, to branch out beyond custom and neighbors, and to establish new land-uses continues to motivate us to find expansion room. And by the 1990s, with the proliferation of think-tanks, institutes, and consultants of every hue-and-cry, STUDY AREAS cropped up on every horizon. All these combine to express endless new forms on The Front.

ACTIVE ZONE

The ambiguity of goings-on at this place guarantees widespread use of its name. The term is usually applied at a distance. It may connote activities either legal or illegal and/or increased law enforcement, depending on context. This allows timid media to identify a neighborhood that is considered to be a "problem area" without using that exact term. Thus "Zone 3 is certainly one of our more active zones," according to Major Frank Echols of the Atlanta Police Department. He was describing a zone that "has traditionally had a soaring crime rate, but now the rate is being reduced." The zone included "the Stewart Avenue business district . . . which has had problems with prostitution and narcotics."[1] In such news accounts, the past tense is used as a handy repository for illegal or marginal activities that still persist. This avoids using the present tense, which while more accurate, would bring criticism to the media, and to the source.

ANNEXATION AREA

This contentious zone of intent is a repository of civic hopes, of a mayor's ambitions, or a city council's anxiety over a shrinking tax base. Annex or die—this is oft perceived as an ambitious city's fate. Thus, among urban doyens of the 1960s—when most U.S. cities were in a growth mode—it was widely accepted that only a major consolidation or merger of a city with its surrounding county could overcome the new suburbanites' resistance to being annexed. "We moved out here to get away from the city!" was a suburban rallying cry. Many such rallies also contained the hidden addendum: ". . . to get away from inner-city minorities."

In spite of such outcry, some two thousand annexations per year occur in the United States, chiefly in states giving their cities jurisdiction over contiguous lands. Such extraterritorial liberality is an exception to the historical distrust of cities written into many state constitutions, making annexation difficult.

Odd-shaped squiggles and goosenecked extensions appear on the map at many a city's edges. These are gerrymanders that extend a city's territory via a narrow annexation pathway, usually along road or utility corridors. They are named for Elbridge Gerry, the Massachusetts governor whose Democratic Party, after his reelection in 1811, redistricted the state into salamander-shaped districts to corral as many friendly voters as possible. Opponents combined Gerry and salamander into the term "gerrymander." This is also called "flagpoling." When flagpoling connects a city with some desired distant tract of tax-yielding territory, e.g., a rich shopping district, it is termed "cherry-picking." The process was outlawed by the U.S. Supreme Court, 1964, requiring districts to be compact and "of contiguous territory."

Another strategy, attempted by Chapel Hill, N.C., in 1988, was to control by annexing "the 500-acre Dubose pasture, considered the town's most beautiful entrance way."[1] An alternate form of cherry-picking is practiced by land developers along the Fronts of competing cities and suburbs. They shop for the "best deal," then petition to be annexed to the competing municipality offering most services for least cost.

Such an attempt was made successfully in 1989 at Colorado Springs, Colo., when a shrewd developer, Frank Aries, assembled 25,060 adjacent acres of the Banning-Lewis Ranchland including a foreign trade zone. He paid more than $200 million for land situated smack in the city's "path of growth."[2] Colorado Springs annexed the whole, and in so doing expanded nearly 50 percent.

States with built-in anti-city biases have allowed many growing cities to be surrounded by tiny but politically potent settlements-turned-into-"cities." These could embrace a scant few hundred residents, and hold Saturday front-yard police court to extract fines from speeders using the single through-street connecting to a nearby shopping center. The frustrated parent city, lacking the politics to annex, then would resort to a complex set of contracts, compacts, agreements, "shotgun marriages"—or resort to government consolidation where possible. Such devices help manage these mosaics of odd forms and contentious jurisdictions in multi-city regions.

At the other extreme, some state laws almost mandate annexation via a routine process of advertising, public hearing, and fait accompli, such as was practiced effectively by Tallahassee, Fla., in the 1990s. As models of city-county consolidations in the late twentieth century, Atlanta, Denver, Jacksonville, Oklahoma City, and Toronto were widely envied, but seldom imitated. An adaptable American city form was taking shape, bringing complex forms of citizenship yet to be perfected into the twenty-first century.

CONVENIENT LOCATION

Variants: opportunity site, sleeper or speculation site, the going-home side of the street.

Millions of research and hunch dollars are spent to choose and to dominate CONVENIENT LOCATIONS—for drive-ins of every sort: liquor stores and grocery stores; cleaners; places to buy snacks, rent films, or make that impulse purchase. Convenience may hover around any location of unappreciated or unsuspected value, a place that sits silently awaiting the insight of an observant and upstart newcomer. Here sits an opportunity site. It is likely to be found in a sleeper situation or sleeper location, its assets under-appraised, its advantages not yet realized or apparent, its accessibility subject to change.

In a highly mobile society, convenience often lies on the going-home side of the street, a hot locale for impulse or "convenience" purchases. Major chains such as Kentucky Fried Chicken traditionally put stores on the "going-home side" to catch hungry buyers flowing homeward. Thousands of new CONVENIENT LOCATIONS sprang up in suburbs along going-to-lunch thoroughfares when the midday traffic grew to equal morning and evening rush hours. But when flows change, when streets are reversed, one-wayed, or otherwise speeded up, convenience may relocate blocks away. Like many a one-time country place called Midway, Half-Mile House, Five Mile Station that has lost its original anchorage, going-home places are at the mercy of modern rerouters and short-cutters. One big change is the shrinkage in flows along old radial routes, and the growth of cross-county commuting. The suburb-to-suburb commute "has become the nation's most prevalent route of daily transportation," according to a demographic study by the Eno Institute for Transportation.[1] Historically, streetcar stops were the CONVENIENT LOCATIONS for retailers in early storefront strips from the 1890s. In the 1920s dealers located their automobile ROWS on going-home streets leading out of DOWNTOWN. By the 1990s these were fast expanding and relocating to suburban strips, clusters, and multi-dealer auto centers.

Convenient candidates lie all around: Here is an old family farm soon to become the site of a highway interchange—with extra space for a shopping center. Over there is a low-rise HOLDOUT building that may be bought cheap to provide new elevator banks for an ageing skyscraper standing forlorn next door and ripe for upgrading. Down by the waterfronts ramshackle docks

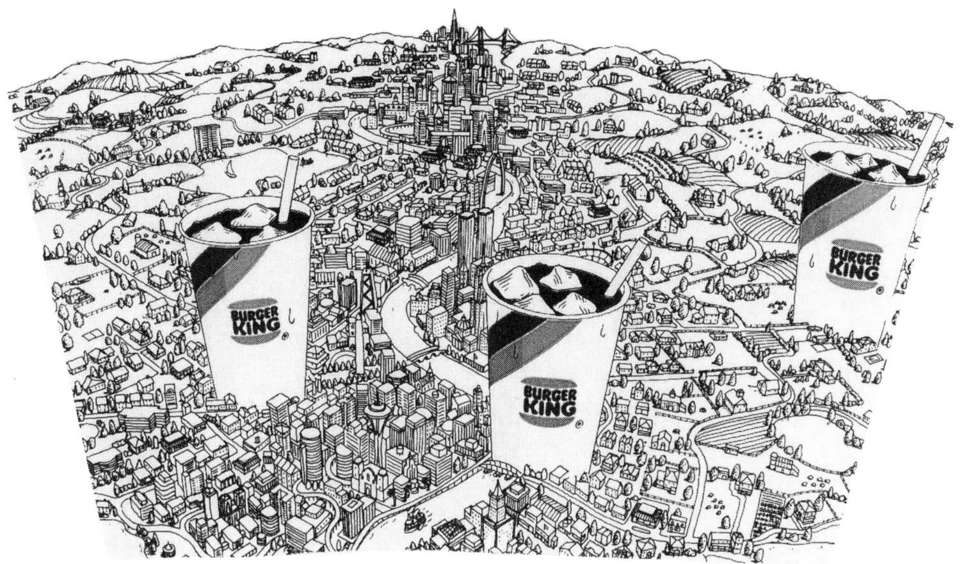

CONVENIENT LOCATION
That most ephemeral of economic assets, "convenience," is forever shifting about in response to traffic, stoplights, one-waying of streets, and the popularization of shortcuts. Among the most assiduous pursuers of CONVENIENT LOCATIONS are the quick-stop-one-stop-fast-food-drive-ins. Sketch courtesy of Burger King Corp., Miami, Fla.

wait to become a festival marketplace. And just yonder lies an innocuous tract that can tie together the scattered holdings of a drunken heir. Then, behold! All may be assembled into the site for a MIXED/MULTI-USE COMPLEX.

On a subtler level, the convenient locale in question may be called a slum, its emotional overtones concealing the locale's other possibilities from most observers—including slum-dwellers. One of the most beautifully bucolic urban tracts in Florida—"Frenchtown," a ten-minute walk from the state capitol in Tallahassee—is locally discounted (1990) by whites because of the DRUG SCENE at a main intersection amid its low-income black population. Class attitudes fixed by the word "slum" are among the most rigid in any pecking order of places. But "slum" says nothing about, while concealing, non-slum conveniences that hover around such places.

See also GENTRIFYING NEIGHBORHOOD, in Chapter 4: Ephemera.

GROWTH AREA

Variants: growth corridor, district, node, pole, sector; development district; expansion district or zone; tourist destination; tomorrowland; power center.

As a replacement for other deities, the great god Growth has been a long-lasting patron to expanding populations whose faith lay in a bigger Gross National Product, in an Ever-Normal Granary, ever-rising incomes—more-of-the-same, whatever that might be.

To live and prosper in a GROWTH AREA was a widespread goal of American colonists. Personal ambitions expanded; pioneer farmers, then townsmen, then city-dwellers learned to dream big, to think

Development. Like Goethe's Dr. Faustus, they expanded their horizons, "from private to public life, from intimacy to activism, from communion to organization."[1] Across the nation they spread their new granges, councils, institutes, chambers, and associations devoted to growth. Early granges, using the older sense of the word, were farmers' organizations, working for the increase of farm crops. Later use expanded to indicate economic and territorial growth: My City vs. All Others.

Inevitably, living in the path of growth came to possess an aura most sought by knowledgeable newcomers. Often they found the choicest sites monopolized by old-timers, and so looked to promote GROWTH AREAS of their own. Well into the 1960s, GROWTH AREA was high on the want-list of most Americans. In Gwinnett County, Ga., northeast of Atlanta, its officials were chagrined in 1988 at the prospect of falling off the "top-ten list" of growing counties. Tijuana, Mexico, with an annual high growth rate of 12 percent in 1986, was touted as part of the new international "Border Metropolis" of San Diego-Tijuana (one that still lacks legal reality).

Growth sector is generally agreed to be that portion of a city aligned along major transportation routes, and thus usually of geometrical, wedge shape, as identified by economist Homer Hoyt.

On the ground, GROWTH AREA may be hard to pin down, its only reality to be found on a map. Out there on the landscape it may consist of a visually confusing and sprawling mix of farmland, woodlots here and there, "For Sale" signs at new subdivisions, and a vast MULTI-USE COMPLEX out by the interchange. Traffic jams mean you're getting close and must read the signs; heavy 7 A.M. and noonday truck traffic means construction zone ahead. Down the highway, past a scatter of new single-story work-and-warehouses, office clusters, and a shopping village or two, some five thousand people may be working at new jobs within a couple square miles of "here" that is called GROWTH AREA. Rush-hour cops speed the flow, to keep GROWTH AREA from suffocating in its own traffic.

In the growth mode, city promoters vie with each other to attract media coverage dripping with press-released phrases like: "anticipated tenants" and "projected population." The rapid increase in traffic through the Newark, N.J., airport (passenger flow up 44 percent in 1983) was expected to cause spillover development "creating a market for first-class offices and retail space."[2] Near Fort Lauderdale, Fla., one twenty-five-year project called Weston, just getting started on drained Everglades land, was described as "a 10,000-acre city . . . a new type of megaproject" under way in 1984. It was designed to contain "more than 45,000 residents, eight schools, several million square feet of office and warehousing space, a firehouse, an airport and 2,000 acres of lakes and waterways."[3] The Route 44 expressway between Virginia Beach and Norfolk was called "the busiest road in one of the nation's fastest-growing areas."[4] In ten years a whole new community north of San Diego, Cal., "has catapulted from less than 10,000 to 100,000," most of the growth occurring after 1988.[5]

In the process, Orange County, once the rural outback for Los Angeles, has become another sort of place—"neither city nor suburb, but with attributes of both

. . . having little to do with Los Angeles," according to *New York Times* critic Paul Goldberger.[6]

GROWTH AREAS are defined by activities. When these slow down or cease, ripples of uncertainty flow outward from the evidence. Following on the heels of local booms comes a lull, a pause, sometimes a depression, a "crash" in real estate prices and wages. By 1990 more than one local GROWTH AREA of earlier decades had run its course. Prices and rents in New York City hit a sharp decline. Economists looked with growing realism at the United States, itself a classic GROWTH AREA. Could the United States keep up with Japan and a combined Europe? Would the prime GROWTH AREA of history—The Great West, as historian Walter Prescott Webb called it—find itself second-rated by other, faster-growing and hence more powerful nations? Or was it not that growth itself was being revalued and redefined?

See also BOOMTOWN, in Chapter 4: Ephemera; ANNEXATION AREA, in this chapter, above; DEPRESSED AREA/REGION, in Chapter 7: Power Vacuum.

GROWTH CONTROL DISTRICT

Only in the late twentieth century did GROWTH CONTROL DISTRICTS emerge under their new and highly politicized name. Growth in America had been taken for granted. But as people watched the spread of population and the exhaustion of various resources, a new mood set in. Growth per se—unexamined and assumed—came to be questioned widely by people co-existing with its byproducts: pollution, gridlock, inflation, tax hikes, and urban unrest.

Among the first to alert the wider public to growth's dangers were nature-lovers and an expanding corps of new-style conservationists. They learned to "think global, act local!" and to flourish statistics of world (and especially Third World) populations out-of-control, threatening future food supplies and well-being. Growth loomed as a menace to newly arrived middle-class families clinging to barely affordable houses—"positional goods in a choice location."[1] Growth produced expressways to uproot whole neighborhoods and to incite a so-called expressway revolt in the sixties. The antagonism toward the new-scale expressways was diverted into broader channels by Lady Bird (Mrs. Lyndon) Johnson's support of "highway beautification" at the 1965 White House Conference on Natural Beauty. This turned into a rallying ground for the new generation of environmentalists and historic preservationists.

An epidemic of bipartisan claustrophobia swept the country by the 1990s. The way had been paved by New Environmentalists in the 1960s, radicalized by the Vietnam War into a deeper look at the erratic workings of an ever-expansive capitalism. Growth Management became a growth-specialty in academia. The Club of Rome's 1972 analysis, *The Limits to Growth: A Search for Suitable Futures,* was widely influential though derided by conservatives. Its project director, Dennis L. Meadows, went on to publish *Alternatives to Growth;* he and his coauthor, Donella H. Meadows, became central figures in a "sustainable growth" movement.

Growth regulation became the crusade

GROWTH CONTROL DISTRICT

New functions hover over No Trespassing signs such as this one on scarce, prime farmland around Vancouver, B.C. After a West Coast dockers' strike left British Columbia with only six weeks' food supply in the 1960s, the government passed laws preserving these nearby Fraser River Valley farmlands, thus concentrating urban growth elsewhere. This site also serves as a hunters' preserve, ensuring double crops.

of an expanding minority, especially in "Ecotopia," the San Francisco-to-Vancouver corridor, and in other beautiful regions. New voices insisted that urbanization be halted, or better regulated. Growth management became a hot political property.

Los Angeles's citizens in 1986 approved Proposition U to decisively limit the city's development. Even Paul Goldberger, architectural critic of the *New York Times,* was finally moved in 1987 to write an occasional column questioning the no-holds-barred overcrowding of Manhattan.

California adopted the continent's most stringent pollution-control laws; these quickly began to alter auto design and to put a crimp in unregulated development. (Meanwhile, all over the continent, real estate developers sought to find, invent, or suborn places that welcomed development in any form, shape, or degree of degradation.)

By the 1980s, a new breed of once-radical journalists were moving up the print-and-TV/radio ladders, "exposing" along the way the extraterritorial costs of growth. They offered visibility to such new models as the late Governor Tom McCall, of Oregon, who urged tourists to "Come—don't stay"; and to Bob Martinez, Florida's first governor of Spanish ancestry, when he repeatedly warned in 1986 that the hundreds of new arrivals each day would put heavy strains on Florida's scrawny tax base.

Some large cities opted for "caps" on the height or presence of skyscrapers, requiring more open space per structure. Many limited their annexation policies, of their own accord or under pressure from resistant neighbors. Some sought to protect themselves with an "urban reserve" (San Diego, Cal.), or "horse farm country" (Lexington, Ky.), or agricultural zone (Vancouver). San Diego in 1987 "became the largest city in the country ever to take action aimed at limiting its population"—capping annual production of housing units at eight thousand.[2] Towns along the Princeton, N.J., corridor on U.S. Highway 1 began in 1987 down-zoning areas intended for office parks. Smaller or resort towns resorted to uniform zoning and subdivision regulations; defining and protecting "sensitive areas"; controlling boat-dock spaces; requiring setbacks, off-road parking and GREENBELTS for large projects; relating density to soil-absorbent capacity; requiring sewers for select large projects; protecting natural areas; and getting new authority to regulate boat traffic in lakes and streams. Land development becomes a tough obstacle course.

Property in Freeport, Maine—home of L. L. Bean and others attracting millions of shoppers—was in 1987 "among the most regulated in Maine." As a headline in the *American Planning Association Journal* put it, "I'll Have My Town Medium-Rural, Please."[3] Buffer zones were in vogue, as were aesthetic control of signs and sometimes architectural styles and colors—these latter rules often encountering opponents shouting "You can't legislate taste!"

Meanwhile, rigid control of city growth had been widely practiced in Communist Europe, while looser controls in France favored a "growth pole" policy. This directed growth along planned corridors that were well supplied with utilities, roads, and rails. Practically no city in the United States would adopt the course followed by Vancouver, which surrounded itself with a protected agricultural GREENBELT in the fertile Fraser River valley. This came from British Columbia's decision to protect local farmland and thus lessen Vancouver's dependence on scarce imported winter vegetables shipped by sea from Seattle. But as U.S. farm/ranchland was called on to produce more food while prime acreage collapsed into urban uses, something had to give. Beyond the wave of U-pick vegetable and fruit farms that sprang up around U.S. cities to meet a demand for freshness, it would take many changes to protect food-land close to cities, as Canadians and the Chinese had done under stronger pressures.

By 1990, new controls across North America had set in motion a screening process: towns known to favor one form of settlement could actively solicit a special class of migrating homeowner or businesses. This was abetted by wide publication of guidebooks (*Places-Rated Almanac* et al.) exposing local quirks, tastes, and rules to a national market of footloose shoppers. Millions of people were "just looking" for a place to live that—once they settled there—would remain forever the same. From such hopes, fond or forlorn, new forms of community, and of community self-control, were aborning and expanding.

OPEN SITE

This contradiction in terms reveals the expectations of those who use it. When using the adjective "open," they may intend to suggest that nothing's going on here. But the addition of the noun "site" hints that "site" is about to be followed by "of"—as in "the site of the crash," "the site of the annual picnic," "the site of a ninety-story skyscraper." Convert that to "construction site" and the expectations come loud and clear. Everybody nods. SITE incites anticipation. OPEN SITE tells us the site is an open target; almost anything can happen here, and probably will.

Attaching "open" to a place you hope will remain truly open and unused invites disappointment. For OPEN SITE is most likely to be what some people will call an opportunity site, a term that lets the cat out of the bag. It suggests open-for-bids, open-for-business, open-for-whatever-comes-down-the-pike. Use that term appropriately, and speculators come a-running, options at the ready.

But the same meanings do not emerge from open land, open space, opening(s), all of which are open to other, wider interpretations. Not all open space will be filled, but an OPEN SITE offers an open invitation.

SHORTCUT

One-lane track between interstate highway lanes provides a handy SHORTCUT for highway police cars, often stationed on radar patrol duty midway between lanes. This offers observant motorists a warning of future surveillance, and gives police a fast getaway for pursuit in either direction. (Interstate 64 from Norfolk, Va., to St. Louis, Mo.)

SHORTCUT

Once upon a time the presence and route of a SHORTCUT was strictly private and often privileged information—a handy part of local folklore. Knowing SHORTCUTS was what separated kids from elders, locals from strangers, escapees from captives, folk knowledge from so-called map truth. Motorists (when that word was still fashionable) traded closely guarded SHORTCUTS to friends. A hilarious British comedy film of the 1950s consisted of a thin story line draped against high-pitched background arguments over the best SHORTCUTS through central London, mostly by way of narrow alleys with excruciating names.

SHORTCUTS are geographic gimmicks, forms of mental leaps, playful hits and misses. They work like shorthand, generic terms, capsules, epitomes, quick sketches, note pads for the next move. For most city folks, SHORTCUTS abound on their mental maps: the least-traffic route between home and job, or among friends and customers; the stop-and-go details of parking lots for errands and shopping; that third stoplight so you avoid traffic off the interstate; the innocent-looking woods trail leading to a favored fishing hole; the way roundabout that afternoon freight that blocks a dozen crossings. All such privately remembered, family-secret sort of SHORTCUTS extend one's personal turf and sense of mastery. Knowing one's place means using its SHORTCUTS.

It hasn't been getting any easier. Hundreds of towns and cities, with federal highway funds to spend, were straightening up their street-grids, cutting out the odd jogs, the hidden turns and narrow squeaks that make up SHORTCUTS. Once rationalized, such geographic quirks could be converted into real estate. The American Automobile Association and local affiliates—with myriad maps and ready-to-assemble Triptiks—offered timesaving advice in multi-foliated versions. But they seldom knew what to advise off the main roads. Backwaters, back alleys, and back roads were off their beaten paths. (To fill the gap, most major bookstores offered a mixed bag of Back-Road-Hideaway travel books.) Without guidebooks, many a seeker after The (short) Way fell back on

SHORTCUT

(Left) Benefitting from the shortcutlery of the U.S. Corps of Engineers, the Pikeville cut-off for the Big Sandy River allows floodwaters to bypass the city of Pikeville, Ky. (off right). This twenty-five-year project provided a multipurpose SHORTCUT for highways and river traffic, as well as flood waters. The old riverbed was filled up to provide the town with new flood-free land. Photo by John C. Lowe.

Talmudic advice: if you don't know where you're going, any road will lead you there.

Along the way, old clues to familiar SHORTCUTS were disappearing under new conformity rules: all corners (according to The Rule from city hall) were to be right-angular; all street lights the same size, shape, color, and proximity; no more oddities at CURBSIDE, no more "informal" homemade street names to jog one's memory. Yet the persistence of formal and informal neighborhoods; the shifts from one historic Old Towne, or one incorporation to the next, from city to county to township and back again—each jurisdiction offered its own oddball selections in names, setbacks, CURBSIDES, and signs as clues to shortcutters: "Don't turn till you see the Merion Township line on the left."

This kind of wayfinding can be expensive. The cost of inefficient route choice has been "estimated as high as $80 billion per year," according to the authors of *Route Choice.*[1]

Few places that help avoid the long-way-round have suffered such trauma as SHORTCUT since computers moved in on it. Early-generation computer-users had penetrated technical positions advising marketing, delivery, and other firms on least-cost routing. Many had learned to solve "The Traveling Salesman Problem," a familiar device among geographers for teaching shortcutting—plotting the most efficient route for a salesman to call on scattered customers. Soon no citywide delivery service could hope to operate efficiently without a computer to plot its drivers' shortcutting routes. Thus by the 1990s, shortcutting was becoming more

(Below) "Eastwood Cut Off": not all road names are so explicit about the shortcutting that led to the present road sign. Here, Eastwood Cut Off (left) supplemented the pioneers' ridge-running road (paved, right foreground) that probably followed an earlier high-and-dry Indian trail. Next came U.S. 60 (invisibly downhill to the left). Most recently came Interstate 64 a mile off to the right—each shortcutting its predecessor.

logical, less happenstantial, and less fun. And more risky. Drivers, hurrying to make deliveries within the X-minutes promised by company ads, were getting into more accidents en route.

The auto-dashboard route-finder arrived on the accessory market, run by computer, having been previsioned by James Bond's Aston Martin giving chase to a golden Rolls-Royce in the 1964 film *Goldfinger.* In 1990 the German firm Blaupunkt began marketing its TravelPilot Vehicle Navigation System device in Germany for $3,700. As your car moves, a video screen shows your position on a digital map. Honda in Japan pondered a similar venture. In the summer of 1990 General Motors began experimenting along a twelve-mile stretch of the Santa Monica Freeway in Los Angeles to find out how cars equipped with electronic navigation systems could avoid congested highways by turning onto SHORTCUTS under the guidance of a dashboard pilot. It's come a long way from folklore and cocktail advisors.

SPECULATIVE SITE

All the great outdoors of North America was, as it remains, a SPECULATIVE SITE in the eyes of the pioneer multitudes. No fixed futures for them; no cut-and-dried plans to stifle their dreams. No sir, it's open territory and whoever gets there first can do whatever . . .

As a statement covering most of American and Canadian history, the foregoing stands unqualified into the present. Frontiers have been passed time and again; homestead laws have come and gone; zoning invented, city-planning applied, taxation imposed and/or perverted. But the itch to speculate, built into the psyche and settlement of two nations, haunts when it does not dominate the shaping of every place, private or public, over all the continent. The history of North American land can be—and has been—told as the history of SPECULATIVE SITES.

Poets and novelists have praised the itch, the drive, the force to get common property and convert it to private uses. When "common property" embraced all of the nearly 9,100,000 square miles of the continent, there was no one to stop all the first-come-first-served from grabbing while grabbing was good—even when grabbing was hard.

Even in the populous 1990s, the drive to convert public lands to private uses continued to fan the fires of development; to "gamble on the rise," to buy on the fringes and wait (as John Jacob Astor advised), to get in on a good thing early; to buy cheap, sell dear. Its defenders are everywhere: they quote the late speculator and presidential advisor Bernard Baruch, who said "To speculate is to observe carefully—and act on what one sees."

Consequently, behind every option, lease, or purchase of a site lies the hope of acquiring what Henry George called "the unearned increment of land value." This is the increase caused by acts of others—by the city, national government, or neighbors—or by their protection, investments, and growth combined. All the canny speculator need do is buy and wait.

But of course it is never that simple. To speculate requires risk: one must make down payment, pay ready cash (once it was gold coin), or promise-to-pay, which makes the promisee a speculator as well. Between option and reward lies many a suicide, bankrupt, or broken life.

Three types of Congressional acts have fueled the flames that light up SPECULATIVE SITES: the various Homestead Acts from 1795 (Western lands for Revolution veterans) and the railroad land-grants after 1861, the Interstate Highway Act of 1954 as variously amended, and the Urban Renewal acts of the 1960s and onward. Each in its own way has created SPECULATIVE SITES by the millions by dumping new ownership and access values on once-remote or unavailable sites. The first homestead acts offered "free land" (from which Indians had barely—or not yet—been driven) for a few cents per acre. In our own times, the Interstate Highway Act dumped free access at the door of suburban and back-country landowners. Suddenly the back forty became an interchange district. Speculators had combed the country, studied (and sometimes sneaked or stolen) the highway plans so as to buy next to a future interchange. And urban renewal agencies continue to buy up slums, blighted, or other earmarked districts for the "new public uses" that were promised by private purchasers.

In each of these great waves, raw or "underused" land was converted into profitable sites for new owners. Much if not all the initial risk had been shouldered by a government. Much if not all whatever profit comes off the deal would accrue to the speculator. How much the new wave of hustle and speculation overseas in the 1990s would follow the American example was a question for the next decades.

STUDY AREA

With the rise of twentieth-century sociology, STUDY AREA became omnipresent on the American scene. When New Deal bureaucrats under President Franklin D. Roosevelt embraced sociology, STUDY AREAS became hot academic and political properties. No legislation could flourish without STUDY AREAS to show the need. The South, identified in a series of famous reports, became "problem [i.e., study] area number one." Urban renewal legislation and federal highway studies in the 1950s added to the stream, as did environmental impact statements in the 1970s. Historic preservationists, a rising breed, marked off their own historic districts for study and salvation. The U.S. Census expanded its geographic scope until abruptly cut back during the Reagan administration. Competing economists mapped market areas, studied what consumers buy by area. Postal zones became handy zip-coded yardsticks for charting buying power, or the lack of it. Geographers were called in to map and analyze areas affected by school desegregation and its bus routing.

No place, it seemed, could survive unstudied. Many newly significant places required many studies: ANNEXATION AREAS, IMPACT AREAS, water-voting-special-purpose districts. Following Supreme Court decisions beginning with Brown vs. Board of Education in 1954, thousands of school districts were re-contoured to reduce racial segregation. Soon, no part of the continent would escape being mapped by satellite, measured by photogrammetry, analyzed by infrared camera, and studied by one and all for odd causes and with unexpected intent. During recurrent shortages of public classroom space, teachers latched onto the nearest outdoors to create field STUDY AREAS.

Along the way, resistance grew. Planners, anxious to get beyond studies, and designers, impatient with a new generation of study-oriented students, lamented "analysis paralysis."

STUDY AREA kept moving through the language, out of its exclusive geographic meaning, and did a turnaround into "area studies." These grew enormously under the impact of World War II. The Office of Strategic Services (OSS) trained sleuths who spawned thousands of dossiers on prospective Allied invasion sites and target areas for bombing and/or sabotage. During the Cold War, U.S. efforts to block USSR expansion generated thousands more area studies using espionage, air reconnaissance, and more standard academic measures of discovery. Almost overnight in the 1990s, the National Security Agency attempted to redirect its huge surveillance capacity so as to target foreign markets for U.S. business and industry. That military refugee, "targeting," looked to have a long linguistic career ahead.

TEMPORARY HOUSING DEVELOPMENT

Vast stretches of American towns and cities first saw daylight as "temporary housing." By preempting Indian hunting clusters, swarming over forlorn prairie crossroads, rushing to the sites of Western gold discoveries, settlers threw up "soddies," wagon beds, portable teepees, shacks, and tent cities as the first stage for structures that later grew to stay. Rooming houses, makeshift hotels, bunkhouses, shotgun houses, workingmen's cottages, tacked-on rooms for rent—all came and went cheap.

In London, it was once widely stated that, "If you build for tenants, you build for tomorrow. But build for horses, and you build for posterity"—a popular tribute to the longevity of those nineteenth-century two-story, brick and stone stables with which upper- and middle-class English neighborhoods were filled. Once converted into mews—apartments-cum-garage—they became a popular form of housing in the 1960s. But by the 1930s in both Britain and in the United States, the term "housing" was coming to suggest, not so much families tucked away in comfy homes, but tenants stacked in barracks-like apartments run by public housing authorities.

The very term "temporary housing" involved a contradiction in terms: all temporary housing overstays its causes. It develops loyalties among tenants, and vested interests among those who do business with, or get votes from, tenants. Housing shortages which generate temporary housing turn out to last longer than expected. Such developments are usually built with government subsidy, against the wishes of neighbors. They undergo recurring waves of neighbor hostility, owner neglect, and official intervention. And then, in times of prosperity, the tenants move on. Depending on local politics and the national economy, they may be replaced by feckless or helpless others, further down the socioeconomic pecking order.

Something akin to magic may occur, however, when office workers get assigned to temporary, makeshift housing. Parkinson's Law, enunciated by C. Northcote Parkinson, asserts that work expands to fill the time and space available. Further, he maintains that work taking place in grand, splendiferous, custom-built headquarters is soon stultified, ossified, inefficient. Such fancy headquarters structures get built at the crest of the wave. From that point onward—as at the Palace of Versailles—the occupant is headed downhill to bankruptcy or oblivion. As Parkinson would assert, "surroundings of dignity" for its planners are the seedbed of military defeat.

But work itself, stimulated by emergency or inspired leadership, can be exciting and innovative in crowded, impromptu, makeshift, easily modified quarters. Such were "The Tempies," temporary offices thrown up along Constitution Avenue on The Mall, Washington, D.C., during World War I. They persisted through scores of remodelings, and generations of come-and-go government agencies. It was nearly a half century past the emergency which set them up that they finally disappeared. And only then under the impetus of a presidential wife, Lady Bird Johnson, whose "Natural Beauty" campaign contributed to clearing the Tempies off The Mall. Once thrown up, they take a long time getting thrown out.

TEMPORARY HOUSING DEVELOPMENT

When Hurricane Andrew struck South Florida in 1992, available space up and down the coast was dragooned for temporary housing, such as this trailer colony at Fort Lauderdale, Fla. A year later, cities and towns throughout the flooding Mississippi and Missouri River Valleys were forced into similar expedients to house flood refugees from all-time record high waters (see TENT CITY). Photo, *Fort Lauderdale Sun-Sentinel.*

CHAPTER 6

Border Zones

Old limits fall, new borders appear in the shape of boundaries, limits, lines, fences, markers, zones, and turf. Many other transitions appear—shifts in density of people or the intensity of land uses—that are temporarily mystifying. The EDGE OF TOWN is swallowed up or bypassed by endless reincarnations of SUBURBIA. Specific place-names get attached to look-alike subdivisions as they leapfrog and overlap. New SPRINGDALES subdivide out of existence the springs and dales for which they are named. Pressures and tensions swirl around THE EDGE, and the line between The Center and Out There grows more permeable as populations and their mobility expand. It is here—as well as in the Testing Grounds we encountered earlier—that modern society's ability or failure to manage its own environment is daily put on exhibit.

BROWSERS' VILLAGE

Variants: Early American, ethnic, shoppers', shopping village.

"Can Westwood Village, the once-elegant shopping area south of the UCLA campus, be restored to its former glory, or is it doomed to remain a 33-acre clutch of movie theaters, pizza parlors, and yogurt dispensaries, lost in the shadows of the new high-rise office buildings along Wilshire Boulevard?"[1]

In words other than those just quoted, can such places be transformed into the newest darling of upscale retailers, a BROWSERS' VILLAGE? Can large swatches of the urban fabric be hurled backward in time, downscaled to become villageated microcosms of long-gone ways and scales of living?

When Charles Darwin observed, in *The Origin of Species,* "fifty-six little trees which had been perpetually browsed down by cattle" he established the image of cattle feeding upon tender young shoots of trees, often at head height.[2] To browse has since wandered away from farm and forest to become a part of shoppers' life. Across the United States in the 1980s, the term BROWSERS' VILLAGE was heavily propagandized in a wave of efforts to turn backward the clock. It described either new or old groupings—in-town, suburban, or exurban—that invited shoppers to come-and-browse, to saunter, to spend half-days or longer strolling or staggering amidst ever-growing clusters of shops, stores, boutiques, kiosks, booths, stands, and standard-brand holes-in-the-wall. It was a by-product of retailers' hopes that shopping could dominate life itself.

Like as not, everything in the new

BROWSERS' VILLAGE was "olded up"—barn siding roughened, self-peeling paint preferred in the Midwest and West; and in the East, rooflines steeply alpine, or trim and bargeboards repainted in Old New England colors. Only in their modest size did most BROWSERS' VILLAGES resemble the ubiquitous small-town village form in which some 9 percent of Americans still lived in 1990. Those face-to-face communities and their internal support systems still rested on old kinships, and sometimes suspicion of outsiders.

The new BROWSERS' VILLAGES often were, or pretended to be, old. They took as their model Early American nineteenth-century country hamlets, still nestled in farming country. They sported leftover pre-auto artifacts. Some, such as Nashville, Ind., had expanded from a tiny, woodworking handicrafters' hamlet of the 1930s. On its festival weekends, traffic backed up for miles. Lenox, Mass., a mountain resort, featured homes and stores remodeled in the 1980s to create fifty new shops and eateries. Long Grove Village, Ill., sprang up in Chicago's far-north exurbia from a modest cluster of farmhouses and converted barns. Near Waterloo, Ontario, a tiny settlement's grist mill set in the midst of farming country was all boutiqued up by the 1980s.

Each new BROWSERS' VILLAGE catered to a highly mobile population of spenders fed up with ubiquitous, crowded, big-scale shopping malls, all of them a product of one generation's tastes and building tactics. These new/old villages, it was hoped, would "give off the warm glow of a seasoned New England town." This, it was claimed, would bring about the "unmalling" of America.[3]

Such a divestiture was unlikely. Not everybody was to be blarneyed into basking in the "warm glow" of another imitation of a long-gone village selling national-brand copies of nineteenth-century goods. These new BROWSERS' VILLAGES were latecomer look-alikes in a long line of old/new devices to capture the fancy and cash of a prosperous mobile middle class.

THE BURN

Variants: burned-out area, burnt-over area.

Only in forested regions do we have terms that tell, not always clearly, how fire has been tamed, domesticated and industrialized as man's dangerous servant, always tugging at its tether. THE BURN, as a verbal noun, indicates a large area of burned-up, burned-over, or burned-off forest or other area. THE BURN follows the tracks of ancient as well as modern man. Pre-human wildfires were caused by lightning, but today 90 percent are caused by man. It was as "keeper of the flame that man first became steward of the land," says Stephen J. Pyne in *Fire in America.*[1]

Deliberate fire was humanity's chief tool in clearing Europe's forests through the Middle Ages. After the Portuguese discovered the island of Madeira in 1470 they began clearance-by-burning, and in seven years destroyed all forests, replacing them with today's orchards and vineyards, the source of Madeira wine.

In North America, THE BURN seldom appears on municipal maps, no matter how drastically fires have reshaped the earlier city or countryside. The term is generally used for forest, rather than city fires. Yet urban fires, like their country cousins, carry long-lasting psychological and physical effects, especially in flammable, wooded SUBURBIA. In dozens of once-Colonial cities, today's historic district is likely to be an unburnt remnant. In Chicago, evidence of an old BURN persists today on the SKYLINE south of the downtown Loop district, a plateau of buildings built to the same height after the Great Fire (1871) with identical fire-towers on top, using gravity flow to supplement piped water. In Seattle the rather uniform old red-brick warehouse-office district near Pioneer Square was built all at once, following the great fire of 1889. Most of Fayetteville, N.C.—600 homes, 125 businesses, some churches, and the state's general assembly hall—all burned in 1831, but today little beyond historical reminders carries the evidence. Better preserved, in the book and film, *Gone with the Wind,* was the military burning of the city of Atlanta by Union troops during the Civil War. This gave rise to the Southern comment that the Union Army commander, General William T. Sherman, was "a mighty fine general—but a little careless with fire!"

Unlike Hollywood movies, urban reforms seldom follow after great city fires, according to historian Christine M. Rosen.[2] She found that civic efforts to follow up massive fires in Chicago (1871), Boston (1872), and Baltimore (1904), by "progressive redevelopment" (better building placement, etc.) generally came to naught.

However, firing-up the landscape has a long history. In 1637 Thomas Morton wrote: "The savages are accustomed to set fire of the country in all places where they can and burn it twice a year."[3] "Indians burned widely for traditional reasons, particularly east of the Continental Divide. Explorers and surveyors often spoke of 'parks' in the Black Hills, Big Horns, and Central and Northern Rockies that [Indian] tribes had created and sustained by fire."[4] Much of the North American Prairie is believed to have been created by repeated natural and Indian burnings. And there is increasing evidence from revisionist historians that New England during the Indians' regime was widely cleared by burning. The decimation of the Indian population of North and Central America—from between four and ten million, down to around one million, due to Spanish and other European diseases—left New England still comparatively thinly populated and open-wooded when the Puritans arrived from 1620 to begin farming old Indian fields.

BURNS, caused by lightning and man, are scattered across the whole of tree-growing North America. Along the borders of old forest BURNS, tourists who know fire history can navigate by visible signs—light from old flames. In such forested regions, old BURNS are well-known to fire rangers, foresters, water district managers, cultural geographers, insurance underwriters, and other realists.

The most infamous of modern BURNS is the Tillamook Burn, created by raging fires that began August 14, 1933. "In a sense, Tillamook was a 300,000 acre fire that burned intermittently for 18 years," observes Stephen J. Pyne.[5] More than ten million board feet of timber was destroyed, over 125,000 acres abandoned. THE BURN became a cause célèbre, was planted and replanted, much of it becoming in 1973 Tillamook State Forest, an innocuous name to conceal its flaming history.

The areas burned in highly publicized "runaway" forest fires of modern times—3,279,000 acres in 121,736 fires in 1978—appear small on the scales of history. For in the years 1925–43 from 21 million to 52 million acres burned each year.[6] The modern fire also mobilizes modern equipment and firefighters. In September 1987 more than twenty thousand firemen were assembled from across the United States to fight 1,877 fires in the Sierra Nevada—nearly 600,000 acres charred in California, Oregon, Idaho.[7] Helicopters, airborne tankers, satellite photos, computer-aided tactics, and international radio nets have joined buckets and brush hooks as fire-fighting tools.

In our nation today, fire is undergoing two historic shifts as it works its way into management practice and managed scenery. The greatest transition in THE BURN, its history and character, shows especially in the expanding pineland plantations of the U.S. South. Here fire has shifted from episodic, erratic, unpredictable occurrence to become scheduled, organized, and predictable. Fire has become just another controlled step in the production line from raw land to woodland plantation to manufactured paper and wood products. One result is towering columns of woodland smoke from controlled BURNS that horrify tourists but increase wood production. Thus fire itself has become an industrial tool and THE BURN a form of manufactured landscape. As "keeper of the flame," man has taken over both the fun and the fury of this flaming weapon.

Fire is also used to "manage wilderness areas" by prescribed or controlled burning to do the job of natural fires, that is, to rid national parks and wilderness areas of excess "fuel"—deadfalls, brush, undergrowth that would boost a totally destructive fire. It was the National Park Service's early refusal to stop natural fires in Yellowstone in 1988 that intensified THE BURN, and caused an uproar among a public ignorant of the essential cleansing function of such fires and unwilling to allow the policy to continue.

The second historic shift occurs in more arid areas, a chief example to be found in the urbanizing of steep lands of California. Here suburbanites, in willful search of the perfect VIEW, settle into scenic wooded canyons and perch on ridgelines, oblivious to their risky surroundings—what firefighters call "The Mixed Chaparral-Eucalyptus-Single-Family-House Fuel Type," a highly volatile mix indeed.[8] New suburbanites continue to build, plant, and remodel homesteads-as-fuel—a replay of the eucalyptus-planting mania of the early 1900s that produced vast new quantities of fuel—"eucalitter."

Only slowly—in the vicinity of each suburban BURN—did the correct term "fuel" begin to replace the older term "trash." Gradually tile and flame-retardant roofing replaced the inflammable cedar-shakes popularized by the Pacific Northwest cedar industry in the 1960s. Reluctantly, the real estate salesman's magic phrase "canyon setting" was being swapped for the fireman's description: "a chimney." Like moths to the flame, many burned-out suburbanites around Laguna Beach, Cal., vowed to move back after widespread fires—some of them set by arsonists—swept over 167,000 acres in November 1993.

Burning to ensure natural plant succession has become riskier as human investment in man-made places continues to take over territory. When forested "wilderness" or undeveloped lands come under pressure, they acquire names, values, prices, and occupants. The protective armor of "property" tightens expensively around places that once burned naturally.

A collision course was noted in the forested counties of California: "Areas that are increasing in population most rapidly are those most prone to wildfires," observed James B. Davis at an international symposium in 1990.[9]

Throughout history, from ancient times THE BURN was widely useful as a weapon of war: Assyrian bas-reliefs have been found dating from the eighth century B.C. that show warriors projecting and extinguishing fires. American Indians used it extensively: to flush enemies from high grass or woods, to harass troops too strong to confront, to deny cover to guerrilla raiders, to conceal movement with smoke, and for long-range messaging.

Beginning in World War II, THE BURN entered an expanded and ominous phase when Allied bombers were directed to create mass fires in civilian areas of German cities. The culminating BURN outgrew its earlier definition when the United States atomic-bombed Hiroshima and Nagasaki, far outdoing an earlier firebombing of Tokyo. For the time being it is widely assumed that THE BURN of "mutually assured destruction" by atomic or nuclear bombing is too horrendous a prospect for Great Powers to contemplate. Meanwhile, equipment, tactics, and language have moved from military usage into civilian fire-fighting: from helipads to message centers to fire lines. Historian Pyne concludes: "Warfare must be considered as an episodic, though potent, means by which fire has been applied to the landscape and distributed around the globe."[10]

THE EDGE

THE EDGE is usually located and identified by people who occupy The Center, looking outward. Tourists are solicited to venture to "the vast, raw, uncrowded edges of the world" of the Canadian Arctic.[1] To encourage you to adventure, a sports outfitter at Banner Elk, N.C., calls his shop, "The Edge of the World."

EDGE-frequenters come in endless variety: beavers and muskrats, crawfish and clams, wading birds, shoaling fish, and hunter-fishers who pursue all of the above. Gamekeepers on South Georgia quail-hunting plantations go to great lengths to

THE EDGE OF TOWN
Only a hookup away from the array of lakes north of Fargo, N.D., and Moorhead, Minn., this boat—poised at the not-for-long edge of Fargo—gives its owners access to such places as Detroit Lakes, described locally as "part of the weekly urban system of Fargo-Moorhead." The fireplug is precisely here because it expects more customers to be built into the background farmland.

manipulate THE EDGE for maximum cover(t), planting food crops to attract game. Landscape designers learn that people prefer to spend time around THE EDGE. A study by S. D. Joardar and J. W. Neill of 6,300 landscape users in Vancouver showed that fewer than 1 percent carried out activities in the open pavement away from physical artifacts such as steps, benches, boundaries, etc.[2] I recall photographing fellow tourists around the columned ruins of the temple of Minerva at Sunion, in Greece. All were clustered or aligned along the temple's EDGES. William H. Whyte, Jr., in his book *City: Rediscovering the Center,* observed that on a busy plaza in New York City, people "show an inclination to station themselves near objects, such as a flagpole or a piece of sculpture . . . steps, or the border of a pool."[3]

To ecologists, THE EDGE is a dynamic encounter zone—called ecotone—between two environments. They study EDGES along that interactive interface between woods and field, between land and marsh, marsh and tidewater. Such EDGES encourage large, varied populations of wild and human life, able to explore and exploit resources in either direction. THE EDGE was the zone that probably produced the human race's first language—when primitive hunters ventured from forest edges into open country, developing grunt-and-shout commands and hand signals to other hunters.

Marshall McLuhan was quite wrong to insist, in *War and Peace in the Global Village,* that "margins cease to exist on this planet."[4] On the contrary, THE EDGE persists, hard at work, on the margins of everyday life. It works as jumping-off place, hang-glider launch, view property, or point of no return. For insular people, it delineates THE EDGE of the universe. It is where one distinguishes here from there—the latter located out in the Great Beyond, another country, a yawning abyss—or simply not-here.

So many millions of tourists jammed right smack up to THE EDGE of Grand Canyon and other famous sights that the National Park Service, from 1966 onward, redesigned such cluttered viewpoints to get rid of structures that obstructed THE VIEW. "The edge of the world" is a resident's term for a deserted twenty-three-mile stretch of California coastline around Honeydew, seventy miles south of Eureka.[5]

But, soon enough, the sharpness of THE EDGE tends to disintegrate, to absorb something of here and of there. THE EDGE tends to become zone rather than sharp line. THE EDGE OF TOWN becomes a border zone, a zone of competition, identified here as part of The Front.

Anthropologists and such students of native cultures flock to the EDGES. In his book *Edges: Human Ecology of the Backcountry,* Ray Raphael describes Mayme Keparisis, living in Trinidad: She "has balanced herself on the seemingly precarious Edge between the Indian and the Christian universe, between traditional artifacts and modern newfangled conveniences."[6] In her great work, *The Edge of the Sea,* Rachel Carson noted that a search for the key to nature's riddle, "sends us back to the edge of the sea, where the drama of life played its first scene on earth and perhaps even its prelude."[7]

THE EDGE is essential to human ability to make decisions, for this is where signals to the human brain intensify. "What is visually significant occurs where the illumination changes. . . . It is along the edge of objects, or where there are variations in the smoothness of a surface, that the intensity of reflected light changes . . . [and thus] . . . can represent physically or visually significant parts of the environment."[8]

EDGES are also packed with cultural cues: contrast the softly convoluted edge of a Victorian-Gothic estate with the hard-edged, walled enclosure of a California modernist; the open, inviting storefront of the shopkeeper with the scowling, bunkered block of the downtown banker's skyscraper; the rickety broken fence of a tenement with bright metal fencing along a new homeowner's property line.

In an ecological sense, all territory consists of EDGES. All places give evidence of the limit of plant and animal species and their territories. A traveler moving eastward from arid Denver into the more humid East will pass many EDGES: the last irrigation ditch, the first broomsedge grass, the first small beech and maple trees, the first insects smearing the car windshield, the first osage orange hedgerows. By this time, around mid-Kansas, he will be east of the twenty-inch rainfall line, a significant historical EDGE.[9]

As a landscape location, THE EDGE is where airborne or flood-borne debris settles on the upwind or upland side of the flow; where birds alight, and quarry (both game and human) hides. Standing at THE EDGE of a woodland patch, an ecologist distinguishes between saum (the perennial herb border), the wooded mantle, and the tree canopy.

EDGE also serves as synonym for end, as in "you fall off the EDGE of the world (or

universe) when you cross the Ohio River going south." Thus it is used loosely (by Northerners) as a synonym for unknown territory. It offers a rich trove of clues to the user's origins, prejudices, and ignorance.

Many an EDGE is being rehabilitated—both in fact and in imagery. As the United States' merchant marine fleet endured a great shrinkage, scores of decrepit industrial waterfronts in the Rust Belt were remodeled in the 1980s into marinas, public promenades, or such inventions as New York's Battery Park City, perched on the built-up EDGE of the Hudson River. A *New York Times* map, March 22, 1987, showed the persistent "Filling in the Edges of Manhattan"—even though that city's Common Council, back in 1789, had directed waterfront landowners to stop their "unlimited extension" into the East River.[10] But developers' pressure on Manhattan's EDGES continues to push them further into its rivers, a contest being played out on many another waterfront, as well.

See also THE EDGE OF TOWN, following.

THE EDGE OF TOWN

This nostalgic term is supposed to indicate that there actually exists a visible point at which a city or town's solid blocks of houses, streets, and utilities come to an abrupt stop, and, suddenly, countryside begins. But in truth and in nature, THE EDGE is a complex encounter zone between two environments or ecological zones. Like a battlefront or weather front, it is a zone of competing and often conflicting energies.

There is seldom to be found a sharp line separating the country and the city. Each is a powerful generator of energies and messages, but those of the city have consistently overwhelmed those of the country as the national majority turned its back on agriculture. Only rarely do the opposing forces reach a truce that produces a sharp-edged, abrupt shift.

For all its evanescence, THE EDGE OF TOWN still works as a sorting-out processor. People who prefer living close to the country move further from the city center. Country residents, anxious to be close—but not too close—to the city, populate

THE EDGE OF TOWN

(Left) Many small towns exhibit a true—although temporary and unmarked—EDGE, such as here at Midway, Ky., where a city street follows the current EDGE of Bluegrass Region farmland on left and the town on right. Meanwhile Midway is becoming part of the commuters' EDGE of Lexington, a half-hour away.

(Right) Here at the outer EDGE of Las Vegas, urbanization comes to an abrupt stop and only dogs, kids, and windblown debris extend much beyond the built-up limits marked by homeowner's fence in foreground.

THE EDGE OF TOWN

(Right) The strong hand of the municipal government of metropolitan Toronto, Ontario, has kept urban sprawl within visible bounds that are rare in the United States. Thus Toronto's western high-rise development comes to a halt at this major thoroughfare, leaving no doubt as to where "the countryside" begins.

THE EDGE. Familiarly called SUBURBIA with such endless variations as exurbia, urban fringe, etc., this zone is in endless tension and flux.

Astutely observant, the nineteenth-century speculator John Jacob Astor, after the year 1812, following his own motto, "Buy on the fringes and wait," made a fortune in real estate. But that was not universal wisdom; not all cities grew as did New York City, the source of Astor family fortunes; not all edges expanded at the same rate. Yet in the profligate space-using society of the last two hundred years, millions of individuals have profited from the geographical shift of urban activities into open land. Buying real estate beyond THE EDGE became a national obsession in the 1950s, fueled by easy federally insured mortgage loans, cheap gasoline, electric lighting, and new highways. Thus emerged an expanded city-form that is still in search of a proper generic name: metro, megaplex, megalopolis, spread city.

The few North American exceptions with sharp city EDGES arise from other forces, natural and political. Desert cities (Las Vegas, Los Angeles, Denver) still possess sharp EDGES where city water lines stop. Toronto, with strong municipal controls, and Vancouver, surrounded by rigidly protected farmland, exhibit marked EDGES of town. But the typical EDGE of a U.S. city is a confusing mosaic of HOLDOUT farms and woodlots, railroad commuting suburbs of the nineteenth century, remnant farm villages, subdivisions from the last few booms, stretches of streets, shops and drive-ins—a mosaic from then and now.

Hundreds of small towns, however, still exhibit sharp EDGES, where there is a precise, identifiable "last-house-in-town," and beyond it lie broad fields or thick forest. Such sharp EDGES show where the energy ran out; the town simply stopped growing.

Now and again, utopians and similar-minded planners look to THE EDGE as a testing ground. "An organized and publicly managed expansion of the urban edge," predicted Thomas E. Jacobson, Chesterfield, Va., "will reduce commuting distances, provide efficient delivery of public services and create predictable development patterns."[1] Hope springs eternal. Only lately, in the 1990s, was hope reinforced by the prospect of rising fuel costs, and early, feeble efforts toward urban growth control.

LAST CHANCE

Variant: "Last-Chance Saloon."

A universal testimony to the persistent differences between liquor laws and their enforcement in one community and its neighbor down the road, LAST CHANCE is marked by urgent roadside admonitions for motorists to buy-now-before-it's-Too-Late! Such LAST CHANCES inevitably occur at state, county, or municipal boundaries where differences between laws and law-enforcement tip the advantage across the border. They mark the end of one form of control and the beginning of another, more lax.

In international travel, such boundaries are often studded with duty-free zones that induce tourists to overload with such price-cut, portable goods as liquor, perfume,

LAST CHANCE/ FIRST CHANCE
Thirst for strong drink heats up locations at county and state lines. Here buyers get their first or last chance at liquor upon leaving or entering "dry territory" where liquor sales are illegal. At Benny's Place, below, for oncoming traffic from left, the sign reads "LAST CHANCE." But thirsty troops from nearby Fort Knox, Ky., line up in the extra waiting lane for their FIRST CHANCE.

and watches. In desert territories, LAST CHANCE may refer more innocently to a filling station with the last water and fuel available for the next hundred miles of anxious travel.

LOVERS' LEAP

Variant: jumping-off place.

LOVERS' LEAP

Not all leaping lovers are as well-memorialized as the Indian maiden Noccalula, who is said to have leapt to her death over Noccalula Falls, now in an Alabama state park. The falls are distinguished by a statue depicting her in the initial stage of ending it all—either for love, or in despair over the treatment of her people by white occupiers of Indian territory, a story still in dispute. From Gadsden-Etowah Tourism Board, Gadsden, Ala.

A special-purpose promontory, LOVERS' LEAP is a place of last recourse for joint-venturing couples who have taken the wrong turns in Lovers' Lane, or for young women determined to End It All after having been led down the garden or primrose path—the latter a nineteenth-century invention to suggest a "Fate Worse Than Death," i.e., premature loss of virginity. Young men, spurned by love or fate, often follow the same trajectory.

Leaping lovers was an image beloved by eighteenth-century cultists in England, followers of the picturesque ideal in landscape design. Their ideal place was often "cool, gloomy, solemn, and sequestered." Its key ingredients were fantastic shapes, twisted oaks, haphazard trails, jutting outcrops, and precipitous cliffs. To leap or fall to one's death from a crumpled crag was to add yet another dramatic element to a landscape studded with man-made RUINS, with sudden, startling views, a "pleasing sensation of gloom," and visual access to spectacular stormy weather. Such disorderly add-ons—these crinkle-crankle oddities—could only have occurred in a society at long last—after nearly two centuries of respite from revolutionary times—able to view RUINS with something beyond remembered horror. These gloom-giving oddities were to be smoothed over in the following nineteenth century by landscape reformers Lancelot ("Capability") Brown and Humphry Repton. But the forms persisted oddly fashionable in the landscape of many Victorian estates in the United States—long after firearms and poisons replaced leaping as a way to End It All.

Quite another set of motivations was involved in the now historic site at Noccalula Falls, Gadsden, Ala., where a bronze statue of Noccalula, "daughter of an Indian chief," is poised at the falls' brink in a state park. Legend (which may have created her name) has it that she leapt to her death after her greedy father promised her to the rich chief of a neighboring tribe. Meanwhile, she was in love with a valorous but poor young warrior, banished by her ambitious father. Her leap is said to have occurred "long ago," in the classic words of a tourist pamphlet available in 1989 at the site.[1] But a revisionist version published in 1989 claimed that Noccalula's romantic leap was in part a fabrication by whites to conceal the genocide behind her legendary suicide. In that alternative version, her death symbolized "the destruction of Indian civilization" via "invasion and enforced exile" from the Cherokees' ancestral homelands in Georgia and North Carolina.[2] They were sent under military escort over the Trail of Tears to Oklahoma, more than a third dying en route.

The Noccalula drop was 90 feet. This is a modest declivity alongside the LOVERS' LEAP at Tallullah Gorge, in north Georgia, into a chasm 750 feet deep. (That was also the locale for the 1972 movie, *Deliverance.*)

At Blowing Rock, N.C., a similar legend adds a novel twist: the occasional high winds at that location, confronting a disap-

pointed Indian maiden attempting her leap, blew her back to her starting point. Modern tourists tempt fate to lesser degree by tossing balls and other retrievables into the howling gusts.

There seems to be no available catalog of LOVERS' LEAPS, although George Stewart in *American Place Names* says the name "is repeated scores of times across the country." But from the few available examples, one may assume yet other, and life-giving, forces are here at work. Many of these locations resemble those of Greek temples to the gods. To anyone having experienced such Greek sites, the suggested presence therein of life and death offers kinship to those North American places where someone—and often many persons—have chosen to put an end to life.

Unless it is discovered that all lovers who leap to their death in pairs or alone are psychotic and/or schizophrenic, the connection between loving and leaping remains more popular myth than medical fact. For non-lovers, high bridges appear to be favored places to take the final plunge.

See also SUICIDE SPOT, in Chapter 7: Power Vacuum.

SPRINGDALE/FIELD/HOUSE/HURST

In transition between their geological origin and their extinction, springs often get attached to the names of early settlements. But few survive urbanization. Adolescence begins when man discovers spring; senility and death when forests are cut and land above the spring is developed for urban purposes. Soon springs live only in place names.

Since history's pre-dawn, springs have attracted animals, hunters, and then worshippers, pilgrims, and settlers. So long as primitive or pioneer homesteaders depended on springs for water or sacred spirits, community rules stopped anyone from polluting or stealing the waters. All that changed with discovery of the pump, and especially gasoline pumps of the nineteenth century. Pumps made it possible to pump domestic and city water from a distance. At once, the spring became old-fashioned and ill-suited to the demands of thirsty cities. Small settlements with names like SPRINGDALE, Springhill, SPRINGHURST, Spring Station, Spring Valley bore testimony either to long-gone local springs, or to pure nostalgia. Settlements called SPRINGFIELD in Massachusetts and Missouri grew into cities.

Few spring-based settlements survived city envelopment. Spring Station, a pioneer Kentucky fort, was subdivided into obscurity. Once the watershed from which a spring flowed became real estate, rain fell on roofs and pavements to be channeled away, often piped underground, mixed with sewage and processed miles away in a sanitary waste facility.

Here and there springs associated with sacred happenings, with holy events, spirits, or a long history of community baptizing, were protected. The Green Springs Valley north of Baltimore, celebrated for its beauty, was earmarked in the 1960s by a noted landscape planner, Ian McHarg, for preservation. In Macon, Ga., the Town Spring, housed in its own masonry building, became a protected landmark, as did the Town Spring of Georgetown, Ky.

Jamestown, Tenn., protects its early Town Spring under a shingled shelter in a small public park, but the flow had shrunk to a trickle by 1991. The Town Spring of Placerville, Cal., becomes Hangman's Creek, which by 1991 was virtually buried amidst motels, parking lots, and an old railroad. Cincinnati's Eden Park has a historic SPRINGHOUSE, built in 1903 to protect the medicinal value of its waters. Cities in limestone country have grown to surround old stone SPRINGHOUSES, many of them dewatered, but protected for sentimental or other values. The Spring Hill Historic District of Mobile, Ala., is on the National Register of Historic Places.

The 1976 edition of The National Register of Historic Places includes:

Spring Branch Butter Factory Site, Iowa
Spring Creek Site, Mich.
Spring Hill, Ohio
Spring Hill College Quadrangle, Ala.
Spring Hill Historic District, Mobile, Ala.
Spring Mill, Ark.
Spring Place, Tenn.
Spring Street Historic District, Maine
Spring Valley Farm, W.Va.
Spring Valley Presbyterian Church, Ore.
Springdale Farm, Pa.
Springfield Armory National Historic Site, Mass.
Springfield Farm, Md.
Springfield, Ky.
Springfield Plantation, Miss.
Springside, N.Y.
Cold Spring, W.Va.
Cold Springs Schoolhouse, S.Dak.
Cold Springs Station Site, Nev.
Warm Springs Bathhouses, Va.
Warm Springs Historic District, Ga.
Clear Springs Plantation, N.C.
Mill Springs Mill, Ky.
Tinkling Spring Presbyterian Church, Va.

In America, urbanization, in effect, becomes desiccation. All the forces of a capitalist society treat water as an economic resource to be mass-produced and distributed. The lowly spring seldom survives such definition. Only when the water supply grows more expensive, scarce, and endangered do local springs get the attention that surrounded their early days.

See also TOWN CREEK, in Chapter 1: Back There.

SUBURBIA

SUBURBIA succeeded upon the American landscape by putting cheap distance between where people work and where they sleep. To protect the Gulf oil that maintained this distance at minimum cost, the United States waged war on Iraq in 1991. Few man-made places can make that claim.

No other man-made place has acquired unto itself a majority of the American population in three generations. In ten years prior to 1958, twelve million people moved to the suburbs—"the greatest migration in the shortest time of the nation's history."[1] After the *Detroit Free Press* in 1989 found more than fifteen thousand derelict buildings in that city, its mayor Coleman Young explained "one million folks have abandoned this city in less than 40 years."[2]

By the 1990s well over half of all continental Americans lived in SUBURBIA. However defined (see variants below), SUBURBIA became the seat of geographic power, the GROWTH AREA in a society putting

SUBURBIA

Spurred by (and stirring up) fears of the urban mob, U.S. periodicals in the 1970s dramatized a besieged and isolated SUBURBIA with cartoons such as this cover of *Newsweek* magazine, 1971. But by the '90s, SUBURBIA, too, was more citified, industrialized, and pocketed with rundown, variegated sectors. It had become a spread-out component of the larger city, sharing its pressures, its woes, and reluctantly, some of its tax obligations. Cover by John Huehnergarth.

SUBURBIA: As it happens, it specializes, acquires names, and is variously, collectively, independently, severally, and occasionally known as

Bedroom Towns
Beyond the Beltway
The Boonies, BOONDOCKS
Border Country, Borderlands, Border Settlements
BOSWASH (the Boston-Washington urbanization, one of several coinages by Herman Kahn)
'Burbs
Commuting Fringe, Suburb, Town
Commuters' Country
Country Club District
Countryside
Daily Movement Systems (Brian Berry)
Dispersed (Dispersal) City
Ecumenopolis (Constantinos Doxiadis)
THE EDGE
Edge City (Joel Garreau)
THE EDGE OF TOWN
Escape Route
Exopolis
Exurb, Exurbia (A. C. Spectorsky)
The Fringe(s), Fringetown
Garden City (Ebenezer Howard, 1898), Suburb
Graduated Settlement
Greenbelt Towns (see GREENBELT)
GROWTH AREA
Inner Suburbs, Inner-Ring Suburbs
Interurbia
Leapfrog Suburb
Low Density City
Megalopolis (Christopher Tunnard, Jean Gottmann)
Metro
Metro(politan) Area, Complex, Fringe, Region, Ring

overwhelming value on physical growth. Having lost population to SUBURBIA, older cities would shrivel and shrink without suburban support, and would-be presidents would lose elections. Without their distant commuters, work stations would go empty, employers would go broke, and oil-dependent industries wither. This North American phenomenon was expanding visibly worldwide.

The key ingredient in all this was distance. Millions of time budgets now included time en route. The distance-from-Back-There that made SUBURBIA work was reinforced by new shopping lists. Behind the lists lay land made cheap by law and valuable by tradition. Land was worth only its speculative price. Its other values had been discarded. With access to cheap land, newcomers bought distance from neighbors old and new. Extravagantly sometimes, they paid for distance in the form of single-family zoning and deed restrictions that yielded large lots, fences, shrubbery—and isolation. They acquired the benefits of distance by passing rules against whatever leaps over it: mass transit, noise, close neighbors, outdoor cooking smells, leaf or trash burning, and highly visible trailers, campers, trucks, or stored goods. In the ideal 'burb, distance is all. Many a suburb extended its distance from city problems by requiring minimum-square-foot house sizes that kept modest-to-lower-income folks out. Some used subtle or raw vigilante tactics to keep minorities at a distance. Others manipulated anti-city attitudes in state legislatures to prevent annexation by the city left behind. It was no accident that newly rich families most successfully distanced themselves from DECLINING AREAS, DOWNTOWNS, PORNO DISTRICTS, and other places left behind Back There in older parts of town. By 1977, the U.S. Census counted more than twenty thousand "suburban communities" and the number has risen steadily.

The resulting segregation was more than racial/ethnic, and more than accidental. The ideal of distancing one's place and self from others—and their problems—had deep cultural roots embedded in postmedieval Europe. It was made to flourish, as never before, in the New World with its subsidies to encourage settlement Out There—railroad land grants, cheap Western land for speculators and settlers. More recently, a study by World Resources Institute showed that the United States "subsidizes motor vehicle transportation by $300 billion a year—more than $1,000 per person above and beyond the direct costs to car users."[3] A computer-assisted survey in Orange County, Cal., found "that developers and related businesses donated 42 percent of campaign funds collected by [county] supervisors and board candidates in the last 14 years. No other group comes close to that figure."[4] Orange County thus turned itself into Commuter Country.

The rise of a subspecies of critical books—*The Organization Man, The Exploding Metropolis, The Split-Level Trap, The Dream Deferred, The Crack in the Picture Window, Suburbia: Its People and Their Politics*—has little immediate effect beyond creating a minor sense of guilt among some refugees. The rest believed they were getting a good deal and a good lifestyle.

A major result of dispersal was widespread loss of propinquity. New job-formation had become distinctly suburban. Between 1970 and 1990 new waves of women had joined the workforce, bringing a formidable increase to family incomes—and to mortgage and second car payments. Critics would rant at how the resulting scatteration looked; at how it distanced families from services, workers from jobs, and raw materials (food, water, fuels) from users; and at how it tied everybody into rigid schedules linked to trips. Economists would count its mounting costs, and sociologists its cultural fragmentation.

But, so long as cheap land was made available, earlier by rail, and later by road, and by cheap mortgage loans, city utilities, fuels, and electricity, the dispersal of households and jobs continued. When the Federal Interstate Highway program was set loose after 1954, it was as though millions of low-priced acres had been loaded onto flatcars and dumped right next door to millions of city customers. There was no turning back the tide. Distance was mass-produced; it was a subsidized land-rush reminiscent of the nineteenth-century Western frontier. SUBURBIA carried out "its familiar role as a land resource for speculators, banks and builders, and as escape hatch for those in the city who could afford it."[5]

That first whiff of The Limits came eventually—the Arabian oil embargo of 1973/74. Millions of Americans found themselves angrily stuck in long queues waiting for scarce gasoline. Its price per gallon multiplied. Thus began the long struggle to strip SUBURBIA and its commuting routes of gas-guzzling big cars. Here began federal programs to improve mass transit among SUBURBIAS and their parent DOWNTOWNS—oft-thwarted by oil lobbyists in and around Congress.

In the interim, SUBURBIA counted its assets, and lobbied to resist centrality. It found spokesmen in "edge-city" promoters and other operators of "the growth machine." When asked why he robbed banks, a noted heister replied, "That's where they keep the money." If asked why they promote suburban development, city bankers today would respond, "That's where they've got the land"—plus subsidized roads to get there.

Ultimately, the city-left-behind—the Back There we examined earlier—intensified its bumpy efforts to survive on a shrinking tax base. It tried to bridge distance by occupational taxes on commuters, to cheapen it by offering mass transit. It struggled to form metro governments, regional coalitions, treaties, and compacts; to develop AIR RIGHTS and other DISTRICTS; to put "caps" on suburban high-rises; and to diminish the subsidies granted to developers in far-out countrysides. (In 1993 Congress put a stop to 5 percent loans to REA [Rural Electric Association] co-ops in wealthy suburbs.) The city played dead-serious games to create new forms of "citizenship" so as to overleap the distance between DOWNTOWN and countryside; and to set up MEETING PLACES in place of ABANDONED AREAS. It would be a last-ditch effort to overcome what economists call "the decay of distance." And it promised to be a long struggle.[6]

See also BROWSERS' VILLAGE and THE VILLAGE, in this chapter; and SECURITY, in Chapter 3: Perks.

The Middle Landscape (Peter Rowe)
Mortgage Manor
The New Frontier
Old County Seat
Outer City, Ring, Suburbs
Outskirts, Out There
Overspill (Great Britain)
Postwar Spread
Railroad (commuting) Suburb
Ribbon Development (Great Britain)
Ruburb, Rurban Area
Rural, Non-farm Settlements
Satellite Settlement, City, Suburb, Towns
Scamscape
Scatteration
Side-of-Town (North, South, East, West)
Slurbs
Small-Town America
Sprawl
Spread City
Streetcar Suburbs (Sam Bass Warner, Jr.)
Stringtowns
The Subdivisions
Suburban Area: Archetypal, Established, Close-in, Fringe, Industrial, Mass-Produced, Middle-Class, Old(er), Other-Directed, Ring, Well-manicured and Upper-Income, Working-Class
The Tracts
The Urban Complex (Robert C. Weaver)
The Urban Field
The Urban Mosaic
Urban Service Area
Yuppie Country

TWILIGHT ZONE

In TWILIGHT ZONES dwell characters and creatures who and which are denizens of two worlds. There is something in the place which attracts or nurtures many such types. In *The Great Chain of Being,* A. O. Lovejoy observed that, "nature loves twilight zones" as places "where forms abide which, if they are to be classified at all, must be assigned to two classes at once."[1] In another context TWILIGHT ZONE is called an ecotone by ecologists—a zone of interaction between two distinctive environments, such as an intertidal beach zone, or the complex edge of a forest abutting an open plain or marsh. Or the deep-water oceanic zone into which surface light barely penetrates.

Like other place-names which begin in the crepuscular dusk between physical realms, TWILIGHT ZONE has migrated—along with its obscurities—into sociology and common usages, each of which specializes in neither-this-nor-thatness. It can denote space occupied by the social demimonde, or a disputed workplace between two labor unions, or a potential DRUG SCENE on a seldom-policed and dimly lit street, or a neighborhood frequented—but how can you be sure?—by refugees from myriad classes or ethnic groups.

So powerful has been the influence of a Rod Serling TV drama, 1959–87, by that name, that TWILIGHT ZONE has picked up an aura of being almost-but-not-quite beyond the touch or taint—a science fiction version—of reality.

Applied to geographic place, twilight has acquired special powers. The planning philosopher E. A. Gutkind asserted that "the twilight now descending upon towns, cities and metropolis marks the end of a perennial revolution that has shaped and reshaped urban communities all over the world for five thousand years." To forestall twilight, his solution was sci-fi indeed: to split up all big cities into "small, meaningful and imaginable units." In his book's sketches, DOWNTOWN and its SKYLINE disappear into a scattered TWILIGHT ZONE of low-rising sub-centers.[2]

Gutkind, for all his lugubriosity, was in good company. Books have been written about twilight at Monticello (home of Thomas Jefferson); twilight in Djakarta, in Italy, South Africa, the Forbidden City (Peking), Vienna ("the capital without a country"); about the twilight of a world, of authority, of capitalism, cities, empire (two books), France, gold, honor, individual liberty, man, parenthood, progressivism, royalty, sailing ships, splendor (i.e., the decline of palaces), steam locomotives, the American mind, the city, the Comintern, the day, the dragon, the elephant, the gods (i.e., the music of the Beatles), the Habsburgs, the Maharajas, the old order (France 1774–78), the Presidency (1970), the sea gods, the souls, the Supreme Court, the tenderfoot, tyrants, white races, and the young; as well as twilight on the Danube and on the range.

But how was TWILIGHT ZONE so seized by negative connotation? Was the absence of THE LIGHT, or the presence of THE DARK, sufficient to make TWILIGHT ZONE a place to avoid? Even a nineteenth-century poem by Thomas Bailey Aldrich sounded in 1877 the same forlorn note:

> *Somewhere in desolate, windswept space*
> *In Twilight Land, in No-man's land*
> *Two hurrying shapes meet face to face*
> *And bade each other stand.*

Taken alone, the word ZONE picks up bad vibes from seedy company. The *Dictionary of Sociology* defines "zone of transition" as a place "in a temporary state of deterioration characterized by a lower grade land use than formerly, and not yet ripe for succeeding more valuable land use."[3] The expression CHANGING NEIGHBORHOOD slinks along under the same assumption—any change is bound to be for the worse.

Shifting our attention from place to time, twilight, the condition, comes in degrees: nautical daylight ends when the sun is twelve degrees below the horizon; civil daylight begins or ends when the sun is six degrees below the horizon; and astronomical daylight ends when the center of the sun is eighteen degrees below the horizon.

It is tempting to conclude, finally, that TWILIGHT ZONE is any place so deprived of either daylight or human sympathy as to offer partial truth in place of the real thing.

See also CHANGING NEIGHBORHOOD, in Chapter 4: Ephemera; THE EDGE, in this chapter, above.

THE VILLAGE

VILLAGE itself, in name and concept, is undergoing the same transmogrification—conversion from place to gimmick—as we found in BROWSERS' VILLAGE. In Old French, the term *ville* had indicated a settlement, usually larger than a hamlet and smaller than a town. The typical European form consisted of a group of adjacent farms that made up a "communal village group." American authors are torn in choosing between town and village, sometimes settling for both. Edmund Wilson once recalled "that world of large old houses . . . in little countrified towns where a rather high degree of civilization flourished against a background of pleasant wildness."[1] More recently John McMullan, writing for Knight-Ridder News Service, asked "Is there room in our urbanized scheme of things for the towns and hamlets we once held dear?"[2]

Situated a notch higher than VILLAGE is small town, and somewhere beyond that is small city. The latter is identified in a 1990 book by G. Scott Thomas, *The Rating Guide to Life in America's Small Cities,* as one of 219 "micropolitan" places of the United States with populations of 15,000 to 50,000—far beyond VILLAGE size, to be sure. In Thomas's view, the best of the lot is San Luis Obispo, Cal., with easygoing charm, mild climate, and a Thursday street fair with the main drag closed off.[3]

Contemporary authors in books such as *Small Town America* (1980), *The American Small Town* (1982), and *The Small Towns Book* (1978), in company with the magazine *Small Town,* pursue a range of smallness. One reminds us that the New England Town has a "Town Hall" facing a "Village Green." The author of *One American Town* recalls that youngsters tore up "the village's venerable bandstand."[4] The term VILLAGE in the United States had never escaped the onus of backwardness (as in Sinclair Lewis's *Main Street*), nor achieved the rich history and seventeenth-century contours of the English village.

In the 1980s, however, the term VILLAGE was seized upon by merchandisers and other land developers—and even by planners anxious to identify with Real Folks settling into places that recalled Back Home. VILLAGE emerged as their favored new urban form, olded-up to resemble its eighteenth-to-nineteenth-century forebears. If a tract development could be somehow contrived to look "villagey," sales might increase.

"Every third planned unit development (PUD) seems to have 'village' attached to its name in an apparently successful attempt to market 19th-century nostalgia in the form of somewhat higher density suburbanization," observed planner Lawrence O. Houstoun, Jr.[5]

In far-off Romania, the reverse process had been at work in the 1980s—the destruction of some eight thousand villages by the then Communist dictator Nicolae Ceauşescu. His goal had been to force villagers into big-city housing blocks under tight sociopolitical control.[6] Off in another direction, Ethiopia by 1986 had collected and resettled some three million peasants into nearly eight thousand new villages.[7]

In the United States, "The Urban Villager" as a useful phrase had been coined in 1962 by sociologist Herbert Gans to describe ageing, predominantly Italian, North Boston, and West End residents forced out of their haunts at great psychic cost to make way for redevelopment.[8]

"For the past couple of years" wrote Ruth Knack, "a small but vocal group of [American] urban designers has been promoting the idea of recreating the old-time village. But while their designs are attractive, they tend to have an artificial Disneyland quality."[9] In 1990, "a condominium complex designed to resemble a 19th-century farm village," and called Battle Road Farm, was taking shape in suburban Lincoln, Mass.[10] Sam Hall Kaplan, the Los Angeles design critic, began the nineties by calling "urban village" the "first cliche of the decade . . . a grab-bag, a shake-and-bake residential historic district, cozy and communal."[11] A popular model on Florida's Panhandle was called "Seaside," and attracted prizes and imitators. Further north, "the concept of the urban village, modeled after a New England town" was alleged to be the starting point for designing "an abstract, geometric state-of-the-art treatment center" on a half-acre that cost $375,000 in New Haven, Conn.[12] Two Californians formulated a new suburban "pedestrian pocket" called Laguna West twelve miles south of Sacramento. They proposed alleys, corner stores, and 3,300 homes clustered around a Village Center with its mass-transit station—a throwback to many a railroad commuter-stop of the 1920s, and to earlier estate clusters of the late 1800s advocated by Andrew Jackson Downing. To give it some semblance of the old village function, its developer said it would even have some jobs within walking distance of homes.[13]

Phoenix, Ariz., followed a wider trend, remapping itself in 1985 into nine urban VILLAGES, and planning a 210-acre annexation, also to be subject to villaging. (But its citizens voted down a 1989 mass-transit plan to connect the VILLAGES.) In Columbus, Ohio, "the idea is to have villages set among greenery in the manner of traditional English or French villages," said New York architect Robert A. M. Stern.[14] He had been imported in 1989 to design housing in a 2,400-acre development at nearby Rocky Ford called The Villages. This was a modern echo of the popular nineteenth-century German Village near downtown Columbus which had survived gentrification in the 1960s (and which later became intensively resettled and upgraded in the 1990s).

By 1989 the process of creating "villages" was so well advanced that a headline writer for the *New York Times* felt called upon to reassure readers that Hingham, Mass.—with its two hundred venerable houses along a village green—is "the real thing, not a re-creation."[15]

None of this did much to slow down the centralizing force of capital-formation that was producing yet bigger cities. Suburbs offered "solutions" that copied village forms, but lacked village jobs and life. Most "villagers" in the United States still shared only remnants of what Robert Redfield (in *Tepoztlan, a Mexican Village,* 1930) called peasant virtues: "a love of the land, a reverence for nature, belief in the intrinsic good of agricultural labor, and a restraint on individual self-seeking in favor of family and community."[16]

Yet in all this, a self-selection process seemed to be at work. Not all city dwellers wished to move outward in space while backward in time to face entrenched village-dwellers unto the fourth generation. Yet those who did move were often grasping for a change-of-life, new housing, easy browsing, along with a change of pace. Thus, many rural villagers found themselves "invaded" by aggressively possessive yuppies importing new lingo (along with Robert's Rules of Order) to casual gatherings that had once passed as a town meeting.

SECTION THREE

Out There

LI

"Out" offers itself in many guises: out there, outside, outdoors, outback, out of bounds, out of town, out of state, out West, out on a limb, out on the town, outpost, out of it (i.e., touch, pocket, reach), out of the way, out and away (one of many variations of out and gone). Such distancing may leave one out of breath, if not out of sorts. Put enough distance between yourself and others and you end up far-out, out of the loop, 'way out or—once spoken in admiring tones—outtasight!

More visibly, the condition of out-ness intrudes upon other places such as out lots, out parcels, outliers, out islands (some seven hundred of them in the Bahamas), outreaches and outskirts. Not to forget the Outer Banks and miscellaneous outers such as islands and suburbs, and Mongolia. Nor New York City's suburbs where "most of [the city's] growth is happening in the outer ring."[1] And out there among the growth sectors, one finds scores if not hundreds of booming outlets, ganged up in outlet centers, on outlet malls, in old converted warehouse rows, or in town clusters.

Auto makers and others go in for "outsourcing" engines and other parts by buying them from outside suppliers, as distinguished from making the parts in-house. During the 1980s, it was noted by sociologist Barbara Ehrenreich that the top brass of American industry "'out-sourced' their manufacturing jobs to the lower-paid and more intimidated work force of the third world."[2] Following a 1991 California crash, the bus company involved was described as "based out of Phoenix," a term which occasionally appears in usage as in "operating out of . . ."

To keep oneself out of reach is deliberate and may be impolite or impolitic, thereby risking being out of touch if not incommunicado. To admit "I'm out of touch" is to confess geographical inferiority, while to be outlandish is to be strange, aberrant, foreign, unbelievable. To end up "on the outs" is to be ostracized, which is bad news indeed. It suggests banishment from society, or being cast out by vote. (Ostracism was the ancient practice using oyster shells or potsherds as counters. The word is from Greek, *ostrakon,* earthen vessel, potsherd, or shell.) However, to enjoy an outside stateroom on a cruise ship is an outright luxury indeed.

The late vituperative columnist Westbrook Pegler had his own set of outcasts: words that he didn't like were "out of town words."[3] Such out-ness has historical precedents: in 1856, the New York *Tribune* observed that "Out of town, which a few years ago meant above Canal-Street, now means across the [Hudson] river or the bay, far down by the seashore, or in the fast receding forests of adjoining counties."[4]

Thus "out" is a designation most often formulated by insiders who are themselves not out—who seldom go out, and who think little of out on its own account—except as being not-in. The term and its current usage in American English most often reflects the view from The Center. Out is a remote location, far removed from whoever makes such designation.

Architectural critic Paul Goldberger of the *New York Times* trotted out his own candidate for coinage—"outtowns." These, he explained, are "outlying urban centers, versions of DOWNTOWN that have sprung up outside conventional urban cores. They are out of town, but they are towns just the same."[5] Another *Times* writer, William Stevens, also attempted a coinage with "outer cities."[6] These were nice tries, but failed to penetrate general usage. In contrast, "edge cities," a coinage by *Washington Post* writer Joel Garreau, made it into a successful and widely reviewed book in 1991.[7] But a so-called edge city is still no more than a component of the larger whole, and the jury is still out as to his entry's linguistic survival.

Conversely, being on the outside is painful if not galling to millions of people relegated, as they may feel, to Out There, remote from and disregarded by The Center. This has been a powerful, alienating force in American politics, as well as the origin of countless booster enterprises—efforts to "put our town on the map." More specifically, "It has only been in the last five to ten years that most national news magazines, networks and newspapers have plopped correspondents in Denver to feed stories to the 'outside' about Colorado and the West," according to Marjie Lundstrom in the *Denver Post,* August 7, 1985.[8]

It was clear from many such accounts, as the twentieth century approached its ending, that being "on-the-outs" was politically poisonous. American society had become deeply divided: rich vs. poor, power elite vs. powerless, drug culture vs. enforcers, America firsters vs. internationalists, "developed" vs. undeveloped.

And, above all, the urban vs. non-urban split continued. This last division concerns us here, for by the 1990s the United States was irreversibly, and still uncomfortably, an urban nation. The "conquest of nature" had succeeded beyond any pioneers' dreams or nightmares. So much of America that was once Out There, in the minds of city-dwellers before World War II, had been settled, subdivided, urbanized. New building lots and separate surveyed tracts were being created at a rate probably exceeding three million per year. (See Epilog.) And in the process, nature itself was being "messed up," once-pristine land and waters despoiled and fouled, in some cases beyond hope or repair, and in most cases at a record-setting pace. There was a new sense of limits abroad in the land. And many of those limits—real or fancied—surrounded the places we examine in this book.

Cities and their regions had got the vote, the voice, the concentrated wealth, the political power. Nobody was going to disinvent urban concentration, even though automobiles and highways had relieved nineteenth-century overcrowding, and even though atomic scares in the Cold War decades (1960s to 1990) had caused flurries of token "decentralization." Not since the earliest days of the republic, when political and economic power snuggled along the Atlantic Coast, had Out There been so powerless.

As a result, populist resentment swelled Out There as more of the empty lands continued to be taken up by refugees, back-to-the-landers, or plain speculators from The Center. When Timothy Egan of the *New York Times* visited Canyon Creek, Mont., he found remote ranchlands being sold, in almost useless forty-acre bites, to city folks—much to the dismay of native Montanans. "Outrage Grows in West over City Slickers as Ranching Neighbors" said the headline over his report.[9] Montana still had more cattle (1,328,000) than people (799,065) in 1991.

Living "In" the System

On the other side of outside, however, 75.2 percent of all Americans were identified by the 1990 U.S. Census as "in"—that is, as "urban" or living in urban areas. No matter that the definition is loose. No matter that it includes wide stretches of Standard Metropolitan Statistical Areas inhabited by farmers, ranchers, or urban refugees and long-distance commuters. Nearly all the nation's millionaires, most of its power centers, much of the eastern United States and of California, and important segments of the rest, lay within the newly coined Daily Urban Systems. The Census Bureau tells us that the proportion of urbanites has risen steadily since 1920, when it was first noted that most Americans even then lived in cities and had begun to cope with traffic jams.

Out There is what's left. A strange creature, Out There, for it reflects both a geographical reality and a state of mind. To many city dwellers, Out There is anything beyond "my block," or "our neighborhood." To others, Out There begins at city limits, or at the first strips, or at some visible reality (or is it a mirage?) called open country.

But there is a still larger Out There, there—the thinly settled and vacant parts of North America: the Great Plains, most of the Rockies, and large remnants of the Great American Desert that loomed so large on nineteenth-century maps of the West. To most Americans Alaska is all Out There in some unknown Great Beyond. To Canadians, their Arctic is so far out as to deserve not provincial but territorial status. The latest of these is self-governing Nunavut, sparsely inhabited by Eskimos but larger (at 770,000 square miles) than Alaska and California combined.

Frank and Deborah Popper, professors from Rutgers University, achieved notoriety Out There by their analysis in the 1990s of the out-ness of the Great Plains. In 1991, 110 of its counties were in distress—losing population, income, soil, votes, and rural infrastructure. Its history of recurrent, endemic, built-in boom-and-bust is notorious. It has set records in overdrawing on its supplies of thin topsoils and underground waters, and in the destruction of fragile grasslands. On the Northern Plains, "drought and population loss are emptying Montana's flat eastern three quarters."[10] And the Plainsmen's resentment of people like the Poppers—and the Poppers' suggestion for restoring much of the 139,000 square miles of depressed, under-used and/or abandoned lands of the Great Plains to a "Buffalo Commons"—is a typical Out There response to experts "in-from-away." The Poppers found that the population of the Great Plains was ageing, departing (over 10 percent loss between 1980 and 1988), and down to four or fewer persons per square mile. At present rates, some two-thirds of Great Plains farms and small communities would vanish by 2020.

The Attitudinal Outback

Anyplace Out There in the United States, one finds old back-country attitudes forever at odds with city-slicker newfangles. Tradition is upset by innovation. Farmers and ranchers carry ungainly role mixtures as wage-earners, pioneers, and capitalists in risky enterprises. Out There, everything happens to disturb predictability, or to bring a-running modern versions of the revenuers: health enforcers, workplace snoopers, or marijuana-busters in helicopters. Not to mention newcomers and officials unaccustomed to life amidst mining, cultivating, ranching, nature's deadlines, or outdoor rambunctiousness. Here is the zone of tension between headquarters and the territory, also known as The Field (the place where people from The Center go out to).

In reality, Out There is where most basic survival heavy work goes on—miners blasting and excavating, farmers raising crops, cattlemen ranching, woodsmen cutting and trimming and felling, pulpwood tenders burning off the underbrush, and front-loaders and fork-lifters stacking raw materials—while forests and WETLANDS do their work quietly purifying air and waters. Out There is largely at the mercy of distant owners, remote markets, and distance itself. Out There is where urban attitudes, values, and practices come up against ancient opponents. Out There, people work with large animals and machines doing heavy, noisy, lonely, dangerously mechanized outdoor work. Their isolation and history tends to breed suspicion of The Center and its agents of change. It is true that parts of Out There burn and stink and stew, but in non-urban configurations. And its remote locations continue to attract so-called survivalists who wait, fully armed and sequestered in their HOLDOUTS, for some future Armageddon.

Other visions of Armageddon flourish concerning the nation's nuclear waste and test sites. Most are located Out There in faraway parts of Idaho, Nevada, and New Mexico. But of the total of 1,275 "Superfund sites" listed in 1992 for eventual cleanup, only 6 percent per year were being cleaned up.[11] It took years of citizen push-and-shove to get the facts revealed: about deaths in the fallout zone, and about future hazards from stored nuclear waste underground. Even the sites' remoteness from urbanization did not insure them against outraged citizen protest.

The Strangers' Impact

Such goings-on Out There infuriate apparatniks from The Center who arrive flourishing their Geiger counters, satellite photos, printouts, and rulebooks assembled in faraway statehouses or offices in Washington, D.C. Meanwhile, new touristic enterprises arise in the great Out There—ski resorts and their condo-dwellers, rented trout streams, hunting preserves, ready-made or remade historic mining villages, off-road camps, day tripper discount centers—all of them erecting "No Trespassing" signs and injecting weekenders, strangers, energy, and cash in episodic flows. The onrush of strangers into formerly out-of-the-way places had never had such impact as in the footloose 1980s, a replay of earlier auto invasions in the 1920s.

Distance is the familiar enemy Out There, and those who cope with it grow accustomed to it. They are more likely than their city cousins to communicate by shortwave radio, and to set up satellite dishes out back. Their ever-visible SKYLINES offer huge flows of non-electronic messages: weather changing, game birds migrating, a barn burning, a dust storm brewing, distant planes homing in along familiar FLIGHT PATHS.

Furthermore, Out There is where ancient as well as high-tech marvels find space for their apparatus and side effects. These occur in, and by means of, such generic places as BLAST SITES, THE BURN, CAMPing places/resorts, DRAWDOWN, HOLDING PATTERN, SOLAR FARMS, TENT CITIES, WATER RANCHES, and WINDFARMS, all of which we examine at closer range in this chapter. But if we backtrack toward The Center, many of these places and their goings-on become illegal or impractical. Among them are EVENT SITES too noisy or unruly closer to town.

Out There the skies offer maneuvering-room for the air races, stunts, parachute jumps, and skydiving that may be politically infeasible closer to The Center. Here are HOLDING PATTERNS for aircraft waiting to land at the HUB closer-in. Far above, often out of sight, are tightly regulated FLIGHT PATHS often so high as to bypass both The Center and Out There. And Out There we find the firing ranges, proving and testing grounds, and other noisy, dangerous places that survive Congressional cutbacks of the military establishment after the United States lost the USSR as its chief antagonist-competitor.

That generic place called EARSHOT all depends: on who's around to hear. There is no "there" unless the transaction is completed—unless somebody's there to hear the sound. "Beyond EARSHOT" describes a person located beyond the reach of specific sound waves—far out indeed. Most sounds are generated at The Center and on The Front—but a sound Out There carries more meaning simply because it is isolated and distinguishable, if not unique. Which is why EARSHOT extends "a fur piece" Out There. And there you frequently find LULUS (locally unwanted land uses)—often noisy—that would encounter supervision or opposition at The Front or The Center.

In-from-Out

Peddlers, solicitors, poll-takers, and reporters are usually made welcome Out There as visitors bearing news from elsewhere. Old-time hospitality still flourishes in much of the outback. No licence or letter of introduction is required. In isolated stretches of Maryland, it is said that "only an Eastern Shoreman fully understands another and beyond that he has small concern for the outside world, although he nods to it in native courtesy as it passes by."[12] But of course there are variations: a local official of the International Brotherhood of Teamsters union was held in contempt of court in Chicago for "Balking at Outside Supervision."[13]

More than once, city planners and others (including the writer), in moments of frustration, have expressed a nostalgic wish for a local Siberia—a remote but easily accessible, uninhabited, unzoned location Out There where all those LULUS could be accommodated with no hassles. But soon enough, as surely as night follows day, LULUS would be followed by hangers-on, suppliers, waste-processors, and workers—all demanding roads, utilities, services, housing, and schools nearby. Soon enough, these newcomers would be joined by others who registered to vote, and would do their political utmost to render illegal those activities that brought their predecessors Out There in the first place. They quickly become variants of exurbanites, who, having made a home Out There in the commutershed, now want the escape hatch closed behind them. It's an oft-told tale.

To restless opportunists at The Center, however, Out There consists of one vast opportunity site. It's speculator's country—fallow, untended, un-urbanized, promising. It is perceived to be inhabited by people who don't know opportunity when it stares them in the face: "settin' on a gold mine and don't know how to hatch." The rivers Out There are yet to be converted to plumbing systems; its workers to militant labor unions, its mailbox numbers to street addresses. Its Indian reservations are not yet absorbed by the "growth machine." Its fistfights, scrapes, and scraps, its shady deals and conniving commonalities not yet redefined in urban terms as crimes.

Out There the struggles continue between "users" and "exchangers": that is, between people who use local assets for domestic or local purposes, and those who measure resources in terms of exchange—strictly goods to be bought, sold, leased, and traded in larger if not distant markets. To the latter, if a "higher-and-better land-use" is available, you should sell out, move off, and let "real achievers" take over that site you've been wasting. Today a cornfield, tomorrow a new town. (If this sounds familiar, it is the same argument used against Indians: they were under-using a resource that white settlers and their stake-holders back in London, New York, and Philadelphia wanted. A similar argument is employed by some Westerners against rural Mexican immigrants Out There in the Southwest.)

The Yin-Yang Embrace

But the future of Out There depends upon The Center—on its politics, its grasp, its information, its dynamic capacity to spread its influence beyond The Front. Out There has lost much of its political grip. Unless ... unless, that is, the environmental migrants of the late twentieth century can continue to instill into Out There a new sense of purpose and validity—a virus so often imported by new outsiders. These new-wave arrivistes read magazines like *Out Back, Country, Homesteader.* They save old *Whole Earth* catalogs. They include hippie homesteaders from the sixties, late-comer exurbanites and weekenders, back-country hikers and bikers, and extensive strains of nature-lovers. They show a capacity to form political coalitions that have begun, like the nineteenth-century Populists, to swing elections in new directions. They fill the power vacuum left in many places by the shrinkage of local farmers, woodsmen, country workers, and repairmen—and their children—who seek the larger markets and allure of The Center.

That is one scenario. But there are others . . . Out There needs The Center's help and participation as badly as The Center needs Out There. Neither is self-sufficient. (Remember the oil-field bumper stickers posted by Texans angered by Northeast efforts to control oil prices during the 1973 shortage: "Let the bastards freeze in THE DARK," they said.) Each place carries part of the load of metropolitan, as well as national, survival. Many who populate Out There continue to resist The Center with vehemence unrestrained. But one can escape only temporarily from common sharing of places and of taxes, political power, and citizen responsibility. Many people try to have it both ways. They stake out their turf along THE EDGE and get stuck in those states of animated suspension that often pervade The Front.

Many political solutions lie in that vague territory known as the offing. These include metro this-and-thats: new forms of neighborhood and regional governments, interstate pacts and multi-county compacts, regional co-ops, and a host of transport strategies and devices to reduce the need, and to cut the costs, of moving from The Center to Out There and back.

It will gradually turn out, perhaps in the longer run, that those who come and go, who opt for The Center or choose a new start Out There, are all parties to the same effort—surviving in a chancy, shifty milieu. The sooner they join hands, politically and otherwise, and learn to navigate in these rough waters, the better their chances to survive.

CHAPTER 7

Power Vacuum

Invisible to many who live there, places where "nothing goes on" jump off the map to trained observers: An ABANDONED FARM reeks not of yesterday but of competitive tomorrows. Who succeeds "Old McDonald"? What's that little white "Zoning Hearing" sign all about? What about the so-called DEPRESSED AREAS, DUMPS, GHOST TOWNS? All of them show promise to speculators and other newcomers not yet held captive by local myopias. THE RUINS and TENT CITIES share anticipations of tomorrows, their rewards and risks. In a society spotted with energy surpluses, only trained eyes and minds can pick out those places where new energies lie in wait to be sprung into visible presence.

ABANDONED FARM/AREA/TOWN

Here lies an object of cyclical public outcry that arises after depressions, or a drop in world commodity prices, or discovery of richer lands or competing products. After the financial Panic of 1819, "from counties everywhere came ominous news of vacated property and wagons headed westward out of the state [Kentucky]."[1] After the Erie Canal opened in 1830, New England was dotted with "run-down farms" and "one-hoss towns" deserted by farmers gone West. In New England in the 1840s, "the hill farmers abandoned their land, selling their hard-won holdings for pittances and moving westward to lands advertised as 'ripe for the plow.' The wilderness returned."[2] Marginal hillside farms were abandoned first. Popular literature after 1900 was dotted with lamentations about "The Abandoned Farmstead," especially in New England; but historian John Stilgoe insists this should be called "wildering"—farms going back to forest or even wilderness—until the next wave of owners takes over.

Farm abandonment is seldom willful or voluntary. Rather, it comes from farmers' dependency on world cycles of boom and bust. It depends on movements of capital over which farmers have no control, on shifting federal farm policies, and especially on the widespread fallacy that food will always arise from land, regardless how badly land and its owner-managers are treated. A cynical federal administration will "let the market decide." In a spectacular case of involuntary abandonment, it was a willful federal government that forced thousands of California's Japanese-ancestry families to abandon farms, homes, businesses in 1942 after the bombing of Pearl Harbor—families "relocated" as World War II prisoners in distant isolated barracks. Their "abandoned" properties were snatched up cheaply by powerful California speculators, then merged into the postwar townscape.

Other "abandonment" is less dramatic, less visible. Schuylerville, in upstate New York, once stabilized by the food and social life supplied by some hundred nearby farms, has withered as "fewer than half remain today, and only 20 of any consequence."[3]

When commodity prices drop, huge farming areas are "retired from production," as the less-competitive farmers quit. Farms enter a new cycle: cropfields and pastures grow up in weeds, then brushland, then forest; get timbered off and revert to woods-pasture, are then cleared again for crops when prices go up. This thirty-to-fifty-year cycle causes the "patchy" and abandoned look of much marginal land in Eastern states, says geographer John Fraser Hart. In the arid West, ABANDONED FARMS revert to semi-desert. Some capital from the sale of abandoned lands flows into fertilizing, irrigating, and high-tilling better lands elsewhere.

Writer Jack Doherty recalls camping outside Portland, Ore., in 1932 in a large deteriorating farm. "Perhaps a bank had repossessed for nonpayment of a loan, or perhaps the county had foreclosed for nonpayment of taxes, both common occurrences in 1932. For whatever reasons, the former owners were gone, and the farm was ours to enjoy, at least for the time being."[4] There the campers dug potatoes and found vegetables, quilts, and an old wash boiler for their temporary residence.

In the eyes of hunters, archaeologists, and historians, clear signs of old ABANDONED FARMS that persist in woodlands include the following: ridged terraces from the wave of field contouring in the 1930s; lone stone chimneys; falling-down orchard trees and plantings of jonquils, narcissus, and other exotics around a former house site; lilacs as reminders of a former privy; a layer of charred firebrands just below the ground surface that once upheld a house; and the stone coping of a disused water well.

How much of present North American territory will qualify for abandonment is much debated. Writing for *High Country News,* Paonia, Colo., Rutgers University professors Frank and Deborah Popper predicted that "despite any conceivable efforts, much of the rural Plains will suffer near-total desertion over the next generation."[5] They aroused protest in the Great Plains, 1990, by advocating that of some 139,000 square miles of depressed, underused, and/or abandoned land revert to buffalo commons.

See also TOADS, in Chapter 4: Ephemera.

ABANDONED FARM/AREA/TOWN

Decrepitude, dereliction, and abandonment all cohabit in large sections of New England, Appalachia, the Great Plains and, above, the knobby tributaries of the Ohio River Valley. Such rough terrain no longer competes with Class A, level, machine-accessible farmland, leaving ramshackle farmhouses to compete with vegetation.

THE BOONDOCKS

So, here we are, out in THE BOONDOCKS —also known as the boonies, the sticks, the bush, the back forty, hinterland, back country, outback, williwags, Siberia, and also the-hell-and-gone. In short, a remote location far removed from those who use the expression.

THE BOONDOCKS probably entered mainland American English as the name of a generic man-made place by 1910, brought back by U.S. Army veterans from the Insurrection in the Philippines. The name, *bundok,* comes from Tagalog, meaning rough country—the interior behind jagged mountains, inaccessible to Americans.

Later, in the 1930s, American sailors came back from Pacific duty talking, not too openly, about a district of whorehouses located, not too inaccessibly, in Shanghai, also known as THE BOONDOCKS. The word migrated stateside among Americans returning from Pacific duty in World War II. By mid-war, it was ensconced in the Marine Corps Reader, 1944, to describe a tough Marine training ground in South Carolina: "the sand and boondocks of Parris Island."

More recently, the American writer William Least Heat Moon set out to write a book about "the three million miles of bent and narrow rural American two-lane, the roads to Podunk and Toonerville. Into the sticks, the boondocks, the burgs, backwaters, jerkwaters, the wide-spots-in-the-road . . . the Middle of Nowhere." It became his best-seller *Blue Highways.*[1]

By the 1960s weather-casters were predicting weather "out in the boonies." It became a favorite term for headline writers to deal with any scene beyond their ken or below their own horizons. Thus a headline in the *New York Times,* describing the new Interstate 78's impact on western New Jersey: "It's Boom Time in What Once Was the Boonies."[2] (But the word *boonies* never appeared in the accompanying article.) Columnist Anna Quindlen tried her own coinage: "The evening commuter rush is getting longer. The 7:45 to West Backofbeyond is more crowded than ever before."[3]

Describing their location in a week-end getaway village between Boston and New York, innkeepers Bob and Penny Nelson of Old Lyme, Conn., tell reporter Berkeley Rice, "It's sure not backwoods New England, but we weren't looking for that. We couldn't run a sophisticated operation like ours out in the boonies."[4] And on National Public Radio, a commentator observed Republican Party efforts, following the Richard Nixon presidency, "to get control of the party away from the East and back out here in the boondocks."[5]

In some Western and Pacific Coast circles, the tules is a synonym for the boonies, referring to swampy or flooded land covered with a form of bulrushes. In common usage, THE BOONDOCKS are invariably distant from users of the term. It's city-dweller's slang for locating, if not denigrating, the non-city Out There. The authors of *Urban Fortunes* refer to the noted Johnson Wax and John Deere firms in, respectively, Racine, Wis., and Moline, Ill., as located in "the corporate hinterland" or BOONDOCKS.[6]

Thus, far out beyond The Center, one finds THE BOONDOCKS lurking in the great out-there, somewhere beyond-the-pale; another colorful trophy piggybacked home from the wars Americans have gone overseas to fight.

THE BOONDOCKS

Seldom immortalized in touristic or other panoramas, THE BOONDOCKS are usually identified only by word of mouth. This Hartfordian view of its neighboring Connecticut towns—with THE BOONDOCKS prudently relegated to somewhere-up-north-New-England-way—appears in a series of widely circulated postcards. Harvey Hutter & Co., Inc., Ossining, N.Y.

THE DARK

Once upon a time THE DARK was an important part of everyday (and night) life. It was the devil's domain where witches bubbled their brew and evil spirits, Beelzebub, criminals, refugees, runaways, prisoners, racketeers, and illicit others all hung out. Where could you sulk or skulk or lurk, if you needed to do so, but in THE DARK? It was an ingredient vital to the working of escape mechanisms. When—in that unlit long ago—you said, "Be sure to get here before dark," you were talking not schedule but survival.

But now we have surrendered THE DARK and all its functions—not all of them illegal or dangerous—to the lighting and utility industries. THE DARK, as a frontier, has been invaded if not overrun. Millions of farmsteads are automatically lit at dusk by barnyard lights turned on and off by photoelectric cells. (In common parlance, these are "barnlot lights," but insurance companies encourage them to protect the barn itself.) In much of the Eastern United States there's little dark left, and in California traces are hard to find.

THE DARK has been, in effect, bought

out, and THE LIGHT moved outdoors at great expense. The Great Outdoors has become a place of round-the-clock activities, mostly floodlighted. Criminals now hide within electronic skullduggery; footpads and cosherers who haunted the eighteenth-century English landscape now do daytime work in what is called "the black economy"—concealing income, evading taxes, fencing stolen goods. The *Los Angeles Times* wrote of "'The Black World' that cloaks military contracts in secrecy and hides multi-million-dollar business from public view."[1]

On a national scale, THE DARK is close to becoming an endangered species, as satellite photos and nighttime travel now reveal. Pressures on the astronomical environment have reached a scale similar to those on the earth itself. In an effort to preserve THE DARK from local extinction, some cities have declared THE LIGHT to be a pollutant. For THE LIGHT generated by a large town or city gets between astronomers and their source material in the high heavens. "Light pollution continues to be as inexorable as urban sprawl. It has noticeably affected virtually every professional observatory in the world."[2] Both the Mount Palomar and Mount Laguna Observatories in California, so as to protect their share of THE DARK, managed in 1985 to get a law passed banning "all nonessential lighting after 11 P.M. in unincorporated areas of San Diego."[3] Mount Wilson Observatory was to close in 1985 "because the lights of Los Angeles have decreased its value."[4] The observatory, after closing, was restored for limited uses.

Some astronomers have looked to the moon, or to orbiting space stations—if they work—as future locations for their scopes, but "if the Hubble Space Telescope is any guide, such facilities will cost at least 1,000 times what they do on the ground."[5]

Yet it appears that THE DARK is a place where modern designers—who like to call themselves "place-makers"—hardly know how to proceed, except to convert THE DARK into THE LIGHT. This they continue to do, so that the entire United States appears on satellite photos to be just another endless constellation, with true darkness a diminishing part of the brightly lit whole.

DEPRESSED AREA/REGION

Here is the flip side of BOOMTOWN, the inevitable sequel to GROWTH AREA; it is the locale, the IMPACT AREA of the last bust. Boom and bust depend upon each other for the system to work; i.e., unemployment here is necessary to employment there, surplus capital here seeks better return elsewhere. That difference in opportunities creates mobility of labor and capital, the yin-yang of capitalist systems.

Along the way, BOOMTOWN had become a focal point for inventors, innovators, risk-takers; a mecca for speculators, hustlers, experimenters, for criminals and pirates, along with simple job-seekers unaware that BOOMTOWN may have already passed its peak and begun its decline. Each boom attracts capital from elsewhere; it uses up and may exhaust local raw materials; it eventually entices fewer customers, wears out its machines and fails to replace them.

Pessimists see DECLINING AREAS as a place to be vacated, fast. Optimists see DECLINING AREAS as Opportunity Sites, once new rules-of-the-game can work in their favor. Thus arises a host of special taxing districts, area development regions, et al., set up to dole out local or regional subsidies to the DEPRESSED AREA from a still-prospering national economy.

Expansion leads inevitably to the limits of that expansion's environment. The ultimate impossibility is an ever-normal granary, a permanent boom, or guaranteed full employment in a free society competing in global markets. That this fact is widely disbelieved in no way affects the way it works.

The term DEPRESSED REGION seldom makes headlines in the DEPRESSED REGION itself. Local media are quick to pick up scraps of evidence pointing to a local "upturn," or to worse conditions elsewhere. (This was known in newspaper circles prior to the 1970s as "Afghanistanism"—a tendency to dramatize how much worse things are in Afghanistan or some other distant and Godforsaken spot.) Readers of the *New York Times* learned that "In Houston, Bottom-Fishing Brings Up Bargains: Out-of-Towners Pursue Deals in Depressed Area." The article explained that "out-of-town investors flocking to this city to scoop up real estate at rock-bottom prices have given rise to the newest sport—'bottom fishing.'" Following the crash of oil prices and savings-and-loan firms, "locals aren't buying because they're broke," observed a local professor. This account recorded "the first wave of out-of-towners" buying up thousands of bankrupt or empty properties in Houston.[1]

Sniffing the wind among booksellers, *Publishers Weekly* published lengthy advice on "What to Do (or Not Do) When the Local Economy Goes Bust." "Amoco is just the latest company to pack up and leave Slidell, LA, taking about 120 families with it. . . . Over the past five years, the town has lost about 3,000 families, as first the oil companies and then aerospace firms relocated." Such an outmigration's impact "is potentially disastrous on bookstores, who depend in large part on educated families that like to read. . . . Despite the dismal economic conditions in New Orleans, Dallas, and Denver [certain key] bookstores are surviving—and even growing. . . . Dallas in particular shows signs of having hit bottom."[2]

Fitchburg, Mass., was described as being "in the midst of a slump that has been worsened by the failure of the bank that made the lion's share of loans for new construction."[3] The crime of "equity skimming" (and thus defrauding U.S. housing agencies) "is all the more insidious because it thrives in areas of severe economic distress—Texas, Oklahoma and Colorado these days."[4] Two New Jersey planners were attacked for comparing the Great Plains of the 1990s to the Dust Bowl of the 1930s.[5]

In a 1975 study of declining metro areas (having less than 1 percent population growth), Edgar Rust found their number rising from 5 in the 1940s to 10 in the next decade, to 26 in the 1970s. Between 1970 and 1988, a total of 120 metro areas had net growth of 2 percent or less, while 17 showed net losses. Kevin Lynch observed that the typical declining city had boomed in the past, thanks to one economic activity in which it specialized. "When that activity faded, or found a more advantageous locale, the city then failed to shift to new enterprise." Original booms that were based on profitable mines, soils, or transport routes, folded up when those resources or routes became exhausted or passé. Lynch concluded that cities which extract from their booms such long-term amenities as a beautiful public environment manage to keep their more choosy citizens longer than do their ugly, amenity-poor competitor cities.[6]

Are geographers and other place-trackers more partial to growth regions than to DECLINING AREAS? The question arises from the "Upper Midwest Economic Study" conducted by noted geographer John Borchert of the University of Minnesota (c. 1961). His maps indicated the growth rate of communities in four categories: "Very Fast" (32 percent annually or faster); "Fast" (25 to 31 percent); and "Moderate" (11 to 25 percent). All others, whether they were being depopulated, growing slowly, or had stabilized, were lumped into a single category, "Slow or decline." Thus it appeared preferable, in the American growth mode, to be four times more precise in classifying growth than in classifying its absence.

Frostbelt or rustbelt cities offered prime examples of depression in the otherwise-wealthy-appearing 1980s. In the coal mining region south of Wheeling, W.Va., employment in mines and in steel, glass, pottery, and paper plants ran to sixty thousand around 1975. By 1990 jobs had been cut in half. Here, as in most DEPRESSED REGIONS, there was talk of "the domino effect"—one lost job impacting others.

A more serious blow to BOOMTOWN mythology came to Houston, Tex., "roaring capital of the oil patch" about 35 percent dependent on the oil industry. Proud of its wealth (thanks to a fivefold jump in world oil prices starting in the 1970s), and of its venturesome spirit, it made much of its lack of zoning controls. The latter contributed to what the *Atlanta Journal-Constitution* called, "a pornographic tide . . . [so that Houston was becoming] the pornographic capital of America."[7] But the *Houston Post* by 1981 was reporting that its own city "ranks near the bottom in [city] services provided" to its citizens, compared with Atlanta, Dallas, Los Angeles, New Orleans, Phoenix, and San Diego. Houston's flood protection was a regional joke. Local planners struggled for a small victory: the first sign control ordinance in 1980.

When oil prices crashed in March 1982 carrying Houston's boom with them, optimism persisted; a local bank, Southwest Bancshares, held onto its plans for an eighty-two-story office tower, to be the region's tallest. But as effects of the oil crisis deepened, long lines of jobless street people were being infiltrated, for the first time, by tough young men preying on the elderly homeless.[8]

After "the chaos of the boom and the despair of the bust," Houston passed its "first substantive planning ordinance" in 1982, as well as its first ordinance requiring street address numbers be posted on all commercial buildings. Zoning was permitted under another name, "development controls," which already had a closet existence under the guise of private deed restrictions. When *New Yorker* magazine writer George W. S. Trow visited in 1988, Houston had forty million square feet of empty unoccupied office space. One observer noted that by 1989 freeways were mostly uncrowded, society women stopped wearing fabulous jewels to charity balls, and Bible-study classes were unusually popular. The devaluation of the Mexican peso in relation to U.S. prices brought a severe drop in Houston's overbuilt hotels' occupancy. Within the decade of Houston's depression, planning and zoning were no longer dirty words. Now they were seen locally not so much as cures for depression but as ways to prepare for the next boom. "Real Texans" viewed this as inevitable.

Above all else, such a DEPRESSED REGION offers its citizens disturbing evidence that their former freedom—from shortages, depression, or outside supervision—was based on temporary advantage, rather than innate virtue. Meanwhile, local media can be counted on to search for optimistic straws in the wind. In the Lakes Region of New Hampshire, brokers were hoping for better sales "after nearly two years of slow sales and plummeting prices."[9] And "two developers of high-end houses in Rhode Island say they are finding buyers despite a 10 percent drop in real estate sales throughout the state since last year."[10]

See also TOADS, in Chapter 4: Ephemera.

DUMP

Variants: trash, rubbish, garbage dump, dumping ground.

No such thing exists as universal trash/rubbish/garbage. Each is socially defined.[1] One man's trash is another's treasure. You discard, or dump, what I keep. One person's garbage is another's compost; one organism excretes what another eats. A major difference between rich and poor is what they throw away. None of us is in touch with universal truth. Trash/rubbish/garbage offers endless variations on the same theme. All may end up in the same DUMP—a place of continuing debate and interurban contention.

Consciousness of garbage disposal came late in the nineteenth century after growth of the post–Civil War industrial United States.[2] DUMP had its linguistic usage as a verb in Dutch and Middle German. Then, as a place, it found a linguistic foothold in the United States when nineteenth-century miners in the West "began calling their waste piles 'dumps.'"[3] We've been detouring around the DUMPS of industrial society ever since.

Garbage has been heaped, hauled, measured, categorized, predicted, analyzed, moved, covered, drowned, incinerated, compressed, processed, recycled, litigated, and discussed, especially in the last century. Mounds of garbage keep growing around us. Our per capita trash is about double that of other developed countries.[4] Hundreds of towns and cities were running out of DUMP space by the 1980s—and seeking distant and less powerful "Siberias" to which they could export wastes.

Among the more innocuous disposal sites are various man-made hills of trash/rubbish/garbage locally entitled Mount Trashmore. Where the first "Mount Trashmore" appeared is not known; one "original" near Chicago in DuPage County is attributed to John R. Sheaffer, an author of *Future Water*.[5] The term is now used generically to designate any large mounded landform composed of urban wastes. Local "Trashmores" were the subject of a design competition by King County, Wash., in July–August, 1979, to determine the best reuse of its old DUMPS. Advised one contestant: slice your old DUMP vertically, cover the slice with glass, and identify each layer to the public: "What we threw away in 1925."

In some GROWTH AREAS, "builders' trash" had to be hauled away to DUMPS—until canny builders learned to cover their mounded leavings with dirt. Thus came instant "earth sculpture."

Since the first New York Dutch burgher's settlement, "several thousand acres of land have been added to the island of Manhattan" as landfill.[6] Dozens of other city sectors rest unknowing on old DUMP sites in Washington, D.C.; Seattle; San Francisco; Norfolk, Va.; Louisville, Ky.; New Orleans; and elsewhere. "Milwaukee today rests on . . . fill, creating severe foundation problems."[7] Under a veneer of topsoil and lush growth, parks from Brooklyn to Burbank, Cal., universities and colleges, subdivisions and cities are sited above yesterday's DUMP.

While pre-industrial landfills are generally benign, DUMPS during the last century became less innocent. They tend to settle unevenly, to catch fire, to generate gases, to stink, and to attract scavenging rodents and humans. Landfills, which have often provided transition from low value to high value real estate, rarely find a welcome anywhere. Even more discouraging, the first blush of efforts to "mine" trash for valuable resources—energy, metal, and reduction of landfill fees—met with unexpected complexities which included deeply polluted strata. Decades later we continue to grapple for "technical" solutions, while any long-term change requires basic shifts in the daily practice of household life.

Currently landfills require "plastic bag" standards including venting, draining, and monitoring for decades. Until the 1960s people didn't "worry about ground water quality. They just basically dug a hole and buried the garbage."[8] And it's not just landfill leachate people worry about. Illegally dumped poisonous or noxious wastes threaten animals, kids, and assorted outbackers who blunder onto old DUMP sites.

Garbage waste disposal was thought in 1992 to be a new recession-proof industry. "Garbage architects" propose secondary uses of reusable items and materials that go into their buildings. By 1992 several U.S. and European car-makers were earmarking certain components as prime recyclables when the vehicle is junked. Daily, and often unaware, we encounter reused items: the gray inside of cereal boxes from newsprint, gypsum board in walls, tar paper on roofs.[9] Discarded tires have been incorporated into new resilient landing strips to prevent cracking.[10] A Japanese firm has designed a trash compactor producing one-ton desk-size cubes to be buried or concrete-coated for construction use.[11] Others develop garbage pellets of burnable trash to mix with high sulfur coal, a process they claim lowers sulfur emissions and is economically practical in the Northeast where DUMPing fees outstrip production costs.[12] DUMPS also serve as efficient locales for producing methane gas, most of which is currently wasted; and to provide archaeologists a splendid record of the habits of earlier throwaway civilizations. To this end, DUMPS have even generated a new field of study—garbology. It depends upon techniques developed by archaeologists who sift through shell mounds and kitchen middens for clues to throwaways of long ago.

In the meantime citizen recycling speeds up. But paper-saving, can-crushing citizens need, and only slowly get, monetary motivation. A Seattle success story began when monthly trash bills increased for nonparticipants. More than 60 percent of the households participated in 1989 with nearly 30 percent of household garbage recycled.[13] Meanwhile, "recycled trash and junk is one of Zimbabwe's major resources."[14]

Along the way, DUMPS are being rechristened as Regional Recycling Centers and Transfer Points. As the place name DUMP itself gets dumped, it suggests that the inheritor of the miners' trailings may be richer in resources and ore streams than many which occur in nature—but only as people relearn how to sort, sift, save (or "mine") them.[15]

GHOST TOWN

In the absence of true ghosts, GHOST TOWNS flourish in the American and Canadian West. The term derives from a mishmash of imagination coupled with tourist promotion. GHOST TOWNS hover in transit, a mix of bravado and decrepitude. Some had emptied out when trains or roads disappeared, leaving semi-deserted whistle-stops. Most are evidence of mining's boom-and-bust cycles; the term GHOST TOWN has become almost a

GHOST TOWN
Western tourism promoters would be lost without GHOST TOWNS as bait for "adventure-tours-into-the-Historic American West." Such de-urbanized skeletons become the venue for staged shootouts, stagecoach robberies, and suchlike contrivances that greet tour buses. Photographer Jill Lachman calls this one—Bodie, Cal.—"one of the best-preserved GHOST TOWNS in the U.S." Photo copyright Jill A. Lachman, Hayward, Cal.

monopoly of the West's writers and its tourist industry.

Barkerville, B.C., prospered from a gold strike in the 1860s, burned to the ground in 1868, was "a genuine ghost town for some years," and then was revived by dredging gold from Antler Creek in the 1920s and from Pine Creek in 1948–50, and by quartz mining in 1933. When restoration began in the 1960s it had a population of 20.[1] Jerome, Ariz., flourished with copper mining in the 1920s. Its Daisy May mine had grossed $125 million when production ended in 1938.[2] Marion, Ga., founded in 1810, grew during the cotton boom of the 1850s, with courthouse, jail, stores, post office, academy, a fine hotel, stage stop, several saloons, and a town ballroom. Its population once numbered three thousand, but it rejected a proposed railroad; its citizens gradually deserted. It was described as a GHOST TOWN in the 1970s and had only one house left standing in the 1980s. Highland County, Ohio, claimed in the 1940s to have ten GHOST TOWNS "whose inhabitants and industries have completely vanished."[3] For a linear variation, consider the summer BOOMTOWN of Ocean City, Md., which after Labor Day "empties and becomes a ghost strip of shuttered boarding houses and padlocked high-rise condominiums."[4]

Over a hundred years ago a Georgia historian, Charles Jones, found enough examples of the end-product of DECLINING AREA for a book, *The Dead Towns of Georgia*.[5] In our own times, tourists snap up such books as *Ghost Towns of the West, Ghost Town Album, Let Us Build Us a City: Eleven Lost Towns,* or *The Lost Towns and Roads of America*.[6] Readers, especially outsiders, continue to exhibit a morbid interest in suitably identified ruins, and in dead and ghostly urban presences, whatever their origins.[7]

THE RUINS

En route toward sacrality or oblivion, RUINS have many beginnings: in human poverty or neglect, in war, windstorm, fashion, or urban renewal. Or in a change of mind. They mean what we choose them to mean; what we want them to mean. In fast times, we slow down to examine and to praise RUINS. In impatient times we get rid of them, or quarry them for building materials. In slow times they sit and wait.

It takes a determined malcontent to find the picturesque in RUINS of his own times, to look about at grand improvements only to fasten his gaze on some waif-and-stray from days long gone. Not until long after the Renaissance did Europeans rediscover the grandeur of Roman and other RUINS. Not until American urban redevelopers of the 1960s had razed thousands of fine nineteenth-century structures did a volatile public rediscover these as "landmarks" and enclose those remaining with legal protections. The "decaying fragility" of RUINS exercises periodic appeal to public concern—these decaying heirlooms of yesterday. As prospective art forms, rubbish and RUINS go through cycles of fashion closely related to building booms.

RUINS can be graded along a rank-order of time: untended, unoccupied, unpainted, run-down, decrepit, ramshackle, decayed, de-roofed, falling-down, disintegrated. At some point in its slow slide into decrepitude, what began as a structure becomes THE RUINS, an object of romantic imaginings.

Speed up the process, surround it with war, multiply THE RUINS, add the stench of rotting human bodies and burned-out wet bedclothes in hundreds of former houses, and THE RUINS undergo horrid transformation. The civilian mind goes into shock. Instincts of looter, vandal, or nurse may come to the fore. Perhaps there is nothing to be done but look and run. On television THE RUINS themselves cease to shock—only the screams of the injured, the grimacing face of the survivor carries the message to an detached, distant audience.

Once RUINS are declared a DISASTER AREA the atmosphere changes; the outside world and its officials move in, bulldozers carve new paths through THE RUINS.

But to protect and preserve THE RUINS against "improvement," as a vivid reminder of the past—that requires extraordinary drive and consensus. Local energies pour into reconstruction; only extraordinary emotions could have protected the bombed-out, ruined church in the center of West Berlin from the rebuilders. Often rage and resentment fuel the preservation of wartime RUINS as-is—such as France's preservation of those Verdun trenches where soldiers were buried by artillery blasts while standing up in their trenches. Only their bayonets still stuck above ground.

Across most of the United States, THE RUINS, as ruins, have long since disappeared. They have been enclosed, modified, renewed, reconstructed. Everything possible and fundable has been done to convert them into something resembling real estate. An active new procession of preservationists can produce maps, models, and computer-images of every stage of ruination and proposed reconstruction. Seldom do RUINS survive all this. Consequently true examples of THE RUINS become more rare. Until the next major earthquake, depression, or other disaster produces a new supply, THE RUINS remain an endangered species.

See also HOLDOUT, in Chapter 1: Back There; DISASTER AREA, in Chapter 2: Patches; PRESIDENTIAL SITE, in Chapter 3: Perks.

SINK

Until humans invented pumps, a SINK was a natural phenomenon—the visible product of land-sinking due to the removal of underlying strata by erosion. SINKS usually had a circular shape, sometimes with water collected at the bottom. But pumps added a new aspect by removing water from aquifers, causing land levels above them to drop. Beginning in Europe in the 1300s, pumps driven by windmills drained water from low-lying fields behind dikes. By the 1600s pumps were draining European underground mines, which made possible more extensive mines. Such expansion by the 1960s, in the case of Pittsburgh, Pa., had caused land above the mines to sink, streets and utility pipes to crack, and houses to collapse. Less spectacular but more widespread was the surface shrinkage in the arid West caused by excessive pumping of underground water for overland irrigation. In 1991 residents of Antelope Valley, Cal., found the ground literally opening beneath their feet. At Edwards Air Force Base, where space shuttles had landed, giant cracks forced the closing of a runway and disfigured prospective subdivision sites at a planned community at nearby Lancaster. According to a 1984 report by UNESCO, California had "the dubious honor" of having the largest area—6,200 square miles—of groundwater-related subsidence in the United States. The report also noted that a 463-square-mile area around Lancaster had sunk three feet between 1955 and 1978.[1] As a *Los Angeles Times* headline noted, March 17, 1991, "Pumping Threatens to Sink High Desert's Future."[2]

SINKS also exist in quite another sense, as "places of last resort into which powerful groups in society shunt, shove, dump, and pour whatever or whomever they do not like or cannot use: auto carcasses, garbage, trash, and minority groups. American society acts as though all these were identical undesirable elements to be pushed over the bank, heaved off the edge, out of sight, out of mind, down into the sink."[3]

American cities possess—although they attempt to conceal—so many SINKS that a first-time observer could conclude that conspiracy was at work. Most SINKS exist in seldom-seen places, or else get screened off by users and neighbors. Consequently they are known chiefly among small boys and other hideaways, dumpers, geographers, and frequent low-altitude fliers. Inevitably they attract the attention of land speculators who perceive all SINKS as prospective turf.

SUICIDE SPOT

Going outdoors to end one's own life is strictly a minority enterprise. Suicides occur mostly indoors, in automobiles, or in transit, with guns or poisons doing the job. The official count: Thirty-two thousand persons per year in the United States are reported as killing themselves, "but the real number is a lot higher."[1] Most who succeed do it with firearms. The most common although less successful means is "self-medication" via poisons or overdosage. Only a tiny minority choose dramatic outdoors spots for their last, or would-be last, act.

In the United States, 870 deaths per year are officially attributed to jumping, but official and medical records are skimpy. One account notes that jumping is frequently executed in public, with site preference being influenced by the "romanticism" of certain bridges.[2] By all odds, however, the favorite outdoor jumping-off-place is San Francisco's Golden Gate Bridge. In its first twenty-eight years of existence, it was the site of over 270 successful suicides[3] and by 1993 the official number had risen to 938 known suicides.[4] "The average was one every three weeks, making this 'the No. 1 suicide shrine in the Western world.'"[5] San Francisco itself is the site of far more suicides per capita than most U.S. cities, and "each year there are more suicides [3832 in 1989] than homicides in California."[6]

By 1987, 134 persons had used the number two suicide venue, the San Diego–Coronado Bay Bridge. It was characterized as "America's second most deadly bridge for suicides," and headlined "The Bridge of No Return."[7]

The Golden Gate Bridge's height of 250 feet (add another ten at low tide) almost guarantees death to the jumper—one hits at about 74 mph. It takes three to four seconds to go the distance; the few survivors say it feels longer. One medical study concludes that a would-be "suicidal leap from either the Golden Gate or the San Francisco–Oakland Bay Bridge is almost invariably fatal. Only one percent of persons jumping from these bridges survives."[8] Also ranking high as overwater SUICIDE SPOTS are the George Washington Bridge, N.Y.; and the Delaware Memorial Bridge, New Castle, Del.[9]

Given the number of honeymooners frequenting Niagara Falls, N.Y., "the Honeymoon CAPITAL of America," its attractions as a SUICIDE SPOT would appear limited. Yet height conquers all, or at least the thirty people a year who end their lives by leaping from either the U.S. or Canadian side. As a suicide site, it was ranked second only to Golden Gate Bridge in November 1987—a distinction held by the San Diego–Coronado Bay Bridge only seven months before.[10]

Most such leapers carry a history of schizophrenia, paranoia, depression, and heavy use of drugs or liquor. Many leaps are preceded by typical pre-suicidal carryings-on: mumbling, crying, threatening.

These leaps do not guarantee death; the few studies available indicate a fair number of survivors, usually with ruinous injuries. As a consequence, many high urban bridges are fenced and/or patrolled, and some favorites are now equipped with telephones so would-be suicides can "reach out and touch someone" to call them back from the brink. The first such "hot line" was set up in 1958 in Los Angeles.[11] So far, suicide barriers—high fences, reinforced by talk-'em-out-of-it guards—have proven effective at the Empire State Building, the Eiffel Tower, Pasadena's Arroyo Seco Bridge, "and other once-notorious plunging platforms."[12] Meanwhile, the new subscience of suicidology is said to be in its infancy, as suicide numbers expand among whites but not among blacks.

See also LOVERS' LEAP, in Chapter 6: Border Zones.

TENT CITY

A sign of breakdown, crisis, disaster, disruption—or sometimes merely innocuous social events—TENT CITY springs up to accompany sudden unexpected surpluses—transients, refugees, homeless victims, pilgrims, guests, or special-events workers. It widely accompanied the increase in homeless persons in the United States, variously estimated at one to four million by 1993.

But its roots dig deep into native ground—probably back to Indian teepee camps set up for summer hunting on migratory game paths and flyways. TENT CITIES dotted the nineteenth-century landscape of camp meetings, chautauquas, mass outings. They were set up in the 1840s at Martha's Vineyard, Mass., to shelter as many as twelve thousand worshippers at religious revivals. In 1910 a madam of Vale, Ore., pitched tents in an isolated clearing, which became known as Whorehouse Meadow, its ladies-of-the-evening doing a thriving business under canvas, catering to range-weary cowboys seeking to saddle-up in unaccustomed luxury. The

meadow figured in a 1968 controversy when a prissy bureaucrat changed the name to Naughty Girl Meadow. The original was restored in 1983.[1]

Later TENT CITY became part of municipal responses to demands for instant shelter—an element in a general loosening-up in the meaning of "shelter." Heavy demand groups include the press/media at political conventions, spillovers from an unexpectedly popular rock concert or state fair, rescue workers at a disaster site, or construction workers at settlements with names like "Dam Village." Often enough TENT CITY goes by other names: it may be called Shacktown, Shantytown, Bogtown, or simply labeled "check-in-station." By 1989, "beneath freeway overpasses, in parking lots and in parks, people around the country are building tent cities that are fast becoming the front lines in political battles over the homeless," wrote a *New York Times* reporter.[2]

TENT CITY is distinguished by its cheap and quick construction. Neighbors object to its laid-on and complained-about sanitation facilities or the lack thereof. TENT CITY is usually sited deliberately in a prominent location to gain political visibility, as was "Resurrection City" on the Washington, D.C., Mall. This was hastily put together in the summer of 1963—a biracial, predominantly black demonstration against civil rights denials and the Vietnam War, following the murder of Dr. Martin Luther King, Jr.

During the era of cheap cotton canvas lasting into the 1950s, TENT CITY was distinctly tan-to-olive drab, its monocolors reflecting the postwar military surpluses of cotton-canvas tarpaulins. This drab era

TENT CITY
Four TENT CITIES such as this were set up following Hurricane Andrew's crushing blow across South Florida on August 24, 1992: three at Homestead, Fla., and one at Florida City. They ranged from 400 beds to 1400; were run by National Guard units and the Federal Emergency Relief Agency; and had a peak population of 3,600 processed by American Red Cross workers. One recalled, "It was like war; three hundred Hueys (helicopters) in and out, nothing green in sight, everything in wreckage." From American Red Cross, Louisville, Ky.

ended by the 1960s. Cotton moved off the no-longer-cheap-labor fields of the South to the irrigated Southwest. Its price went up, and cheaper sheet plastic filled the gap.

New-generation TENT CITIES became multicolored and garishly visible, sometimes aping the yellow-and-white pavilions of elite debut parties and receptions, and of such elegant golf matches as the 117th British Open at Royal Latham. Far more than plastic, however, the rise of outdoor protest groups and the mass-production of Special Events for the growing trade convention industry brought on a revolution in TENT CITY. Each larger and gaudier special event—national political conventions and booming trade shows, or rock concerts attracting fifty thousand and more attendees and hangers-on—involved lavish media coverage and policing. Thus was created a specialized form of TENT CITY, chiefly inhabited by convention-based workers, television apparatus, and crews, with their collective demands for V.I.P. treatment and set-up space.

At the 1984 Republican National Convention, Dallas, Tex., TENT CITY exhibited its traditional function as a protest site for anti-nuclear, anti-poverty, pro-gay, and generally anti-Reagan-administration groups. "For nearly a year the city battled to keep protesters from camping in city parks. In the end city officials compromised, making available the dry stream bed of the Trinity River for a TENT CITY—and even spraying for insects and helping to remove the poison ivy. Perhaps not so coincidentally, the riverbed was also selected as the site of a daily cattle drive."[3] This particular TENT CITY was "provided for opponents of the Republican Party or the Reagan Administration, as well as for those who prefer to camp out at the convention."[4]

At Kentucky Derby time in May, the University of Louisville at one time set up TENT CITY on hockey and soccer fields for visiting students, $5 per night. In Memphis, Tenn., the Press Tent set up at Auction Avenue and Front Street was a small example at a big event: a ground-breaking, September 15, 1989, for the thirty-two-story Great American Pyramid, a stainless steel "signature landmark," later opened in 1991 at a projected cost of $200 million.

TENT CITY also became a highly publicized gimmick used by protest groups to call attention to their goals or plight. In 1987, Los Angeles sponsored a TENT CITY as an alternative to street camps in the skid row area. This led to the formation of two groups to help the homeless.[5]

Expanding public interest became evident in the 1990s as the homeless ended their political invisibility. Once assembled in eye-catching campsites and TENT CITIES, they made well-housed people disgusted, and uncomfortable with their own feelings of disgust. Such feelings arose, not only from the fact that many homeless were addicts to drugs and/or liquor—but that increasing numbers were, to the middle class, "just like us." This was a reminder that, even in so-called good times, many DEPRESSED AREAS generate waves of homeless refugees.

Christmas brings another variation: a TENT CITY for the poor and homeless such as that opened in Los Angeles in 1985 as a seasonal refuge "on the abandoned old State Building site across from City Hall. [It] housed about 300 in true Southern California style—concealing misery beneath blue and white tents resembling canvas pavilions at a Beverly Hills garden party. The shutting down of Tent City returned its inhabitants to downtown streets."[6] By 1989, when jails and prisons had become widely and illegally overcrowded, TENT CITY offered a temporary solution at Secaucus, N.J. The overcrowded Hudson County Jail got a tented annex on a patch of Hackensack Meadowland. It had a mess tent, recreation tent, and ten brown Army tents with wooden floors, all in a row, heavily fenced and cheaper than barracks. While it lasted, it was popular with prisoners especially for its outdoor sports.

Having experienced TENT CITY life on the deserts of Iraq, and having "noticed during Desert Storm that the Air Force had much better living conditions," the U.S. Army developed a prototype TENT CITY in 1992 on the outskirts of Fort Bragg, N.C. Designed to be set up ten to fifteen miles behind a military front (with presumed air supremacy), five more 550-bed compounds were scheduled for production.[7]

But TENT CITY faces yet another transformation, into trailer city. The price of secondhand house trailers and mobile homes dropped steadily in the late twentieth century, so that prefab trailers, often with widespread awnings and pop-out storage spaces, became essential features of TENT CITY and other temporary settlements. The "instant city" created at the Woodstock (N.Y.) Rock Festival of 1969 offered an early model of society's growing capacity for overnight settlements. Trailer city does the quickstep in that continuing process.

See also THE SETUP, in Chapter 4: Ephemera.

CHAPTER 8

Opportunity Sites

European investors stand amazed at how still-underdeveloped much United States territory appears. Here lie further energy sources to be tapped, and outdated fixations to be reconsidered. Here may be AIRSPACE containing potential AIR RIGHTS AREAS to be identified and structured; old BLAST SITES ripe for reconstruction, FLEA MARKETS that can extend their reach and season, MAILBOXES that can be structured to become shelters; THE ICE ready to have its season stretched and its uses multiplied; SAND CASTLES waiting to become all-season attractions; and those FURROWS, SOLAR FARMS, and WINDFARMS that reach out to snare untapped sources of energy.

Every new BYPASS opens up access for site-seekers. Here a surplus Army post is offered to private business; there a run-down street gets converted into a ROW. Quiet coastal waters are "discovered" by hordes of weekend new-boat owners. Spooky, photogenic mansions get listed in *Location Update* magazine to attract movie producers. Native Indians from Ecuador convert street corners from Manhattan to Amsterdam into sales pitches for colorful cotton/wool weavings. And Long Beach, Cal., begins (1993) spending $542.5 million to expand its harbor for PACIFIC RIM trade. Opportunity sites large and small, routine or dramatic—all upset old time-and-distance and cost-benefit ratios, and put the status quo up for grabs.

AIRSPACE

There once was a time when English common law assured the owner of fee title to land that his ownership extended to the High Heavens above, and into the bowels, if not the very center, of the Earth below. That geometric icosahedron of ownership, even though it could be exploited only at or close to the Earth's surface, gave landowners a mighty sense of control. It linked Man, the Gods, and the Devil himself, and Earth became their BATTLEGROUND.

That was before airplanes, rockets, and satellites. That was before the so-called Death of God and before the birth of legal abstractions called FLIGHT PATHS and air routes. Once these latter had been established in practice and in law, AIRSPACE would never again be the same.

In the generation after airmail flights began in the 1920s, AIRSPACE became a complex, crafted webbing invisible to the naked eye. Once fancy-free, the plaything of gods and poets, it became the stock in trade of the aviation industry.

Consider: AIRSPACE above the continental United States has been converted into a network of 185,637 nautical miles of exclusive commercial air routes, with jet routes accounting for 150,496 of the total; many square miles of "no-fly" military restricted AIRSPACE, most of which is requested from the Federal Aviation Administration on an hourly or daily basis; and scores of global routes of satellites.[1] AIRSPACE is loaded with debris from dead or exploded rockets. It also includes that shrinking realm of personal space above back yard into which the modern landowner may still intrude with kites—but not too highflying. Model planes under power are mostly restricted to approved locations.

Consider as well: the intrusions into the once-private realm of back yard and other properties by noise and sight of aircraft; the intrusions of light and electronic signals, some of which may not be legally intercepted by groundlings. And finally the invasion of what once was private space by the all-seeing eyes of satellites and highflying photographic planes. AIRSPACE has become a grab-bag of competing interests.

See also EARSHOT, in Chapter 4: Ephemera.

BATTLEFIELD/GROUND

Throughout the wars of history, the BATTLEFIELD is the ultimate place of decision; the testing ground of men and materiel, the site of ongoing disciplined or other violent action. "The battle's the thing," said military historian S. L. A. Marshall. The site itself comes in endless variations, and once hostilities cease, it degrades quickly into a hazardous source of scavenged civilian supplies and souvenirs. Exactly what happened during and after large encounters may continue to be in dispute for centuries. The history of great battles continues to be rewritten.

In celebrated cases, the site becomes a historic BATTLEFIELD—memorialized, managed, mapped, and the object of touristic and patriotic attention, usually under supervision by the U.S. National Park Service or regional counterparts. Along the way it produces income for tour guides, case studies for historians, and casualties among careless dealers in salvaged explosives and firearms. Few nations outside the United States set aside entire BATTLEFIELDS as historic sites, although Europe has long been noted for its memorials, cemeteries, ossuaries, unknown-soldier tombs, and historic markers going back to Greek and Roman battles. (Near

AIRSPACE

(Overleaf) Large areas of North America are under wraps, overlaid by various degrees of warnings, prohibitions, and other military restrictions, shown on this map of military air-training areas along the Mississippi-Alabama border. "National security" is in control here. Its laxity or its strictures vary with international tensions. U.S. Department of Commerce, Washington, D.C.

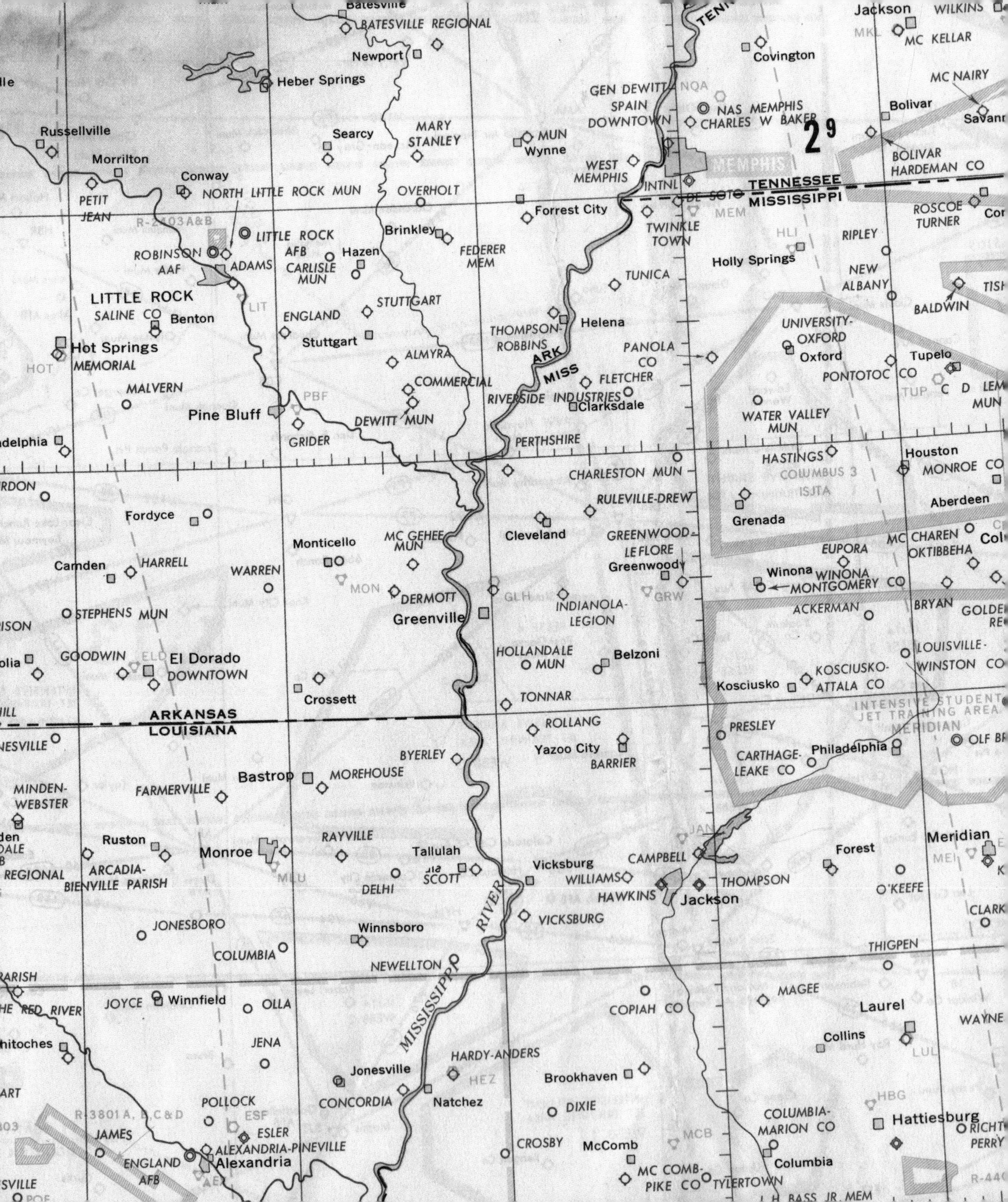
Batesville
BATESVILLE REGIONAL
Newport
Heber Springs
Russellville
Searcy
MARY STANLEY
Morrilton
Conway
NORTH LITTLE ROCK MUN
OVERHOLT
PETIT JEAN
R-2403A&B
LITTLE ROCK AFB
ROBINSON AAF
ADAMS
Hazen
CARLISLE MUN
Brinkley
FEDERER MEM
LITTLE ROCK
SALINE CO
Benton
LIT
ENGLAND
STUTTGART
Stuttgart
Hot Springs
MEMORIAL
HOT
ALMYRA
COMMERCIAL
MALVERN
PBF
Pine Bluff
DEWITT MUN
GRIDER
adelphia
RDON
Fordyce
Monticello
MC GEHEE MUN
Camden
HARRELL
WARREN
MON
DERMOTT
STEPHENS MUN
RISON
GOODWIN
ELD
El Dorado
DOWNTOWN
olia
Crossett
HILL
ARKANSAS
LOUISIANA
NESVILLE
Bastrop
MOREHOUSE
MINDEN-WEBSTER
FARMERVILLE
Ruston
Monroe
RAYVILLE
REGIONAL
ARCADIA-BIENVILLE PARISH
MLU
DELHI
Tallulah
SCOTT
JONESBORO
COLUMBIA
Winnsboro
NEWELLTON
RARISH
JOYCE
Winnfield
OLLA
HE RED RIVER
nitoches
JENA
Jonesville
CONCORDIA
Natchez
ART
POLLOCK
R-3801 A, B, C & D
ESF
JAMES
ESLER
ALEXANDRIA-PINEVILLE
Alexandria
ENGLAND AFB
SVILLE
POE
TENN
GEN DEWITT SPAIN
DOWNTOWN
NQA
NAS MEMPHIS
CHARLES W BAKER
MUN
Wynne
WEST MEMPHIS
MEMPHIS
INTNL
DE SOTO
MEM
TENNESSEE
MISSISSIPPI
Forrest City
TWINKLE TOWN
HLI
Holly Springs
TUNICA
Helena
THOMPSON-ROBBINS
ARK
MISS
PANOLA CO
FLETCHER
RIVERSIDE INDUSTRIES
Clarksdale
PERTHSHIRE
CHARLESTON MUN
RULEVILLE-DREW
Cleveland
GREENWOOD-LE FLORE
Greenwood
GRW
GLH
INDIANOLA-LEGION
Greenville
HOLLANDALE MUN
Belzoni
TONNAR
ROLLANG
Yazoo City
BARRIER
BYERLEY
JAN
CAMPBELL
Vicksburg
WILLIAMS
HAWKINS
Jackson
THOMPSON
VICKSBURG
RIVER
MISSISSIPPI
COPIAH CO
HARDY-ANDERS
HEZ
Brookhaven
DIXIE
CROSBY
McComb
MC COMB-PIKE CO
TYLERTOWN
Jackson
WILKINS
MKL
MC KELLAR
Covington
MC NAIRY
Bolivar
Savann
2 9
BOLIVAR HARDEMAN CO
ROSCOE TURNER
Cor
RIPLEY
NEW ALBANY
BALDWIN
TISH
UNIVERSITY-OXFORD
Oxford
Tupelo
PONTOTOC CO
TUP
C D LEM MUN
WATER VALLEY MUN
HASTINGS
Houston
COLUMBUS 3
ISJTA
MONROE CO
Grenada
Aberdeen
EUPORA
MC CHAREN
OKTIBBEHA
Col
Winona
WINONA MONTGOMERY CO
ACKERMAN
BRYAN
GOLDE
LOUISVILLE-WINSTON CO
Kosciusko
KOSCIUSKO-ATTALA CO
INTENSIVE STUDENT JET TRAINING AREA
MERIDIAN
PRESLEY
OLF BF
CARTHAGE-LEAKE CO
Philadelphia
Meridian
Forest
MEI
O'KEEFE
CLARK
THIGPEN
MAGEE
Laurel
WAYNE
Collins
LUL
HBG
COLUMBIA-MARION CO
Hattiesburg
RICHT
PERRY
Columbia
MCB
L H BASS JR MEM
R-440

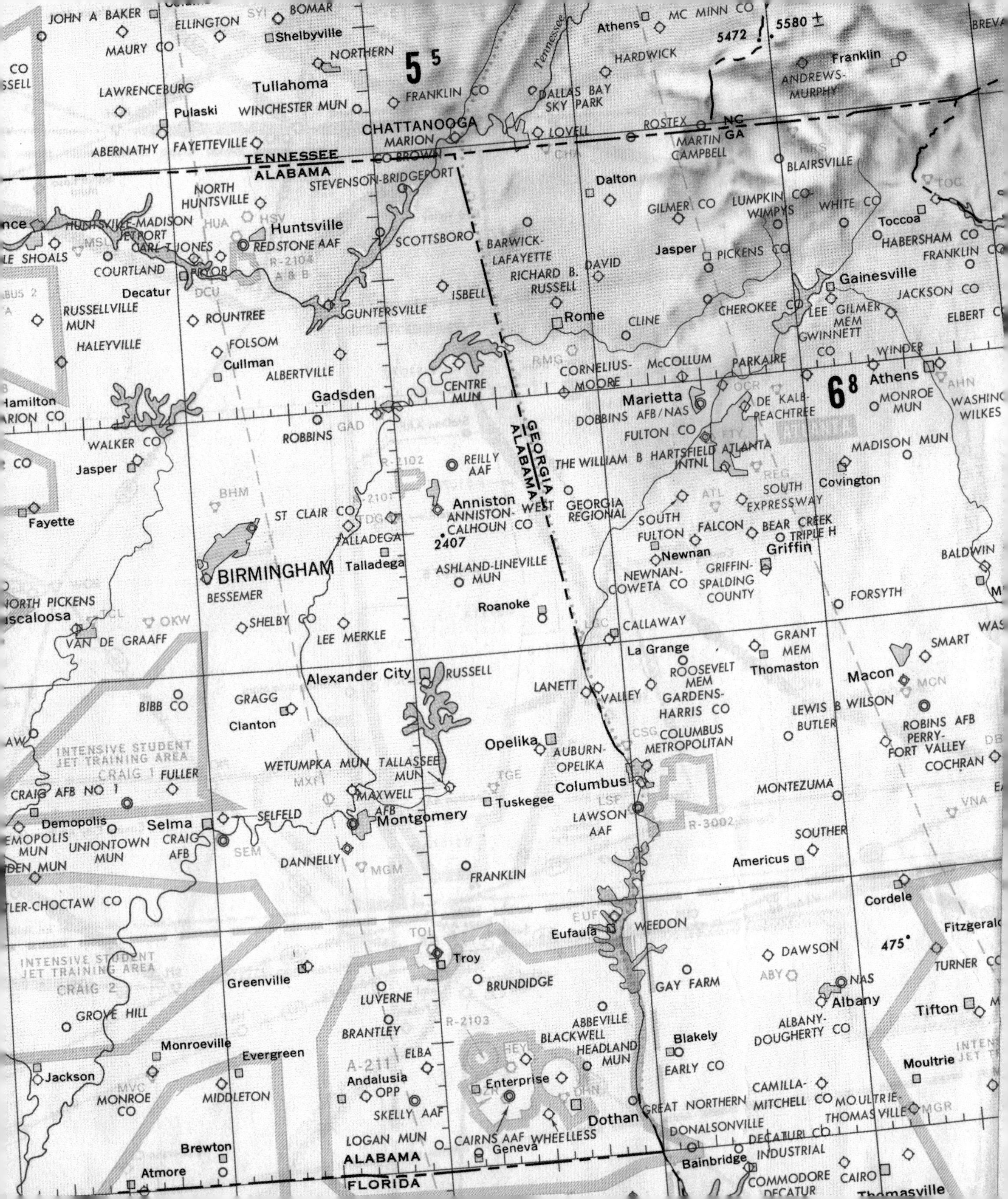

JOHN A BAKER
ELLINGTON
BOMAR
Shelbyville
MAURY CO
NORTHERN
5 5
Tullahoma
LAWRENCEBURG
Pulaski
WINCHESTER MUN
FRANKLIN CO
Tennessee
CHATTANOOGA
MARION
ABERNATHY
FAYETTEVILLE
TENNESSEE
ALABAMA
CO BROWN
STEVENSON-BRIDGEPORT
Athens
MC MINN CO
5472
5580 ±
HARDWICK
DALLAS BAY
SKY PARK
LOVELL
ROSTEX
NC
GA
MARTIN CAMPBELL
Franklin
ANDREWS-MURPHY
BLAIRSVILLE
NORTH HUNTSVILLE
HUNTSVILLE-MADISON JETPORT
CARL T JONES
Huntsville
REDSTONE AAF
R-2104 A & B
SHOALS
COURTLAND
PRYOR
Decatur
DCU
RUSSELLVILLE MUN
ROUNTREE
HALEYVILLE
FOLSOM
Cullman
ALBERTVILLE
Hamilton
MARION CO
SCOTTSBORO
GUNTERSVILLE
ISBELL
BARWICK-LAFAYETTE
RICHARD B. RUSSELL
DAVID
Rome
CLINE
Dalton
GILMER CO
LUMPKIN CO-WIMPYS
WHITE CO
Toccoa
HABERSHAM CO
FRANKLIN CO
Jasper
PICKENS CO
Gainesville
JACKSON CO
CHEROKEE CO
LEE GILMER MEM
GWINNETT CO
ELBERT CO
WINDER
CENTRE MUN
Gadsden
CORNELIUS-MOORE
McCOLLUM
PARKAIRE
Marietta
DOBBINS AFB/NAS
DE KALB-PEACHTREE
6 8
Athens
MONROE MUN
WASHINGTON WILKES
WALKER CO
Jasper
ROBBINS
GAD
R-2102
REILLY AAF
GEORGIA
ALABAMA
FULTON CO
THE WILLIAM B HARTSFIELD ATLANTA INTL
ATLANTA
MADISON MUN
Covington
BHM
R-2101
ST CLAIR CO
TALLADEGA
Anniston
ANNISTON-CALHOUN CO
WEST GEORGIA REGIONAL
2407
SOUTH FULTON
FALCON
SOUTH EXPRESSWAY
BEAR CREEK
TRIPLE H
Griffin
Newnan
Fayette
BIRMINGHAM
BESSEMER
Talladega
ASHLAND-LINEVILLE MUN
NEWNAN-COWETA CO
GRIFFIN-SPALDING COUNTY
BALDWIN
NORTH PICKENS
Tuscaloosa
TCL
OKW
VAN DE GRAAFF
SHELBY
LEE MERKLE
Roanoke
CALLAWAY
FORSYTH
La Grange
GRANT MEM
Thomaston
SMART
Alexander City
RUSSELL
GRAGG
BIBB CO
Clanton
LANETT
VALLEY
ROOSEVELT MEM GARDENS-HARRIS CO
Macon
LEWIS B WILSON
ROBINS AFB
INTENSIVE STUDENT JET TRAINING AREA
CRAIG 1
FULLER
CRAIG AFB NO 1
WETUMPKA MUN
TALLASSEE MUN
Opelika
AUBURN-OPELIKA
CSG
COLUMBUS METROPOLITAN
Columbus
BUTLER
PERRY-FORT VALLEY
COCHRAN
MXF
MAXWELL AFB
Montgomery
TGE
Tuskegee
LSF
LAWSON AAF
R-3002
MONTEZUMA
VNA
Demopolis
DEMOPOLIS MUN
UNIONTOWN MUN
Selma
SELFELD
CRAIG AFB
SEM
DANNELLY
MGM
FRANKLIN
SOUTHER
Americus
BUTLER-CHOCTAW CO
Cordele
EUF
Eufaula
WEEDON
Fitzgerald
INTENSIVE STUDENT JET TRAINING AREA
CRAIG 2
TOI
Troy
BRUNDIDGE
Greenville
LUVERNE
GAY FARM
DAWSON
ABY
NAS
Albany
475
TURNER CO
GROVE HILL
BRANTLEY
R-2103
ABBEVILLE
BLACKWELL
HEADLAND MUN
Blakely
ALBANY-DOUGHERTY CO
Tifton
Monroeville
Evergreen
ELBA
A-211
HEY
EARLY CO
Moultrie
Jackson
MONROE CO
MVC
MIDDLETON
Andalusia
OPP
Enterprise
DHN
SKELLY AAF
CAMILLA-MITCHELL CO
MOULTRIE-THOMASVILLE
MGR
GREAT NORTHERN
Dothan
LOGAN MUN
CAIRNS AAF
WHEELLESS
Geneva
DONALSONVILLE
DECATUR
INDUSTRIAL
Brewton
ALABAMA
FLORIDA
Atmore
Bainbridge
COMMODORE DECATUR
CAIRO
Thomasville

Verdun, France, tourists can still perceive traces of upright bayonets of soldiers who had been buried in a sudden giant shell-burst in one of World War I's great battles.)

The battles celebrated on North American sites came in many forms: pitched, running, last-ditch, long-range, or hand-to-hand. They were duels between geometricians armed with mortars and cannon; skirmishes between horsemen at full gallop, devastating charges across open fields in the face of point-blank artillery fire, or ambushes conducted from behind field-stone walls, or after a nighttime sneak approach. Much is also missing. Nowhere in North America can tourists see firsthand the devastation of twentieth-century war, the scars of pitched battles lasting for months, or of nuclear blasts, of saturation bombing, or minefields still buried in square miles of former battle turf. Long after the fact, Americans could visit European BATTLEGROUNDS where Americans fought in two World Wars, but few could ever share the shocking reality of the original article. Thus insulated from the real thing, Americans continued to view BATTLEFIELDS as something as far removed from life as death.

BLAST SITE

Behind this definition lies the recognition that more BLASTS occur these days, doing more complex and extensive damage, because of the widening presence of explosive and/or poisonous gases, liquids, and combinations—in pipes, plants, transports. New ordinances are adopted to regulate transportation of explosives. Yet these move daily along our streets and highways. BLAST SITE is not to be confused with blasting site, which is a place of ongoing work where carefully controlled blasts, usually with dynamite, do the work of thousands of men. In contrast, BLAST SITE undoes the works of man. Here is where place-substitution occurs by means of a special form of violence. Before The Blast, this was once a school, factory, or dock. Now it is a BLAST SITE, with little or no evidence of the former place. The site is most likely flattened, with a crater marking its center. For blocks or miles around, blackened or blasted bits of buildings, houses, cars, and bodies are scattered. Fragments and debris may be distributed far and wide, pawed over until little identifiable remains.

BLAST and storage SITES often coincide; ill-treated dynamite tends to blow up where it is stored, sometimes where it is made, and often while being carelessly moved. After a dynamite plant exploded, described in the 1916 book *Dynamite Stories* by Hudson Maxim, the plant foreman observed of two missing employees, "I guess you'll have to plow the ground if you want to bury them."[1] The blast of April 16, 1947, in Texas City, Tex., was in fact a chain reaction set off at dockside by an exploding ship filled with ammonium nitrate. It killed 576 persons; 178 were missing. Pieces of the freighter Grandcamp were found fifteen miles away.[2]

Oddities among BLAST SITES abound. In Louisville, Ky., explosive gases seeping in 1981 from a Ralston-Purina soybean-oil plant sewer blew up linearly.[3] The site of the linear blast extended underground miles away from Purina, blowing up street pavements, manhole covers, cars both parked and moving, and structures en route.

BYPASS

This is the byproduct of a temporary, makeshift detour that has graduated into the next higher dimension. It shortens so many trips, it makes so much sense as to attract not only more traffic but official budgeteers. This leads to new signs, street widening, and perhaps a new name: Outer Loop, Beltway, Shortway—namely BYPASS.

Some detours keep their names and become landmarked: The Atcheson, Topeka, and Santa Fe Railroad was an early booster of New Mexico and northern Arizona. Its off-line bus tours, such as the Indian Detour, "gave the visitor a glimpse of many of the pueblos and ruins in the Santa Fe-Taos area."[1]

Early road BYPASSES around hamlets and villages, often built in the 1930s, followed simple new rights-of-way carved off the back lots, or cut through the pastures of nineteenth-century settlements. By the 1950s these had been paved and widened to become many a New Main Street that circumvented the original. BYPASS was further expensively written into the American landscape by the Interstate Highway Act of 1949 (as variously amended), which created thousands of miles of superhighways around, as well as through, towns and cities.[2]

Where major traffic has been diverted to the BYPASS, it may be jeeringly called "buy-pass" after its conversion into a commercial strip, known locally as Strip City. Often "buy-pass" is as ephemeral as the BYPASS which gave it birth. Marginally located from the start, it seldom moves up the economic ladder, and often turns into a low-rent mecca for cheap traders.

Other BYPASSES, however, became instant generators of fast money for speculators and landowners around new regional interchanges. They built office parks, research centers, shopping complexes. Some qualified to be dubbed as "edge cities" by Joel Garreau; most of them solicited customers from new suburbs, those other by-products of access via the new belts, loops, etc.[3] All together, they accelerated the drain-off that pulled people, cars, and cash out of The Center.

In handling floodwaters, BYPASS plays an even more dramatic role. The most alarming example is Atchafalaya Slough Bypass in Louisiana, a series of old bayous along which the oft-flooding Mississippi River is expected to force its way into a new and permanent path. This will effectively bypass (and "backwater") the port and city of New Orleans. It is expected to drastically alter the economic and political geography of the Mississippi Valley, and affect most directly those Southern and Gulf Coast States dependent on the current navigation channel, as modified by the flood of 1993. The distributive flow of the river, and of goods moving on the Ohio, Missouri, and Mississippi River systems will never be the same. There is no known method at this writing for the Great Flood's power to be diverted from its rerouting. Calling the FLOODWAY by its occasionally official name of "inundation area" or "overflow area" will not lessen its future impact.

See also FLOODWAY, in Chapter 4: Ephemera.

CAMP

Variants: camping place, camp spot, campsite, camp-out, campgrounds.

"To find a camping place" suggests that one is choosing any bucolic or vegetated location where one can pitch CAMP, spread mats or sleeping bags, and get away with it overnight. These turn out most often to be happenstantial rather than generally recognized or designated places.

In Middle English the term was picchen or pycchen to indicate fixing, thrusting (as a tent peg into the ground), erecting, or securing a tent or CAMP, hence to plant or to fix in place. In the 1960s the popularity of verbal nouns such as "sit-in, be-in, cook-out, and hookup" included "a camp-out" as a place as well as an activity. "Over thirty years ago the Spanish Steps [Rome, Italy] were the campsite of hundreds of scruffy self-proclaiming 'Existentialists,' not all Italian, of course, waiting for something, anything to happen."[1] Campsites dot our historic landscape. Indian households, once their harvesting was done, "struck their wigwams, stored the bulk of their corn and beans, and moved to Campsites for fall hunting."[2] All this traipsing about set oddly with the Puritan settlers, who expected Indians to settle down as well.

There is a well-recognized order in places for camping, an order made more rigid by the examples of Boy/Girl Scout CAMPS, by extensive military training, and by the growing popularity of KOA (Kampgrounds of America) and other national chains. These more highly structured places—KOA especially—offer uniform signs, layouts, regional guidebooks, full-time staffs, and health certificates. The key is predictability. Travelers, tired and exhausted, know what to expect, and get it.

Along the way "tourist camps" had flourished in the 1920s, then fell away under competition from motels after the 1940s.

But would-be campers, undeterred by all this, carved out a new realm for "real" camping in remote locations. National and state parks came under great pressure from backpackers and outdoors types enspirited by out-backers of the 1960s. Increasing conflicts, as determined campers encounter the fencing-off of vast tracts of back country, are a likely prospect for the future.

FALL COLOR COUNTRY

Now that its manufacture and distribution is completed, now that the product has been accepted in the marketplace, the result is an idealized, reconstituted countryside, complete with picturesque villages, lovely streams, flashing waters, hillsides inviting exploration. Originally carved out of wilderness to be settled, it has been redesigned to be seen. It is dotted with panoramas that give photographers itchy trigger fingers and cause even seasoned journalists to lower their guard, drop their objectivity, spread-eagle their prose.

This lovely world barely existed two generations ago, except in isolated spots. Today its innermost secrets spread from friend to friend, and its range has expanded from New England to the Rockies. News networks track its changes day by day. Sunday and regional magazines proclaim its virtues. It generates huge traffic jams, and millions of dollars of revenue per day.

FALL COLOR COUNTRY
Compressing the key ingredients of FALL COLOR COUNTRY into one painting, this Georgia mail-order house masters the stereotypes: rich colors (in the original); a painter's easel to frame THE SCENE; a footloose family with playful dog; a traditional, besteepled white church and village—i.e., nostalgic ingredients of a once-agricultural landscape of farmers/villagers now penetrated or overrun by weekenders and tourists. Cover courtesy of Charles Keath, Ltd., Norcross, Ga.

This manufactured creature, this product of mass-marketing, this by-product of color photography, this creature of modern tourism and a new regional economic base is FALL COLOR COUNTRY, also known as Fall Foliage Country. Once thought to be a natural local phenomenon, it has been expanded to become the essence of contrived reality: you must be there on schedule, preferably for Opening Day. Merchants depend on it—like Christmas shopping—to "make their season," and to generate year-end profit. Local harvest festivals capitalize on it. Breathless TV commercials entice travelers with accounts of "peak color locations." FALL COLOR COUNTRY has its growing infrastructure of motels, tour buses, convention bureaus. In

Chattanooga, Tenn., an enterprising boat owner offers Fall Foliage Tours on the Tennessee River, which itself offers incomparable views of colorful mountainsides.

Planning for these affairs begins early. New York State is a leader in late-summer ads enticing mobilizable viewers. The photography industry, capitalizing on the chance to sell millions of rolls and cassettes of film and tape, is happy when a *Denver Post* writer observes, "You automatically reach for the camera every time you approach" a gleaming hillside of yellow aspen. Many states have set up versions of New Hampshire's network of "leaf watchers" who send in twice-weekly accounts of peaking color. Various states offer a free long-distance 1-800 number, to give up-to-the-minute accounts of where the color is peaking. Vermont has "foliage spotters" who help stir up the $80 million that tourists spend to confront a biological phenomenon that involves minimum human conflict.

Fashions in tourist attractions change. Agricultural harvests, once small-scale and visible on every country road, are more than ever mechanized, twenty-four-hour big-scale operations. Will corporate farmers seize on the harvest scene as a generator of tourism with the arrival of video-camera-toting families, munching seasonal goodies while they sing-along at the new agrimotel complex? Some family U-pick farms point the way, with their peak-season U-pick days, organized bus tours, ladies' specials, and Thanksgiving feasts. One-day horse-races have become week-long Derby and Preakness Festivals, the Indy 500 a week-long racing celebration. FALL COLOR COUNTRY is joining this host of coming attractions.

FLEA MARKET

Straight from the Middle Ages: add automobiles and the ubiquitous forklift truck, and FLEA MARKET emerges in modern guise (and sometimes mod gear). It arises from VACANT LOTS in the height of impromptu, and appears along highways in larger forms. Locations favor easy access, but also respond to local legends and associations.

Not to be confused with yard sale, FLEA MARKET is at least one step up the managerial ladder—with a known (and often advertised) location, laid out in public view, and run for a profit.

Mostly FLEA MARKETS just appear, open to all comers; no blueprints, few preliminaries. Seldom does a change-of-zoning cloud the sky. Behind the apparently chaotic clutter of collectibles and whatnots verging towards junkdom, basic systems are at work here.

The modern FLEA MARKET lives off surplus material-goods accumulations of an industrial society. Roadside and front-door variations have a long history. Today's FLEA MARKET will feature outgrown children's clothing, the clutter from abandoned barns, shops, woodsheds, or cellars. It recirculates returns, rejects, and throwaways; lifetime hobby collections at life's ending, via middlemen along the way. And its sales goods reflect local economies; outside Detroit, retired auto workers in the 1960s set up a heavy-metal FLEA MARKET to which they brought lifetime hoards of tools, dies, and auto parts.

frontages. Its vendors, often unsupervised and unmonitered, may sell distressed or damaged goods, firearms or other fugitive stuffs. Often the site itself is property in distress, a condition which a FLEA MARKET is expected to remedy.

Indoorsing caught up with the FLEA MARKET in the 1970s. Along the way, the FLEA MARKET had been picked up by hotshot entrepreneurs able to assemble wide tracts of land close to highway interchanges, adding hardstand parking, and indoor stalls rented by the hundreds. It was becoming a warehousing operation, complete with forklift trucks, walkie-talkie supervisors, weekend traffic jams, and off-duty cops patrolling the grounds. Here some new hustlers-turned-developers actu-

FLEA MARKET

Off-brand locations, odd lots, and roadside waif-and-stray-sites attract thousands of yard-sale sellers and buyers in season. North America's longest FLEA MARKET may be the 450 miles of U.S. Highway 127 from the outskirts of Cincinnati, Ohio, to Gadsden, Ala. For those shopping both sides of the road on one long August weekend, it's called "The Whole 900-Mile Yard Sale." Action concentrates around Jamestown, Fentress County, Tenn., where a newly elected county judge started the venture in 1986—to attract travelers "off the Interstate" (I-75) and onto local roads, yards, and markets.

In good times and bad, the FLEA MARKET survives. But its twentieth-century extravaganzas—the five-to-hundred-acre display grounds with parking for a thousand cars—arise out of a national consumer economy built on high turnover of short-lived goods, plus the wide ownership of family vans and pickup trucks. Once arisen from the margins of the economy, the FLEA MARKET flourishes in the geography of cheap leftovers—odd-shaped land parcels to which other uses make no current claim. It can sneak overnight onto high-visibility and under-used highway

ally built permanent structures with stalls for rent to a new breed of pickup-trucked buyers and sellers. Many of these operated out of converted vans.

Further upscale and indoors, state fair boards entered the game, opening their fairgrounds, and huge airconditioned exhibit halls, to spring, fall, and post-Christmas flea-market sales, advertised throughout the region, attracting shoppers in predictable waves. As mass attractions, FLEA MARKETS came to life many steps below and before the more hifalutin festival market places. But both types shared an

FLEA MARKET

This expanding FLEA MARKET with 1,500 spaces for dealers near Attalla, in central Alabama (1993), has made the shift from an eighty-acre family crop farm to a U-pick vegetable operation, and then to a small FLEA MARKET, now thrice enlarged and catering to regional cruiser-shoppers a few miles off U.S. 127. Such locations form the end of international production lines. Worn-out bucks and family discards stop here for recycling. Photo by Jamie Terrell.

indispensable proximity: more cheap merchandise jampacked into one place than most shoppers see in a lifetime.

FLIGHT PATH

Barely conceivable to modern air travelers is the concept of an early FLIGHT PATH, say in the 1920s. It hugged the ground. Goggled pilots in those open-cockpit days had to keep visual contact with Mother Earth, ever ready for rough landings in hayfields and rocky pastures. Barnstormers, stunt fliers, and experimental rocket-launchers considered all outdoors as their FLIGHT PATH of choice. But no longer.

FLIGHT PATH has moved upward into rigidly prescribed AIRSPACE. As a designated zone in the sky, it asserts its presence to groundlings only by the noise and sight of planes and their contrails. For travelers anxious to find an obscurely marked local airport, scanning distant horizons offers locational clues in the size, shape, and direction of planes ascending or descending.

But FLIGHT PATHS near airports turn into approach zones, which become an environmental IMPACT AREA to those living and working underneath the overwhelming noisy rush of landings and takeoffs. Such are noise levels that only thirteen U.S. airports would, in 1985, permit the Concorde supersonic (240,000 horsepower) jet to use their runways. FLIGHT PATH carries on in many political guises: noise IMPACT AREA proved to be the most divisive from the 1960s into the 1990s—an era of vast expansion and runway-lengthening at U.S. airports.

Up there in the sky, AIRSPACE is that realm below space itself (also known as outer space) and is subdivided into a precisely defined and complicated network of FLIGHT PATHS. These are invisible highways and superhighways "complete with aerial equivalents of underpasses, overpasses, access lanes, dangerous intersections, bottlenecks, detours and morning and evening rush hours."[1] These are three-deckers, stacked up to 45,000 feet; only the Concorde, military, and spycraft fly higher. Designated FLIGHT PATHS are officially eight miles wide. Since the air-controllers' strike of 1981, fewer controllers are available; hence planes that once flew as close as five miles apart were kept to 10–20, and sometimes 40 miles, apart. Pilots were supposed to maintain 1,000 feet vertical separation; closer than that was an "operational error." If too close, pilots were expected to report a "near-midair collision." Eastbound planes fly at even altitudes, westbound at odd elevations (say 21,000 feet).

Some nine hundred routes have been numbered by the Federal Aviation Administration. Many were revised to handle new traffic volumes originating from HUBS after the airlines were deregulated in 1978. Twenty-three federal airway route-control centers (in 1989) manage the flow, which logged a record of 33.6 million flights in 1986. The New York Tracon (Terminal Radar Approach Control) on Long Island had 150 controllers in 1988 monitoring 1.71 million flights a year through 15,000 square miles of AIRSPACE over four states. Amid all this, Air Force One, the military jet carrying President Ronald Reagan, had a near-miss (1.58 miles) on October 12, 1988, after a string of operational screw-ups. By 1992 the NYTracon was expected to be able to double its capacity, and handle 3,400 planes per hour.

By its nature, FLIGHT PATH transcends political borders. Yet border politics intrudes anyhow into FLIGHT PATH. This was evident when the United States attempted to test cruise missiles from the Beaufort Sea southward into Alberta, arousing opposition from both the USSR and Canada's New Democratic Party. Meanwhile, Soviet bombers' intrusions into Canadian AIRSPACE almost doubled in 1986–87. Such flights crossed what Soviet leader Mikhail Gorbachev was proposing—before his expulsion—as a demilitarized "northern peace zone."[2]

Far above all this, the myriad FLIGHT PATHS of orbiting satellites are widely published while, even as late as 1990, the purposes and payloads of military rockets to watch Russia were still being kept secret by the U.S. Defense Department, pending further downsizing of U.S. stockpiles.

See also EARSHOT, in Chapter 4: Ephemera.

THE FURROW

What began as a slight, primitive scratch on the earth's surface behind a forked stick drawn by human beings has expanded to become a prime mover of civilized landscapes today. Possibly ten thousand years ago in Asia, or two thousand years ago in North America, THE FURROW became the progenitor of field, farm, plantation, the trail, and finally the highway. Each of these man-made places depends upon energy applied to a blade, a knife, a plow, or scoop drawn across the ground in a continuous and often straight line.

THE FURROW'S forcible entry prized open the earth's surface to make sustenance—one step up from the digger stick to make holes in the ground for seeds or tubers.[1] THE FURROW came after the hole-and-stick, the "triumph of the continuous" in historian Fernand Braudel's penetrating phrase.

That magical summarizing phrase, the "triumph of the continuous," came to Braudel in his Johns Hopkins University lectures, 1976.[2] He used it to describe the conversion of French seasonal street fairs to year-round shops with owners, addresses, fixed hours, stock-in-trade, and regular shipments. It was a landmark jump-start for expanding local, regional, and global markets—and capitalism itself.

From THE FURROW came surpluses to be stored, so that, once human beings could count on food from THE FURROW, they could shift from hunting and foraging to crop-raising. And this called for new habits, longer commitment to place. It tied together man and site. No more going off chasing game when it was time to plant

THE FURROW

Superseded, sidetracked, outmoded, this nostalgified artifact is a reminder of times when a well-turned FURROW was the mark of a good farmer who kept his plow, mule, or horse in good shape. Now it's often the sign of a retired farmer—or his son, grandson—or has migrated through a local yard sale. As farmers are fewer, the number of farm-nostalgia items dwindles, and a new generation of artifacts is taking their places: front-yard tackle and antiques with no farming connections at all.

and time to reap. Sticking to THE FURROW converted random planters to organized croppers, the predecessor of today's corporate farmer. And from one's FURROWS came surpluses to trade, more food for larger families, bigger tribes, more power in trade and combat. These first food surpluses made it possible to settle down in extended families, to form villages, to cooperate, to trade work, to have feasts. "Food production made it possible for more people in more places to form larger and more complex groupings."[3] THE FURROW meant a new division of labor. It became man's orbit, a mark of possession and ownership. Before THE FURROW intervened, women tended the fields, men went off hunting. Once THE FURROW made its demands—for organized teams of oxen and later horses to pull heavier plows—men dominated THE FURROW and later the machines that dug it.

The first FURROW made by plows may have appeared in Egypt and Mesopotamia by 3000 B.C. and on Britain's lighter soils by 1000 B.C. FURROWS made with iron plowshares were being turned by 400 B.C. To early speculative minds, THE FURROW and its products hinted at what seemed to be the infinitely expandable nature of food production.

By the Middle Ages, plows were being pulled through Europe's heavy soils by teams of six to eight oxen, too expensive for one farmer to afford. Hence much plowing and animal-tending was communal, and field sizes dictated by the size and length of THE FURROW. Plowing strips were standardized in England by Edward I (1272–1307) at 40 rods (i.e., "one-furrow-long" or one furlong, still a familiar measure in horse-racing; it equals 220 yards, or 201.17 meters). A standard measure for a day's plowing was 40 rods long, and 4 rods wide—our present acre.

Visible traces of old FURROWS remain: in the furrowed terraces of old contour-plowed Southern hillsides now thick with plantation pine; in the swell and swale of mounded-up terraces that help drain wet soils; and behind the heavy horse-drawn plows of Amish and Mennonite farmers who, as they scorn powered machines, carve out the ancient ritual of horse-drawn plow and THE FURROW across their own fields.

By the time THE FURROW reached soils of the American Midwest, it was produced by heavier, horse-drawn plows with multi-purpose plowshares of metal, capable of producing FURROWS deeper and off to the far horizons. John Deere had the first slab of "plow steel" cast in 1846, and when efficient, mass-produced Farmall gasoline tractors arrived after 1920, field size expanded even more.[4] Plows were supplemented by add-on tools for planting, watering, cultivating, and fertilizing. Today it has become the $75,000 all-wheel-drive tractor, an all-day mobile home for the cultivator, with piped-in music, cooled air, food, and lights for working THE DARK. THE FURROW has been widened far beyond its single-line form to become another large, temporary element in a mass-produced capital-intensive landscape.

THE ICE

In all water there is the possibility of ice—and vice versa, since ice and the water from which it emerges are both pure potential. Given the proper latitude, ice is inevitable—lots of it (ice pack), a continent full of it (polar ice cap), or a glass of it (bourbon and soda). Natural ice occurs at the meeting point for life in suspension and life in action; in one form (as a broken pack in the sea lanes) it is a threat to navigation; but in its smoothest natural forms it offers life-giving mobility via dogsled, snowshoe, ski plane, or snowmobile to Eskimos, Aleuts, Mongolians, and extreme latitudinarians the world over. For them, hard ice means easy movement over great distances.

Long after ice-making became mechanical, the ability to apply such knowledge to the sky unlocked the present rain-making era around 1946. Not until men learned to

THE FURROW
Now mass-produced by heavy machinery, THE FURROW can be deep-plowed, disked, smoothed, and seeded all in a single-pass operation. For working still-wet fields, late-model machines are four-wheel powered or tracked, and may stand as tall as a two-story house, with powerful headlights for all-night work. Photo, John Deere & Co., Moline, Ill.

"seed" rain-bearing clouds by sprinkling them from a plane with artificial dry-ice crystals or with silver iodide crystals—not until then did man's effort to make rain go beyond prayer and spell-casting.

But back on earth, THE ICE remains an environment which transforms all habits of living and moving. And with the expanding search for fish and other seafoods, petroleum, and undersea minerals, operations under THE ICE are expanding via the construction of artificial islands, using techniques that expand from Arctic experiences with man-made ice-roads and seagoing icebreaker ships. Thus THE ICE has been identified as a new frontier, chiefly by Northern Hemisphere firms equipped with high-tech, ice-breaking, and under-ice exploiting potential. On a different level, there's a historic marker near Wind River Indian Reservation, Wyo., proclaiming the spot where settlers "broke open the sod under which they found natural deposits of ice" for mixing drinks.[1]

THE MAILBOXES

Roadside evidence of centralization is offered by the U.S. Post Office. This photo is from its ongoing campaign to make one-stop pickup easier for mail collectors, as at this rusticated community kiosk. Photo, U.S. Postal Service, Washington, D.C.

THE MAILBOXES

Sticking up along thousands of miles of North American roadsides, THE MAILBOXES are vital to every suburban or country dwelling cluster. They unite outback with upfront, BOONDOCKS with county seats, here with there. We are talking not so much of that single metal or plastic box out in front of one address, or perched on old milk cans and one-hoss plows, welded to a rusty crankshaft, teetering on rotten two-by-fours, or on beautiful slender shafts of "post-rock" in central Kansas . . . but of THE MAILBOXES, a collectivity. They gang up at intersections, they cluster along roadsides, they come uniform or variegated, brand-new or decrepit. Along crowded city blocks, in tacked-up clusters in entryways, they reveal the presence of families legally and illegally jammed into converted flats.

Outside the built-up cities, they nestle up side by side as the daily port-of-call for millions of folks. On many a country road, they're the only signal a stranger gets that back up the holler, off down the valley, there's a cluster of families who share, if nothing more, THE MAILBOXES.

This new collectivity has helped, while being abetted by, the spread of trailer CAMPS, mobile-home parks, and new settlements of folks here-today and gone-tomorrow. Their MAILBOXES are organized into clumps, single rows, double rows, two-level rows, sometimes housed in miniature make-believe cottages, and requiring a pull-out lane to accommodate residents stopping for mail. Analyzing new suburbs in the 1950s, William H. Whyte, Jr., discovered that many a cluster of THE MAILBOXES is an inevitable place for neighbors to meet—an observation easily applied to MAILBOXES and garbage racks.[1]

Here is where private life goes public, even if only to receive junk mail and L. L. Bean catalogs. Here is where the long arm of the law reaches out to protect little packets of paper stuffed into metal boxes, coming and going. "Rural Free Delivery" (RFD)—that new device which guaranteed a Sears Roebuck catalog into every willing farmhouse—began its long reach around Charleston, W.Va., on Oct. 1, 1896. Soon it replaced little fourth-class post offices, and wiped out star routes that had been "the backbone of the old rural mail service." MAILBOXES had begun their migration into the outdoors. By 1900, RFD covered

THE MAILBOXES
In the hills above Redwood City, Cal., standard black MAILBOXES form a roadside lineup along suburban Palomar Drive. Here, residents of adjacent streets set up their own sub-clusters, a focus for mail as well as for socializing.

29,000 miles, and by 1959 extended to 1,733,000 miles. Each route's length had expanded from 23 to 56 miles.

But RFD is a term now abandoned by the U.S. Post Office. Often photographed, and collected, THE MAILBOXES frequently appear in slide-shows and books on the artifacts of folk culture. Perhaps to city folks they already represented a disappearing way of life.

In Santa Fe, N.M., there's an old joke about remittance men, as they were called in the British Empire. "If the mail stopped and the mailboxes fell down, all these Easterners depending on their remittance checks from Back East would be broke within two weeks."

In some new exurbs THE MAILBOXES get organized out of folk culture. Like Cinderella, they are transformed ... into a drive-through "local postal facility." You drive in under cover, reach out of the car window, dial your combination into your prefab stainless steel lockbox, in-the-wall or on a free-standing pedestal, and pick up or drop your mail. In more intimate purlieus, THE MAILBOXES often become a daily focus for socializing. And in some exclusive enclaves where the zip code signifies high status, THE MAILBOXES, in a historic or snazzy building, serve as a just-for-us community rendezvous.

The U.S. Post Office promotes clustering. It publishes a handsome brochure showing chic-looking little gazebos and boutiquey-boxy shelters, lighted at night, to protect you and your mailboxes. These constitute the actual "address," small privately owned facilities endowed with the protection of the U.S. Post Office for postage-paid mail only. Private, non-stamped messages to neighbors are likely to be discarded by zealous or new-on-the-route postmen.

They're also a signal to folks who don't or can't trust their neighbors not to tamper with the U.S. Mail. But they further signify that the Post Office wants us to cooperate with one-stop pickup, and thus the Post Office becomes a party to the development of clustering in SUBURBIA.

Already in some cities there has arrived

on THE SCENE a new version of the old small-town-crossroads-store-pump-island-MAILBOXES. This time around it has varied generic names like multi-use-community-facility postage center, or a franchised jiffy mailer-copier-lockbox. As more functions are loaded onto such new versions of THE MAILBOXES—the doling out of income-tax forms, posting of "Wanted" and other notices—they will continue to migrate under roof, becoming part of what the Material Culture school calls "Built America"—sometimes scaled-down variations or components of the ubiquitous MULTI-USE COMPLEX.

But material objects continue to be dematerialized by the onset of telephone-answering devices. "This is my voice mailbox," says the recorded voice of a woman who advertises her "800" telephone number for enlisting co-housing members.

By the time we all are receiving electronic mail, THE MAILBOXES will have been enlarged to handle bulk distributions. The sign-at-the-door system for home delivery used by the U.S. Post Office, United Parcel Service, Federal Express, et al., may well be supplemented (or supplanted) by electronic check-in. Your super-mailbox would open electronically to any vendors you pre-select.

PACIFIC RIM

There does exist, undeniably, a Pacific Ocean. It does have an acknowledged, though debatable, RIM. That RIM, it has become increasingly apparent, is developing into the major GROWTH AREA of the known world. That GROWTH AREA includes the U.S. West Coast, notably Los Angeles, which wants to be the RIM "CAPITAL," whatever that comes to mean. And President Bill Clinton made it clear in 1993 that he intended for the U.S. to be a senior partner in PACIFIC RIM development.

Before RIM there was "The Ring of Fire," denoting the chain of faults, quake zones, and volcanic areas circling the Pacific. PACIFIC RIM is a geopolitical concept assembled, by and large, in Los Angeles, which knows a big market when it lies next door—even if across a fourteen-hour jet flight.

Until 1993, PACIFIC RIM'S existence was acknowledged and debated chiefly on the American West Coast. Its boundaries are set by the shores of Asia, North America, and, less firmly engaged, South America. Its existence had been vigorously promulgated, chiefly during the cross-Pacific trade boom of the 1970s, by expansion-minded capitalists and their allies in and around Los Angeles. As movie CAPITAL and America's chief dream factory, Los Angeles knew that naming is a first step toward cooption and capture.

The PACIFIC RIM most Americans read about began to assume new shape in the 1990s. During the Cold War, U.S. military forces in the Pacific had been outnumbered four to one by those in Europe. But by 1994 U.S. military PRESENCE in Europe had shrunk, and the Europe-Pacific ratio was expected to be one-to-one. A Singapore military analyst saw this as evidence of "a strategic contest for leadership in the Asia-Pacific."[1]

Los Angeles typically was picking up old European ideas. Geopolitician Karl Haushofer had fanned the flames of Ger-

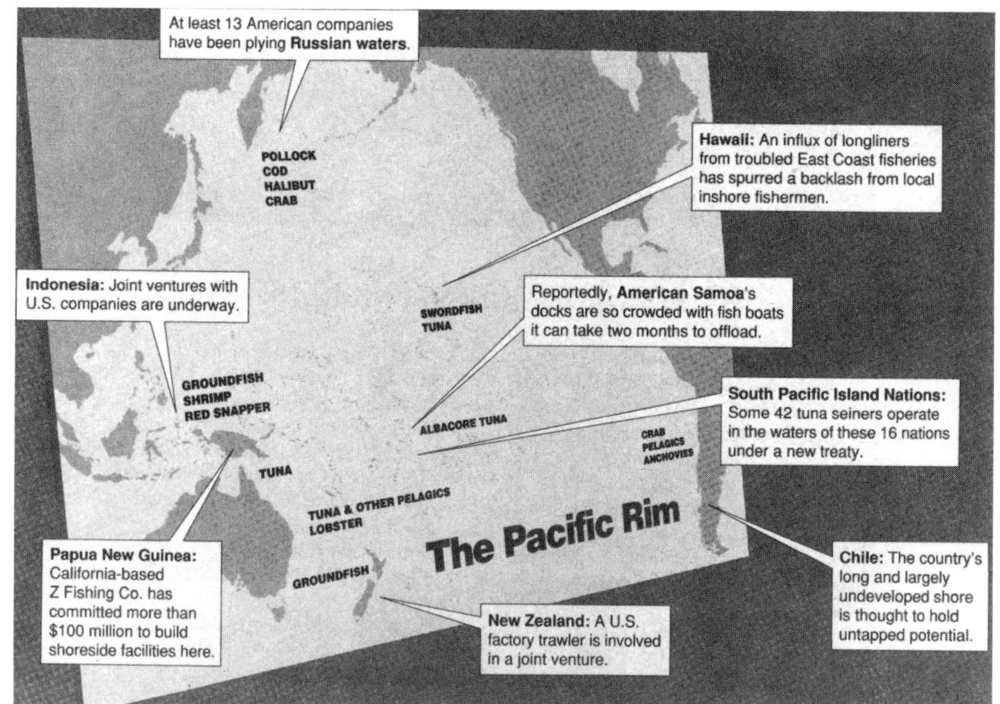

PACIFIC RIM

Fishing in distant waters—the world FISHERY—is old stuff for global fleets of all nations, as indicated by this March 1993 map of the PACIFIC RIM, published by the *National Fisherman.* But U.S. businessmen are now having to learn the fishermen's trade—angling for new business on the PACIFIC RIM, far from home base and familiar territory, competing with "Rimsters" from Japan, China, Taiwan, Korea, et al.

man nationalism in stressing the centrality of "The Heartland" (Eurasia). Geographer Robert Strausz-Hupe had asserted in 1942 that "The U.S. in its quest for pan-Pacific domination faces conflict with the British Empire, Japan and Russia."[2] But that was before the Empire had shrunk into Commonwealth, and China bespoke its rightful place on the RIM.

As leading Los Angelenos saw it, their city in the 1980s had become "the financial capital of the Pacific Rim." West Coast newspapers missed few opportunities to include RIM reactions in news of trade and currency. Already, it was believed by "rimsters," RIM nations made up the world's next GROWTH AREA. "From Southern California to Canada, the booming West Coast port cities have embraced investment from Asia as their economic future."[3] The Pacific was expected soon to exceed the Atlantic in carrying tonnage of world trade. "U.S. exports to Japan now [1993] surpass those to Germany, France, and Italy combined."[4] Soon, it was believed by West Coast optimists, the RIM would outproduce and out-trade the new twelve-nation European Economic Community brought to life in 1992. "For Frank Gibney [director, the Pacific Basin Institute, Santa Barbara, Cal.], all the evidence insists that the center of the world has shifted significantly."[5] Some claimed this was already happening; and that the twenty-first will be "the Asian Century." Asserted John A. Alwin in 1992 "the nexus of world trade and investment is shifting from the Atlantic Community of nations to the Pacific and its rim of countries."[6]

The drama of instant, round-the-world

communications, along with the rise of Hong Kong as a twenty-four-hour world electronic banking CAPITAL, has fostered the spread of a global electronic network, over which data and money flow instantly, worldwide. Yet over land and water, there are still time-and-space differentials between Los Angeles and Hong Kong—a sixteen-hour time difference and fourteen hours of flying time; whereas the New York–Paris difference is six hours by the clock, three hours of flying via Concorde, and seven via TWA. Freighters enjoy similar advantages in the cross-Atlantic trade. But on TV, an ocean span is a second's jump: by 1992, the Pacific RIM'S nascent TV ad market had already reached $14 billion.[7]

Thus, PACIFIC RIM and a new Europe both offer economic competition to the United States. The decades into the twenty-first century will reveal how much of the big-RIM talk is reality and how much is hype. "Rhetoric notwithstanding," wrote an international trade advisor, "this is not manifest destiny with water wings."[8]

PRESENCE

Within a generation after World War II, PRESENCE assumed a new and important place in the public eye. It determines, if it does not actually define, territory. It implies possession or, at the least, influence. It takes-place; it occupies and can dominate space. Always in the offing is the possibility of armed enforcement or intervention by outsiders.

If PRESENCE constitutes "that portion of space within one's ken, call, or influence" (as *Webster III* has it), it cuts a broad swath, indeed, across world geography and in international politics: Britain had acquired a PRESENCE throughout its Empire; the Nazi PRESENCE in Latin America during World War II was much feared by the United States. The U.S. PRESENCE in the Philippines and Hawaii was highly visible via its military bases and anchorages. The United States continued to dicker in 1991 for "a continuing PRESENCE" in Singapore and in the Philippines. "U.S. Aid to a Troubled Philippines Refocuses Debate on Military Presence," said a 1990 headline.[1] And after open combat ceased, the "Contras [were] Maintaining Presence in Nicaragua."[2] After the Gulf War, Defense Secretary Cheney and others at the Pentagon "reportedly want to use the current opportunity to establish something closer to a permanent presence in the [Persian] Gulf than others in the Administration think politically wise."[3]

As physical interventions by the United States penetrated other national AIRSPACE, lands, and waters, so did its PRESENCE. Around the world, American military bases, embassies, and legations became known as "Little Americas . . . frequent adjuncts to the American diplomatic presence . . . hermetic enclaves of residences, shops, schools, recreational, and other facilities that inhibit serious dealings with the surrounding land and population."[4]

The U.S. PRESENCE (sometimes referred to as "the U.N. coalition PRESENCE") in the Middle East before and after the brief Gulf War, 1991, assumed its ultimate, complex form. It included the physical reality of refueling stations and docks, air and ground military bases, observation posts, command posts, minefields, defensive perimeters, staging areas, fleet anchorages, listening stations (including mobile antenna farms), supply depots, lines of communication, fallback positions, and debarkation zones. In sufficient numbers, such man-made places constitute a powerful PRESENCE in either combat or demilitarized zones.

To maintain PRESENCE as a form of control may be simple and direct—a "line in the sand," a barbed-wire fence, automatic gates, or minefields. Once open, declared war is ended, PRESENCE may be maintained by armed occupation, by manned outposts, observation stations, and by inspections. Or it may be maintained in more tenuous form: recorded warnings via loudspeakers or car radio, local propaganda, international broadcast, or by warning fly-overs, airborne leafletting, or even skywriting.

In common with weapons, military lingo, and public policy, PRESENCE has maintained and expanded its foothold on domestic language: "Federal Presence in Alabama Town Helps Cut Crime, Police Chief Says."[5] Further, planner Lawrence O. Houstoun, Jr., in *Architecture* magazine, December 1989, noted that, "While it isn't Pittsburgh's answer to the Empire State Building, PPG [a downtown skyscraper] is a major skyline presence."

A newer meaning had permeated the language of the 1990s. Increasingly, PRESENCE was being viewed as local, physical evidence of distant, ever-expanding authority, especially of the federal government "a visible encoding within our public environment."[6]

It was inevitable that this prejudicial view should be adopted by architectural critics, such as Kenneth L. Warriner, Jr., in the Summer 1990 issue of *Avant Garde.* He viewed PRESENCE as qualities of a plan or structure that asserts itself, and intrudes into the public view—"a stable foil" to all that swirls around it, a quality that can be "read." He attributes PRESENCE to the noted Villa Savoye—a building that appears to be "floating above the landscape." It "can be taken up as an entity" by the observer.[7]

To proceed further: such a place gives off signals. It is set off, distinct from its surroundings. It demands attention to its formal structure and high visibility. It has become what popular usage terms a Prominent Feature with a generic name acknowledged by the public: "headquarters skyscraper" or "showcase property." It occupies or, in fact, may in itself be a high-visibility site, a GOOD ADDRESS, a hundred percent location, and has become a visible asset, a part of the SKYLINE, perhaps a "view property," and no doubt a key part of the urban infrastructure. PRESENCE, indeed, has a long reach.

To disregard PRESENCE is to abandon forefront in favor of outback, and to prefer the obscurity of places having no known address. This points to what Warriner dislikes about a building with PRESENCE. According to the Warriner view, such a place is offensively supportive of "Western values." By stressing visibility, clarity, organization of space, etc., such a place or structure neglects—if it does not destroy—other, non-Western values.

It would appear that the not-so-hidden agenda of such an argument is an attempt to undermine clarity of expression in the language of the built environment. Yet "clarity," far from being the tyrannical device of Western culture, is a basic necessity for understanding one's surroundings.

SAND CASTLES

Since "sand-castling" became an international spectator sport with annual contests around the Atlantic and Pacific Coasts, it has attracted full-time teams of competitors. San Diego's Imperial Beach rode the boom with more than thirty creations (shown here in 1992). Photo, John Nelson, *San Diego Union-Tribune.*

Too many man-made places fail in just that quality of explaining themselves.

Warriner appears to be pushing PRESENCES into the background, so as to "allow meaning to emerge out of what people can do together." His argument, fuzzy as it is, appears to define Good Architecture as that which emerges from a sort of informal consensus arising from communal wishes and activities.

But consensus necessarily extends to the language itself. Without widespread agreement on the names for places and directions a society quickly loses its bearings, while the physical environment loses its PRESENCE.

Furthermore, all man-made places base their existence on some form of consensus. All offer testimony to forms of human agreement about the environment. They do so, not only in language, but by virtue of frontage on a public way, by their connection to a public utility, or their dependence upon public access. Without these accessibilities, places have only the most marginal existence. Without public access there is no such thing as neighbor, passerby, drive-ins, drop-in, drop-off, or meaningful adjacency. Access forms connections between places—and people who use them.

The term PRESENCE, in its non-combative usage, implies full access. And that is not a bad quality for practitioners of public professions such as architecture, planning, and landscape architecture to keep as their goal. To disregard PRESENCE is to disregard community.

THE SAND CASTLES

Whether Leonardo Da Vinci made sand models of his work is still debated among a growing breed of North American "sandcastlers." To them, building castles-in-the-sand has become, not a dream, not a sculptor's step toward the Real Thing, but a full-time, competitive fun-business. They create entire landscapes, some extending for miles along an ocean beach, then disappearing overnight. These sand castles and other constructions became centerpieces for annual festivals and competitions. "Sandcastling" moved beyond family games for the kiddies to become a verb. It acquired media personalities, in-group publications, and international competitions. These pull spectators by the thousands. A leading figure, sandcastler Todd Vander Pluym, got $120,000 for a 120-foot row of sand buildings at Albuquerque, N.M.

The handmade products vary in size from head-high replicas to a record-breaking sand castle (made with machinery) at Kaseda, Japan, 56' 2" high. A former record was held by "The Lost City of Atlantis," 52' high, at Treasure Island, Fla. But U.S. competition rules usually set 14 feet as a ceiling, and four hours as a time limit. Records for length are still being broken: a group called Totally in Sand with some four hundred volunteers set the 1989 record with a sculpture more than three miles long.[1]

Sandcastling origins, far upstream in time, may go back to Egyptians who, four thousand years ago, made sand replicas of pyramids and monuments. More recently, sand artists flourished along Atlantic City's

Boardwalk, carving scenes in sand, and also mass-producing quick portraits of tourists from 1897 until they were banned from that beach in 1944.[2]

In the 1960s, "dirt artists" (a.k.a. earth artists or landscape sculptors) began dramatizing the plastic possibilities of the open landscape. In its present (but still evolving) form, sandcastling dates back to 1952 when a landmarking contest was held at Fort Lauderdale, Fla., at a Fourth of July festival. Active sandcastling sites have been concentrated along the beachy margins of North America, principally on the West Coast where the sport caught on in a big way. There were contests in the 1960s at Newport Beach, Alameda, Cannon Beach, and Carmel, Cal. Cannon Beach's sandcastling started with a washout after a storm which isolated the town. Merchants backed the contest to attract trade. It eventually attracted some 1,000 contestants and some 250,000 people—more than the town could handle.[3]

Sandcastling now joins a host of other occupations which, increasingly in the 1990s, enliven varied sites with once limited but now expanding artistic activities. One-day horse and auto races have expanded into week-long festivals. EVENT SITES mushroom to accommodate civic events, fireworks displays, conventions, and reunions.

Highly organized with rules of the game, sandcastling is very much a spectator sport. Prizes are large, sponsors varied, and competition keen. Friends compete, and race around to scope out a unique piece before it collapses. In addition to huge crowds, they attract many artists and architects: one-fifth of the competitions were sponsored by chapters of the American Institute of Architects. The internationals include one at Hong Kong. More than ninety competitions and organizers were included in a 1990 reference book.[4]

Among the more popular non-castle subjects are fortifications, cathedrals, dragons, snakes, pyramids, Gulliver, Mt. Rushmore, the VW bug, the Seven Dwarfs, Alice in Wonderland characters, reclining mermaids, and other fabulous figures.

California dominated the early play. But its steep beaches, starved by silt-stopping inland dams, made casting difficult. The rules were relaxed to allow casters to mix their own precise silt and clay content to stabilize the moist sand. Rules often specified a four-hour construction limit. Team play became essential for the exhausting rush to set and brace wooden forms, and to move, moisturize, hard-pack, shape, and carve the sand—all by hand and hand-tools. When the tide rolls in, or the site lease expires, it all comes tumbling down.

This fits artist George Roualt's observation, "art is never finished, only abandoned." But sand castles could also become semipermanent pieces of new-built environments. They could expand into untapped markets along the Great and other Lakes and inland waterways as magnets to reanimate decrepit waterfronts. Watching sandcastlers at work has become an attraction in itself, so sandcastlers get hired to do their thing indoors to magnetize customers into shopping malls. If this continues, sandcastling will become yet another outdoors sport put under a roof in the interest of crowd control and cash flow.

SOLAR FARM

To tap the ultimate powerhouse—the sun from which all earthly fuels are descended—has long been the stuff of dreams. It took the Arabian oil embargo of 1973–74 and a jump in oil prices to add new meaning to SOLAR FARM. This is an evolving term for an outdoor array of solar panels, the use of which is undergoing a second wave of expansion. The newer FARMS increasingly use photovoltaic cells for converting sunshine into electrical energy. In this new search for diverse sources of energy, the United States had begun moving away from earlier forms of "stored sunshine"; the peak years for using firewood for heating were prior to 1850; for coal, around 1919; for natural gas, 1973; and for petroleum, about 1975. The nation was clearly into a multi-source struggle over energy.

European, U.S., and Japanese firms jumped into photovoltaic development and competition in the 1970s. Sales by U.S. firms were "increasing at an average rate of 25 percent per year, from $75 million in 1980 to $150 million in 1983. If measured by the peak power capacity of photovoltaic cells, the growth rate is 50 percent per year. System conversion efficiency has increased from 5 percent to nearly 10 percent. Costs have dropped by a factor of 10 since 1974, as rapid a pace as that achieved in the case of calculators. . . . [Foreign competition is increasing.] Evidence suggests that the U.S. market share in photovoltaics will level off at 50 to 55 percent."[1] With continued expansion, SOLAR FARMS could occupy many an empty site or desert location in coming years.

Favored sites for SOLAR FARMS and associated power stations are in the Southwest, notably southern California and Arizona, where some spots get 4,000 annual hours of sunshine. (Parts of Washington and New Hampshire get as little as 1,600 hours.)

But once the SOLAR FARM goes urban, the scene changes; the farm down-sizes to a set of panels in someone's backyard, or on the roof. A single solar collector measuring 6 × 8 feet would appear to need no more space than its footing. But its "place" requires access to the sun's full light, summer and winter. Do one's "sun rights" prevent a neighbor from building a garage that blocks sunlight from reaching one's personal collector? The legal question comes to be: can my "right" to sunlight override my neighbor's right to full use of his property (i.e., a new garage)? The answer was no, in one particular case. The Florida court of appeals, in the Eden Roc case, 1959, held that the Fontainbleau Hotel of Miami Beach had the right to build a fourteen-story addition, even though it shadowed Eden Roc's swimming pool next door. That is, the property owner did not have a lawful right to the free flow of light across the neighbor's property. (If you did so, then you—and not your neighbor—would be, for practical purposes, its owner.) Thus city folks who "farm" sunlight must adapt their crops to city sites. (If you are determined to ensure sunlight, in spite of your neighbor, go to Colorado. Its laws let you buy an easement over your neighbor's lot to ensure your own access to sunlight.)

WINDFARM

Variants: windmill farm or windpower field.

American tourists go to great lengths to visit remnant windmills lining canals in Holland, and dotting Cape Cod—or even further to see the world's oldest remnant windmill farm overlooking the highland Plains of Lassithi on the island of Crete. Here an estimated ten thousand windmills, using ancient jib-sail designs, once pumped water from a giant underground aquifer to irrigate what has been called the breadbasket of Crete.[1]

The windmill appears to have been invented in Turkey around 200 B.C. and by the eleventh century was extensively used in the Middle East. Returning Crusaders brought it to Europe, where the Dutch put it to work draining Holland. By 1586 the first windpowered paper mill came

WINDFARM

Biggest array of WINDFARMS in the U.S. clusters along fifty-four square miles of breezy ridges and windgaps around Altamont Pass near Livermore, Cal., picking up saleable energy for distribution by Pacific Gas & Electric Co. By 1993, more than 7,500 wind turbines, one per every five acres, had been installed in recent years across this mostly cattle-grazing landscape. (Others cluster around Tehachapi Pass near the Mojave Desert.) Some neighbors living within EARSHOT of these and other WINDFARMS have gone to court or to legislatures to cut the noise. Photo by Michael Brohm.

onstream, and sawmills around 1700. Back in Lassithi, many hundreds still spin today on the plains; but up at the mountain notch that faces strong prevailing winds, only stone foundations remain where windmills once powered huge grain mills by the road over the pass.

In the United States, WINDFARMS are expanding into spectacular and sometimes noisy sources of electrical power generated by large modern turbines. Their blades, rotors, and other catchment devices are rotated by the force of passing winds. Prior to the 1970s the rare visible examples were single and often rickety farm windmills, mostly in the Midwest and Great Plains, producing a modest 1 hp, pumping up underground water for farm/ranch homes and gardens, and for livestock on remote fields.

The flurry to find new energy sources following the 1973–74 Arab oil embargo caused a rush of inventors and speculators to buy up windy sites. Governments, federal and especially Californian, offered subsidies to turbine manufacturers and to WINDFARM developers. (A counterpart rush began to revive old water mills and renovate their sites.)

It was a time of turbulence, not all of it wind-generated. "Most of the driving forces seem[ed] to be beyond the control of local authority."[2] A new federal program (Comprehensive Community Energy Management) in 1978 set out to show how power could be generated in sixteen locales.

The new law, in Section 210 of the Federal Public Utilities Regulatory Policies Act (1978), required utility companies to accept into their networks wind-and-water-generated power from these new sources—from cogeneration and small power producers. It required "just and reasonable" prices for ten years, making investment profitable, until President Reagan ended the boom. By 1992 California had "85 percent of the world's installed wind-power capacity."[3]

Good wind sites are rare; some parts of North America seldom offer enough wind to raise a dust devil. But in Livermore, Cal., "it's a sight to daunt even the dauntless Don Quixote, the world's original and most revered tilter at windmills. . . . On both sides of Interstate 580 east of San Francisco, marching along the ridgelines of the fat-hilled Altamont Pass, are what seem to be battalions of exoskeletal outer space creatures . . . more windmills than even the man of La Mancha would care to shake a lance at."[4] By 1990, there were 4,000 wind-catchers on the gusty ridges and slopes around Altamont, Cal., "and 12,000 more in other parts of the state."[5] On a breezy day, their output can exceed that of one large nuclear power plant. Altamont, Tehachapi, and San Gorgonio passes combined are the sites of 16,000 turbines costing some $3 billion, producing up to 1,500 megawatts. Ninety percent of the nation's wind production is in California. Wind turbines' efficiency tripled in the 1980s, while California's cost of producing wind-electricity dropped as much as 300 percent. Fifteen percent of the island of Hawaii's electricity was wind-generated in 1989.

A typical California WINDFARM occupies a prominent location. Its spinning turbines on eighty-foot steel towers are visible for miles. They occupy only 10 to 15 percent of the land, which is typically used for pasture.

An early cause célèbre took place a hundred miles west of Los Angeles near the plush resort of Palm Springs, Cal. There, at

WINDFARM

Where it all began: Some of the world's earliest and greatest WINDFARMS flourish on the Lassithi Plain, in the uplands of Crete. Hundreds of these windmills pump underground water up to irrigate grain crops. Grains were once ground in giant stone mills flanking a nearby mountain pass and wind gap.

a junction of low and high deserts, lies San Gorgonio Pass, said to be "the best wind resource area in the United States."[6] Local residents had been angered by a SoCal-Edison proposal for hefty electric rate increases in '81. After a flurry of local opposition-people saying they needed an alternative source, the utility firm set up two wind turbines on the pass in 1980–81. One soon collapsed, the other never got going. Local residents were unimpressed. Yet entrepreneurs, seeing lower-priced turbines coming soon, and sniffing the volume of untapped wind, filed applications in 1981 for thirteen WINDFARMS. Most were turned down. Planning historian James A. Throgmorton concludes: "Local planners impeded development because the existing political structure of electric power supply did not permit the risks and benefits of local energy development to be allocated in a manner that local residents considered fair."[7]

In a larger national context in the 1980s, this seemed to be small stuff. But it contained a large lesson for the future: at 1990 prices of competing electricity, wind still "doesn't quite pay."[8] If all conditions were right, wind power would still generate only 5 percent of needed electricity in the

United States. Yet conservation has worked wonders: the U.S. economy grew "by 35 percent from 1973 to 1986 with no extra energy use."[9] Wind, sun, and geothermal energies combined, plus determined conservation in the nineties, still offered the United States relative independence from imported fuels.

Earlier, enthusiasm for wind power had gone-with-the-wind when oil prices stabilized and federal subsidies disappeared. "The coup de grace came in 1986 when the last of the [federal] tax credits that had propelled the industry to such an auspicious start expired."[10] Meanwhile, many neighbors raised hell over wind noise. A proposal to build Los Angeles County's first WINDFARM near Gorham was killed by neighbors insisting the 458 turbines would "hurt property values." In a celebrated case, the city of Palm Springs, neighbor to some 4,000 ridgetop turbines, asked the landowning U.S. Bureau of Land Management to take the sails out of the wind. Iraq's invasion of Kuwait in 1990 set the cycle in motion again, soaring oil prices spurring re-interest in WINDFARMS.

An exotic new type of WINDFARM was proposed by Mexico City's mayor Manuel Camacho Solis: one hundred giant fan complexes, each covering thirty-three acres, to blow polluted air away from the city, and beyond its surrounding ring of mountains. These will take, not generate, energy, in a desperate effort to pump out bad air.[11]

Back in the traditional mode, after three years' testing, the Green Mountain Power Corporation in Vermont expects new wind turbines atop Little Equinox Mountain (elevation 3,300 feet) to produce 10 percent of the power needs of its region by 1995.[12]

While some neighbors resist, the wind persists. Future fuel shortages and the inevitable shift toward non-fossil-fuel economies will send surveyors and speculators back to the wind gaps. And on such high grounds they will continue to seek what the Greeks thought to be the eternal gift of the gods.

See also DOWNWIND, in Chapter 9: The Limits.

CHAPTER 9

The Limits

Many places that appear to be Opportunity Sites—IRRIGATED AREA, THE PUBLIC DOMAIN, WETLANDS—at some point reach practicable limits. At some time in the past, barely and imperceptibly, there began a slow run-on-the-bank of capital supplies: land, water, sunlight, and air. The publication in 1962 of Rachel Carson's book, *Silent Spring*, was a turning point. Another was the disastrous air-pollution inversion at Donora, Pa., in 1948. No longer could Americans view nature as a bottomless pit, an inexhaustible gold mine, an untapped reservoir, the guarantor of unlimited credit.

The Limits, as now perceived, have no known, fixed, or predictable location. They may crop up where we least expect them. Who foresaw that super-rich, urbanized California would—in the eyes of its many unemployed in 1993—appear to have run out of steam, of opportunity, of cheap water, and pure air? Or that Philadelphia, with its shrunken industry, would find itself too big for its population? The Limits may crop up where they have been long expected—in those border zones-of-confrontation between competing populations and limited resources. Often they appear at the terminus of service-delivery pipelines—out where rural free delivery ran out, where United Parcel Service seldom runs, where "city services" are exorbitant, and only the airwaves regularly deliver their rich menu of the faraway.

In what follows on these pages, many apparently ever-expandable places will reach their limits. These include places that people invent and sometimes lose control over. Entering the twenty-first century, we must stay alert to the limits that nature and exploding world populations continue to inject into competitive landscapes.

AVALANCHE ZONE

This murky linguistic term has descended to us from the dangers of the Swiss Alps where, at any winter moment an avalanche may roar downward. "It is difficult for plainsmen to appreciate the extent of the danger," observed mountain-geographer Roderick Peattie.[1]

If a non-mountaineer looks upwards, he/she will be awed by the steepness of slopes. Conditions for avalanches occur when the mountainside slopes at 22 degrees and carries 40 to 50 cm of snow; or slopes 30 degrees carrying 15 cm; or 50 degrees ("precipitous") carrying 5 cm of snow. If the slopes are steeply glaciated, chances are good for avalanches. Snow may move when it rests on ground, but more likely when it rests on ice. It follows known routes down chutes, ravines, or slide corridors, some of them man-enhanced to guide the moving precipice of snow along a less-damaging path.

Only in the northern United States, Canada, and Alaska do conditions approach those of well-studied Switzerland, which has 9,368 avalanche corridors, including 5,294 that carry their load several times a year.[2] What people do in order to coexist with AVALANCHE ZONES is similar around the globe: they (1) move elsewhere; (2) keep a sharp and now electronic watch for early signs; (3) build their houses and brace their villages to survive the sudden weighty blow of dry or wet snow at high speed; (4) erect chutes, barriers, diversions, walls, and sheds to deflect it, or tunnels to get under it; and (5) set up elaborate radio nets and snow-removal and road-clearance systems. Coexistence is expensive.

AVALANCHE ZONE'S most modest cousin-once-removed is marked "Falling Rock Zone," but on mountainsides more stringent warnings appear where roads themselves may suddenly fall away and disappear.

THE BEACH

A braided, multimedia, fluid, restless river of sand, wet and dry, mixed with wind and water in endless, ceaseless tidal mutation. This amphibious, shifty footing for human fun and work, this neighbor to all coastal peoples is a constant stream existing both above and below water level—extending for miles, not only along coastlines, but from far inland out to an ever-shifting sea bottom thirty to forty feet deep, three or four miles offshore. It changes out of all recognition over time, and requires endless supplies of air and water to maintain its flowing self.

The physical BEACH may have a bottom that drops off suddenly into depths which, in California, supply endless upwellings of chilled water. Or, as in Sanibel Island, Fla., BEACHES may be part of a gently sloping ramp extending for miles offshore. This gentle slope assures that sea shells, driven inland by winter storms, arrive in good condition onshore, creating one of the world's premium shelling BEACHES, a magnet for elderly "shellers," especially after a causeway linked BEACH to mainland in 1963.

Mainland BEACHES and offshore ("barrier island") BEACHES vary greatly, the latter more dominated by energy stresses coming from several directions in waves, winds, tides, storms, and overwash. Barrier islands, their dunes and their BEACHES—stretching from New England to Mexico—are in slow retreat, one that would become a rout with the expected rise in ocean levels. Between islands, the geography changes endlessly: Fire Island's Inlet, N.Y., moved six miles westward between 1694 and 1958.[1] "Emerald Isle [on Bogue Bank, N.C.] is now so highly developed [by resorters] that whenever the inlet moves it will be condemned as a 'natural disaster.'"[2] Captiva and Little Captiva Islands, Fla., were segregated not by the Supreme Court but by hurricane.

By occupying BEACHES and converting them to real estate, man engages in endless warfare to control BEACHES and their natural components. No "erosion" exists in nature; "erosion" is what happens to nature converted to asset. In short terms, mankind "wins" an occasional battle and asserts ownership; but in longer terms that restless dynamic event called BEACH regains control. "Being close to the beach . . . means becoming part of the beach—something no house survives satisfactorily."[3] On such occasions, short-fused landowners take it as a personal insult. Only mankind's short-term memory conceals the astronomical costs of calling BEACH real property, and attempting to "restore" it to some prior condition described in a legal deed. Forces that make BEACHES do not read. Forces that unmake BEACHES were causing Florida to lose three hundred acres per year in 1957, according to Dr. Per Brunn, University of Florida oceanographer.[4]

All this commotion has brought an intensity of gaze to the modern tax assessors trying to strike a balance between land values going up and land itself going away.

Yet the mystique of owning, occupying, or living near THE BEACH retains its power, even as natural forces undermine many a locale. (Erosion increases with nearly every beach structure built.) A metaphorical aura extends inland a flexible distance from THE BEACH. "Close to the beach" may mean: (1) actually directly upon THE BEACH—often a stilted house open to attack from the next hurricane; (2) on the dunes, a location almost as precarious as the open BEACH; (3) across the beach highway; (4) close enough to walk to in one's swimsuit; (5) close enough to walk to, carrying a cooler and beach gear; (6) close enough to hear breakers; (7) close enough to feel the sea breeze; (8) close enough to see the ocean [for buyers, the rule here is: investigate before investing—does one see THE BEACH, the breakers, or merely the far horizon?]; or finally, (9) "within driving distance of the beach." This stretches THE BEACH some two hundred miles inland. Yet even this distant penumbra embraces the contact zone for weekenders, summer cottagers, and city folk anxious to maintain contact, however tenuous, with THE BEACH.

This contact has been long and contentious. A "saturnalia of moist wetness" was said to have been inspired by the first form-fitting swimsuits (then "bathing attire") in the 1870s. That was when a narrow-gage railroad began hauling Bostonians out to Revere Beach to watch fireworks and boxing, to eat potato chips (which Revere Beach claimed to have invented), to frequent seaside shanties and their "hostesses," and to drink a lot. But that was only a taste of what were to come—by the millions a century later. Revere also became the scene of intense black-white competition for turf following desegregation after the 1950s.

New York's secretary of state in 1990

lamented that "increasingly, the beachfront is being developed and walled off from the public" by many hands, public and private. Hundreds of beachfront towns responded to thronging beachgoers by incorporating. The island of Sanibel, Fla., did the same in 1974.

Beach histories of the late twentieth century are filled with similar scenarios: a once-secluded ("pristine") BEACH is discovered by speculators, the bay-side swamp filled, dunes bulldozed, high-rise hotel planned, local residents outraged. Old-timers scream of "the lack of planning," public hearings are conducted at courthouse or state capitol, and—so long as local interests gang up to solicit outside capital—development of some sort follows.

Into this predictable scenario the Coastal Zone Management Act, passed by Congress in 1972, added weight to its protectors, if not permanence to THE BEACH itself. It brought new players into the game. It also brought federally insured flood insurance. Its higher premiums in high-risk areas help to protect the government more than beach property or its owners.

Meanwhile, large stretches of THE BEACH were being sequestered by towns and cities to forestall inundation by strangers. Where they could get away with it, local officials limited parking, beach permits, and, in New England, lobstering, to local residents. Or they heavily charged outsiders. When Belmar, N.J., its population swelling from six thousand to sixty thousand on summer weekends, was ordered by a judge not to charge more than $3.25 per beach-user, the town complained this wasn't nearly enough to cover beach-maintenance costs.

The Connecticut Civil Liberties Union advocated first-come, first-serve for all beach-goers. But townsmen, arguing that they pay local taxes to maintain BEACHES, insisted their limited resources should not be swamped by outsiders. On a larger scale, conservationists manage to buy up and otherwise protect from the onslaught many stretches of "endangered habitat."

By the 1990s, popular California BEACHES were accommodating a million visitors in a summer's day, while Jones Beach State Park on Long Island, N.Y., handled 7.5 million throughout the year 1989. At Pismo Beach State Park, California created a "state vehicular area"—five miles of drive-in BEACHES open to vehicles—to cruise, to compete, to get stuck. California itself faced a dwindling of its total BEACH area owing to a century of major dam building which trapped upstream those sediments which once helped stabilize BEACHES.

Meanwhile, BEACHES within the reach of populations have developed their own social orders, sometimes invisible to strangers. Social life takes endless forms, usually arrayed in lines parallel to the surf. Within the surf, gradations of surfers, boaters, and swimmers keep their distances. In the shallows tiny paddlers and timid swimmers with parents dart in and out. Along the wet strand stroll shellers, exercisers, paraders, partner-seekers. Sitting and lying close to the strolling flow are explorational teen-agers and sexually active pairs and groups. Strenuous volleyball players scatter sand, balls, and shouts. Further back, families spread blankets and towels, open umbrellas, and picnic for the day. Still further away from the action, older folk may snooze. The linear order may change with the days of the week, with school and other holidays, and where caterers supply cabanas and wind shelters. The array shifts drastically on BEACHES open to autos and bikers.

Once populated mostly by "beach people," California BEACHES by the 1950s had generated a lifestyle which spread all around the continent: complex, photogenic, and often competitive. "Spring break" crowds of collegians had swelled to almost one million by 1990, reaching far to Caribbean BEACHES. Fort Lauderdale, Fla., was the early spring-break mecca for East Coasters, 350,000 in a fortnight, until local regulations in the late 1980s diverted the hordes to Daytona Beach (400,000 spring-breakers in 1990).

Fears of skin cancer became widespread by the 1980s, but failed to stop the lemming-like human flow seaward. Such fears did, however, create a market for sun-shielding clothing, shelters, and cosmetics.

In the 1950s, the old social order of American BEACHES, based on an acceptable code-of-conduct among coherent groupings of like-minded BEACH lovers, began to be challenged. New populations moved into, high-rises looked down upon, and strangers abounded at BEACHES that attracted huge holiday crowds often carrying water-sports gear and noisemakers. Segregation became more pronounced: some BEACHES were "taken over" by homosexuals, muscle-builders, surfers, fishermen, air crews, or ethnic groups. Such takeovers varied from simple dominance by numbers to overt intimidation and physical force. Nudists retreated to more isolated venues.

But one 1979 study of a heavily urban California ("Southland") BEACH concluded that "remarkably little serious trouble occurs"—especially as compared with racing events and rock concerts. For the most part, beach-goers had adopted "a common definition of the beach situation as well as a shared set of routine behaviors." So long as such a social contract exists, THE BEACH and beach-goers will carry on their restless routines. So long as the American population continues its glacial migration toward the coasts, those social contracts will continue to be renegotiated.

BIOREGION

This is a discovered locale, a "rediscovery" of the 1960s popularized by the so-called Ecology Movement with its roots in biology. Anybody may become a "discoverer" by identifying a grouping of place-related and biologically influenced goings-on. But the concept's deeper roots rest in Greek geography three centuries before Christ. The Latin *regere,* meaning "to direct or to rule," carries the word into its present usage. Closer to our times, Charles Darwin's *Origin of Species* (1859) set off a wave of diggings into bio-geographic influences upon man.

"Regional studies" first came to the fore with the works of the German researchers Alexander von Humboldt and Karl Ritter. The term "region" with its Latin meanings survived attackers, defenders, and redefinitions. Loosely and widely used, it became whatever geographic area suited the convenience of whoever used the term. But among bureaucrats and governors, it connotes an area over which we exercise control.

The geographer Ritter defined regions as watersheds, or whatever was enclosed by mountain ranges. He called them "natural units" of various orders, into which the earth could be divided. In the early twentieth century, the Scotch biologist-teacher, Patrick Geddes (mentor of historian Lewis Mumford), taught from his Outlook Tower in Edinburgh as a vantage from which one could unify "the striking landscape, near and far." Geddes used a technological gimmick, the camera obscura, to see the region whole; and regional surveys to diagnose social ailments. "Town and county interests are commonly treated separately with injury to both," he said. All human sciences focused on the great trinity—Place, Folk, and Work—as he interpreted it.

The region has long been seen as the ideal area within which to view a city, and to plan for environmental improvements. Whether one calls what Geddes then saw a functional area or natural area, whether one stands inside or outside the geographic complex called BIOREGION, it has proven hard to pin down. Geddes, like Aristotle, preferred standing on his promontory to best view a city's BIOREGION. Geddes considered urban water sources as "precisely the most important, the ultimate and determinant condition of population, and the inexorable limit of their growth." So did the Romans who used aqueducts, as did the later Californians, to extend their grasp over far-off sources.

Most experts see the "there" there in BIOREGION as consisting of "spatial groupings." Occasionally they will put the regional there-ness onto local maps, but only after much dispute. The simplest measure used to define a BIOREGION is its watershed. But modern analysis of complex factors by computer has offered BIOREGION as a plaything for number-crunchers.

No matter how well-analyzed, however, BIOREGION seldom makes headlines or cocktail talk—except when locals try to fix a political border between here and there, such as voting on a legally defined metropolitan area, an improvement district, or a planning region. Many a local attempt to create metropolitan government rises or falls around a sense of regional identity, and competing definitions of same.

Other and less city-centered definers see BIOREGION in more radical terms—as distinct from, and antagonistic to, the big American city, its voracity, and its capacity to damage or to destroy environments. Instead, these reformers envision a total reorientation of American life around BIOREGIONS of villages, small towns, and revived rural landscapes. They picked up New Deal friends and funds during the Depression Thirties, and again during a utopian era of "postwar planning" in the 1940s.

Peter Van Dresser, a radical New Mexico planner, in *A Landscape for Humans* (1972) and other books, saw the northern New Mexico Uplands as a BIOREGION—a "specific regional community" needing assistance but not urban dominance. Back of many another BIOREGION plan was a similar distrust of ever-expanding cities which seemed to threaten the BIOREGION. For such thinkers, Lewis Mumford's critique of urban-industrial culture became their Bible; and Benton MacKaye's Appalachian Trail served as their favorite escape route or relief valve. Ian McHarg's later teaching at the University of Pennsylvania and in his book, *Design with Nature* (1969), applied his own post-Geddes concepts to BIOREGIONS of varied sizes—from a Maryland valley to Staten Island, N.Y., to the Earth. His evangelistic fervor was in the great Geddes-MacKaye tradition.

Amidst trends and countertrends, regions are ever more variously defined. Once a BIOREGION is mapped, however, knowledgeable locals will nod approvingly and say "Of course!" What they nod about and may know about are those aspects of local climate, topography, geology, geography, flora-and-fauna, and traits of human culture—Geddes's Folk—which combine to distinguish their very own BIOREGION visibly, notably, and provably from others. Such are the core region of Appalachia, Van Dresser's New Mexico–Colorado uplands, the Bluegrass of Kentucky, Kansas's Post Rock Country, the cutover region of northern Michigan, et al.

But sometimes it's hard to tell. Many a BIOREGION is so cut up by political borders as to render difficult any cross-border perception, much less cooperation. Many North American valleys are split by rivers (Mississippi, Missouri, Ohio) and by political boundaries that divide competing states and market territories. Meanwhile, folklorists and historical romancers seek and sometimes find mythic cultural unity in "Riverine Peoples" along the riverbanks.

Prefaced by the magic modifier "metropolitan," the word "region" picked up voltage when "Standard Metropolitan Areas" were first defined in 1949 by the Bureau of the Budget and then redefined as "Standard Metropolitan Statistical Areas" in 1959. Such a "region" became a promotional tool for competing cities, and was lampooned by planner Frederick A. Bair, Jr., in the 1960s as being defined by "whatever there are federal grants-in-aid-of."

As it gets identified and studied, a BIOREGION begins to preach to its citizens: about what their place is best and worst suited for; about the proper crops to raise; about which direction(s) to expand one's city or market; about where to go, or whom to fight for more water or fresh air. To know one's BIOREGION means to understand the geographical, biological, and cultural forces at work upon it and its citizens. "You have to enter into a dialog with your own bioregion before you tell the world how to change," said eco-solutionist Donald Conroy.[1] Such local understandings, rooted in the nuts-and-bolts of everyday real life, can form the basis for all broader knowledge.

Gradually, the nineteenth-century dominance of proximity—of things going on next to each other—has given way. Fewer people needed or demanded everything close-at-hand. The right-next-door-ness of urban life fell apart when railroad commuting, and then autos, cheap gas, and mobility intervened. Here spread into strange places, away off into nearby counties. And in this process many an intact BIOREGION has become turf for competing towns, cities, and corporations seeking to take it over.

Yet even as it grew, here persisted in its hereness. The science of ecology expanded from studying the where-ness of plant and animal life to the study of human cultures, and the limits of their support systems. The Chicago School (Amos Hawley, sociologist Louis Wirth, Otis Dudley Duncan, et al.) swung its emphasis onto growing cities, while European scholars stayed attuned to a broader biological field. Only in the environmentally conscious 1960s did U.S. ecologists move from biological studies into cultural front lines, defining ecosystems with eco-awareness, and adopting eco-politics. The Green political movement in northern Europe found allies in North America.

But as commutersheds expanded forty-five minutes outward from jobs, as cities grabbed and pumped water to or from far counties or states, as the dry West looked to reluctant Canada for future water, as decentralized plants shipped parts and parcels hither and yon across enlarged market areas, many a BIOREGION no longer felt, acted, or looked like its ancestor. Its grand complexity, some claimed, could only be grasped by computer-imagery. Chances are, Sir Patrick Geddes would have jumped at the chance—and then recoiled from what he found.

Regional planning, once broadly based (as in the Tennessee Valley Authority) but rarely carried out in the United States, in the 1980s had become a tool for marketing managers and few others. It remained on the agenda of planning schools, but of few politicians outside California. If anything could move it into mainstream politics, it would be the continuing environmental crises—falling water tables in the West, acid rain from the Ohio Valley denuding Appalachian and New England forests, oceanside pollution fouling East Coast BEACHES and shrinking the North Atlantic FISHERY, DDT in Antarctic penguins, and global warming. Gradually, the crisis was spreading through and beyond BIOREGION as environmentalists continued to preach cooperation across city, state, regional, and national boundaries.

See also DISTRICT, in Chapter 3: Perks.

DOWNWIND

Sniffing the wind has been a survival technique among animals, savages, hunters, and warriors for millennia—long before the term moved into politics. Gunpowder in warfare gave new importance to knowing which way the wind was blowing. Before smokeless gunpowder came into use, BATTLEFIELDS were "clouded with a thick, choking pall of smoke" which commanders had to anticipate, hoping the enemy rather than themselves would be choked and blinded by wind-borne smoke.[1]

Hostesses and hunters alike specialize in that risky zone known as DOWNWIND. For nothing disturbs a party or spooks wild game quicker than a sudden, strange, harsh, or unintended odor. Rarely on social occasions are persons reeking of garlic, manure, sweat, or other body odor reminded by others DOWNWIND that they smell, stink, and give offense. Rarely are they invited back.

But the plot, as well as the atmosphere, thickens when the offending source is a chemical complex covering two thousand acres, paying millions of dollars in local taxes, employing a thousand local workers, while exuding an olfactory plume DOWNWIND across two populous counties. "It smells like money," was the typical earlier reaction. Smoke and smell in the nineteenth century connoted a job at the other

DOWNWIND

Locations around Lake Michigan receive both benefits and losses from being "DOWNWIND." Early settlers arrayed their cherry orchards along the eastern shore (shaded) to take advantage of cold winds blowing off frigid lake waters. This delays early springtime budding until all danger of late frosts is past—and favors an expanding strip of summer resorters. In the 1960s, the same prevailing westerlies caught up smoky outpourings of Gary's steel mills and wafted them eastward to produce "the LaPorte Effect": dirty rain falling on LaPorte, Ind. (diagonal hatching). Map redrawn by Joan Sommers.

end. Black smoke signaled men-at-work. "Smokestack-chasing" was widely used to boost a home town's economy.

In some cities, notably Los Angeles, Denver, and Pittsburgh, DOWNWIND is part of a pollution-removal system that only works part-time. As chemical plants expanded, DOWNWIND locations became a major point of contact between human beings and sometimes distant sources of smoke, chemicals, and other airborne nastiness. Watching and testing for windborne gases had been a preoccupation among World War I troops after soldiers were first exposed to mustard gas attacks. Fears of gas attacks permeated U.S. preparations for invading Iraq in 1991. Sniffing, or otherwise testing, the wind remains a universal practice, in and out of war and politics.

For centuries, critics and health specialists had known there was more to smoke than jobs. John Evelyn, the English silviculturist, in 1661 had written a still-reprinted pamphlet attack, *Fumifugium: Or the Smoake of London,* in which he coined the phrase "an infernal nuisance." Yet in the nineteenth century, any chemical fumes loosed from factories were dismissed as an episode in the local cost of living, unworthy of official recognition. Some instances did, however, penetrate the local language: New York's notorious Gashouse District was named for the source of its nearby stink. By the 1960s, those living DOWNWIND from a string of ageing chemical plants, nuclear test sites, or leaking DUMPS had vastly expanded the DOWNWIND population and politicized the discussion. The rigors of mortis spoke loud to the crowds.

What once were dismissed as "short-lived odor incidents" expanded to become geopolitical realities. An early example was the Monongahela River town of Donora, Pa., where pollutants built up locally from plants processing steel, wire, and zinc. During an atmospheric inversion, with no escape route DOWNWIND, pollutants stayed where they were generated. In the last week of October 1948 the pollutants affected 5,910 persons, 49.7 percent of the residents, of whom 18 died in three days. Hundreds more died prematurely but later. It was a well-documented landmark case. In another, some 3,500 persons died in Greater London after a smog episode in 1952; and another thousand in 1956.

Editor John Lair in the *Saturday Review of Literature* expanded the subject by his disclosure of "the LaPorte Effect." This consisted of dirty rain falling on LaPorte, Ind., directly after heavy steel and smoke production at the upwind steel mills of Gary, Ind., to the west. After an "episode" at Texas City, Tex., in 1987, four thousand persons were evacuated. From Louisville's West End in Kentucky, twenty-two thousand persons were evacuated after a barge loaded with chlorine jammed onto an Ohio River dam; and up to twenty-seven thousand were evacuated DOWNWIND to escape a huge toxic cloud from a City of Commerce chemical plant in Los Angeles County, September 3, 1988. Torrance, Cal., alerted itself to the dangers of its neighboring Mobile refinery after explosions and fires in 1987. A reporter later wrote that "Torrance is now known for good schools, good air—when the wind doesn't come from the refinery."[2] DOWNWIND locations in all such cases offered risk if not death. By 1988, the air in fifty-eight U.S. cities had been classified by the Environmental Protection Agency by a less noxious term: "substandard."

Under pressure from new laws in the 1970s, many power plants and other smoke-generators added tall smokestacks to reduce local complaints. This also distributed pollution more widely. Nor did all DOWNWIND contaminants stay airborne, as their producers hoped. Figures in 1986 showed that "approximately 95 percent of all toxic pollutants in Lake Superior have drifted there through the air." For other Great Lakes, the percentages were the following: 75 in Michigan and Huron; 48–50 in Ontario; and 25–48 in Lake Erie.[3]

As a result of chemical spills and nuclear diffusions, "downwinders" became newly conscious of their rights not-to-be-downwinded. This was true especially near the Nevada Proving Ground, where nuclear test explosions were begun in 1954. Federal officials argued that the tests—though risky—were an alternative to the "risk of annihilation" from a Russian nuclear attack. As it turned out, more than twenty-five test atomic blasts in Nevada during the 1950s produced "clouds filled with radioactive debris . . . carried by high-altitude currents over the downwind valleys, deserts, and small towns of Nevada, Arizona, and Utah."[4] Efforts by federal officials to conceal the results opened the way for widespreading distrust of nuclear agencies. Not until 1990 did federal officials confirm the fact that most people living near the Hanford, Wash., nuclear project, 1944–47, had been exposed to low doses of radiation. The downwinders called a news conference to protest in 1991, nearly a half-century after the fact.

The city of Stuttgart, Germany, pioneered in airshed management in the 1960s by clearing all houses and other smoke-producers from an upwind valley to improve its own air quality. Canada, aggrieved at being used as a downwind DUMP for airborne U.S. pollutants, protested for years before achieving a small victory with a transboundary air pollution agreement in 1991. Similar disputes crossed NATIONAL BORDERS in Europe as forests in Holland, Germany, and Sweden showed acute acid-rain poisoning.

The political effects were widespread. No longer in the United States after 1990 was it possible to use DOWNWIND as a supine SINK for pollutants. Eastern states and cities living DOWNWIND from polluting sources supported new anti-pollution laws of the 1980s. Eastern U.S. and Canadian forests (similar to those in Germany) were under duress; lakes and woodlands were stressed. If the Ohio Valley—a major pollutant source—had been a separate nation, downwinders in New England might have been prompted to threaten war. As it was, eight Northeastern states formed a compact to control their own as well as upwind pollutants.

Residents DOWNWIND from chemical plants in California's Silicon Valley and elsewhere became active politically to reduce their risks. California adopted the most stringent anti-pollution laws in U.S. history, partly aimed at automobile sources—leading to the redesigning of cars sold in that state.

Since scientific measurements grew more precise, a large industry began to analyze what it means to live DOWNWIND, and how blame and costs can be allocated. By the summer of 1988 one-quarter of all Americans were said to be breathing bad air for part of the year. "Smog Alerts" became commonplace in many regions. Living DOWNWIND thus served to promote "broad coalitions between environmentalists and business groups who fear that poor air will hinder prosperity in tourism, retirement, and agriculture."[5]

Beyond all this, modern weather discoveries point to global distances over which airborne changes are carried, to affect distant downwinders half a world away. By 1992 it was fairly well established that the source of odd weather patterns in North America—record wintertime warmth, plus torrential Gulf Coast rains—was El Niño, the warming of the distant Southwest Pacific Ocean occurring every three to seven years.

See also NO SMOKING AREA, in Chapter 2: Patches.

DRAWDOWN

Note the difference: technically, DRAWDOWN (sometimes hyphenated) is that volume of water that is withdrawn or "drawn down" when released or pumped from a flood-control or other reservoir, and/or when a drought reduces the incoming water supply. High-water mark, on the other hand, is that mark, either natural or man-made, which points to the height of the last flood, or overfilled reservoir.

But the term DRAWDOWN has, by extension and extensive use, come to indicate the surrounding expanse of banks, mud flats, deltas, outwash, and bottoms left exposed when reservoir water is drawn down. This happens normally when winter and spring high waters are stored, then released in dry months to keep downstreams flowing.

Such a vast spread of exposed DRAWDOWN becomes highly visible, a distinctive part of the scene. In the drought summer of 1986, Lake Lanier, which supplies Atlanta, shrunk by nearly seven square miles, exposing wide stretches of dried shorelines. On any such DRAWDOWN horizontal bands of debris and wash indicate previous levels. It becomes a place to be avoided except by the most avid hunters, fishermen, scavengers, or bird-watchers, and other naturalists. At times of extreme low water, the DRAWDOWN expands to cover most of the so-called lake area. It acquires long-range visibility, especially in infrared air photos where its exposed bottom reflects huge amounts of solar heat. Little clouds, that in wetter days formed above reservoirs, hover no longer up there in the sky.

Consequently DRAWDOWN has generated an array of expedients in water-scarce regions; raising the level of old dams became a familiar effort in the 1980s. Pumping water from distant sources has become an effective though expensive tactic. Neighbors along many a water's EDGE erect makeshift structures to render THE EDGE more useful: catwalks; duckboards; portable or floatable docks; boat-houses maneuvered to perch at the new, low-water line; stones hauled down to form little hardstands for fishermen. Long strings of temporary steps now connect the lower shore with boathouses perched high and dry at the old high-water line.

Once DRAWDOWN dries out, it becomes a scavenging-ground for hunters of souvenirs, beer cans, snagged fishing lures, dropped coins, and equipment. During droughts, DRAWDOWN becomes the photographic venue for television crews looking for crisis footage. From surrounding thirsty lands, ranchers, farmers, and water-haulers move to the edges, stake out sometimes hotly contested positions for their irrigation pumps. As DRAWDOWN expands, distant cities dependent on reservoirs issue rules and reminders: "No watering of lawns," and "It takes three glasses of water to wash one glass."

As a political presence, DRAWDOWN attracts local critics who far outnumber its supporters. These are chiefly resorters, or owners and patrons of fishing docks and ramps, and marine suppliers who see their sphere of influence dwindling as DRAWDOWN expands. But in a major drought, the controversy shifts from local to federal agencies. They have long been under pressure to allocate more water for expanding cities, less for irrigating farmers/ranchers. The low-water line at DRAWDOWN signals a political showdown at the statehouse door and on the floor of the U.S. Congress.

See also GROWTH AREA, in Chapter 5: Testing Grounds.

DRAWDOWN
All waterside activities around reservoirs stop, go, shift, or move, depending on the state of the DRAWDOWN—the shifting water levels, especially in Western multipurpose reservoirs such as Friant Reservoir, Cal. For boaters, "dockside" may be here when the level is "drawn down," or far up the bank, depending on the water level, which fluctuates to accommodate floods and downstream irrigators.

DROUGHT AREA

Here is a marginal but occupied farming area, between abnormally rainy years. It was originally located in a district of small or erratic rainfall in the foolish expectation that man's settlements could increase the basic climatic source of rainfall, or that irrigation water could be imported great distances, in perpetuity, and at someone else's expense.

Here is also a major city (New York, Los Angeles, Phoenix) dependent on distant watersheds, competing with other states for rivers of limited volume. Its pavements flush away rainfall, its springs long since have dried up, its streams are culverted, its citizens now and again rationed: lawn-watering on alternate weeks, bricks in the toilet tank, exhortations in the media.

Into most DROUGHT AREAS are poured life's savings, risk capital, and national taxes. If the drought is seen as a temporary one, various flows of "disaster relief" may be tapped: local, state, and national.

At the onset of these recurring episodes, the usual remedies are trotted out: conservation of water supplies, rationing, trading, down-sizing. Volunteer rationing soon breaks down, and old disputes heat up. As residential water-users all over the United States keep up their incessant demands—far exceeding one hundred gallons per day per household—once-humid cities and regions begin to feel the first effects of man-induced drought.[1] Irrigation had already made its move from the West to the East as local water-use exceeded local supplies. By the 1960s there were legal battles among Eastern cities competing for local waters: New York vs. Philadelphia for the Delaware River; Boston vs. upstate towns for the Mystic and Charles Rivers; Truro, Mass., vs. its neighbors on Cape Cod for underground waters.

In the spring of 1988, a vast river of air a mile thick, some sixty miles wide, ten miles above ground and moving eastward at 200 mph departed from its normal course over the Great Plains. Called the jet stream, it usually flows from both the Pacific Ocean and Gulf of Mexico, incorporating storms with their rains. But within its altered flow were ingredients of a major drought. Normal jet streams supply rainfall to the Midwest and eastern Great Plains. But the 1988 stream split. Its two branches trapped a great ridge of high pressure over the central United States. It blocked storms and rain, and brought on a devastating drought. One explanation was a "teleconnection" between the event itself and El Niño—midwinter changes in Pacific Ocean atmosphere that brings cold water welling toward the surface. This Pacific coolant affects storms which move stateside, but now helped lock dry air into place over central states.

While causes were still being investigated, DROUGHT AREAS suffered as of old. In the 1988 drought, cattle ranges dried out, corn crops withered, farms in the thirsty land turned brittle, farmers desperate. The governors of ten dried-out states lobbied for federal drought relief. It brought money, and bought time, but no rain.

In the historically dry states, a new land ethic has taken hold. Drip irrigation via tiny nozzles in the root zone began in the 1970s to replace wasteful sprays and leaky canals. Western homeowners, no longer acting as though they lived in drizzly New

England, gradually began to replace water-dependent lawn grass with desert plants. In the North Platte River Valley, Nebraska enforced "voluntary" limits to the withdrawal of scarce underground waters. From West to East the lessons flowed as the rivers diminished; water once seen as a God-given right, was now a litigious, negotiable asset. Once so defined, all the poetry in the world could hardly bring it back to its original, pristine fluidity. Meanwhile, many a dry-country diehard spurned the word "drought," preferring the temporizing term, "extended dry spell."

Yet dry spells had become more inevitable, though less predictable in the usual day-to-day sense. Climatic change seemed to be under way.

Above all else loomed the possibility of political changes in the pecking order of water-users. Western farmer/ranchers stood first at the traditional irrigation ditch gate—"first in use, first in line." But at the downstream end-of-the-line, large businesses took the lion's share: "250 customers out of 600,000 used 18 percent of L.A.'s Water," said a local headline in 1990. The big drinkers were both private and public: from Anheuser-Busch beer makers to Caltrans, the highway agency.

See also IRRIGATED AREA/LANDS, in this chapter, below.

THE FISHERY

Global view

The world's best-organized hunters are showing early signs of becoming farmers. They are the fish hunters: a worldwide array of fresh- and salt-water fisherfolk whose tools range from multi-million-dollar oceangoing factory ships to primitive spears used by Native Americans on inland whitewaters. Their numbers expand as their quarry diminishes. There were signs in the 1990s that fishing—in every sense implied by the term—is headed the way of cattle-raising. THE FISHERY is in the early stages of conversion from open range to managed "pastures," with quotas and caps on the kill and the catch.

All fisherfolk, from ocean trawlers to fly-casters, depend upon that watery resource known as THE FISHERY. This term is now used to describe all marine hunting grounds—from oceans fished under international treaties to inland lakes, ponds, and streams. This widespreading place—little known to the general public under that name—has emerged into international politics as world population continues to grow by approximately 480,000 per day, and consumes more fish foods than ever before.

To conceive of THE FISHERY on today's terms would have been unthinkable to Izaak Walton (*The Compleat Angler,* 1653). For it has become the negotiable, identifiable, real workplace of a worldwide industry, as well as the haunt of international sport fishers. While 96 percent of the world's food still comes from land, yet three-fourths of the world's surface is water. As population expands, THE FISHERY is under pressure to produce—predictably and on demand—more fish tonnage per annum. By 1987, 67 percent of America's fish diet was imported. By 1990 we were eating fifteen pounds of fish products per capita per year, an all-time high. And sports fishermen were demanding their right to ever-normal (i.e., bigger) catches, both offshore and on domestic streams and lakes.

Great changes are to come. Barely under way is a history-making conversion of the world's hunting grounds of open-ended waters into three-dimensional managed pastures and feed lots. An international treaty of 1982 and later modifications created Exclusive Economic Zones around all coastlined nations. The shift—probably under new forms of ownership and control as hunting grounds become managed pastures—will preoccupy generations yet unborn.

Historical recap

Local contentions over fishing rights have long histories: disputes between Canada and France have continued for over 225 years, lately centered on the right to fish for cod around the tiny St. Pierre and Miquelon Islands off Newfoundland. Alaskans contended with Russians on both sides of the Bering Strait long before Alaskan statehood.

In the process, such fishing grounds as the fertile Georges Bank off Cape Cod–Nova Scotia have become significant parts of the Atlantic FISHERIES. A shallow (60 meter) area of some 100 × 200 kilometers, its swirling currents a food chain rich with deepwater bounty, it was identified by European explorers as early as the 1520s. In the late 1600s European fishermen nicknamed the North Atlantic their "Herring Pond" and were actively fishing off the Atlantic coast by the 1700s. The United States and Canada, friendly along both land and sea borders, finally agreed in 1989 to split fishing rights across Georges Bank.

Modern technique: Rise of oceanic capture

The place-term FISHERY ascended into wider currency after fishing went big-scale in the 1950s. THE FISHERY'S content is evasive, its medium fluid. The quarry cross and recross national waters and time zones. Prior to radar, satellites, and computers, no human mind could grasp THE FISHERY'S full flux. Information about it—some closely guarded—has gradually moved into international data banks. New satellite monitors and computer analysis made it possible to identify what once were called fishing grounds, now bigger and more complex. Satellite images from the Coastal Zone Color Scanner, bought by oceanic fishermen, moved via fax machines to boats in 1990, and were soon to be beamed directly to scanners on shipboard, enabling fishermen to spot quarry more quickly.

Huge fishing fleets from Japan, Eastern Europe, Russia, the United States, et al., began in the 1950s their own months-long "world cruises" to fish and to overfish international waters far from home port. Whales were among the largest quarry to be caught, cut up, and frozen in floating fish-processing factories. One-time "moon-watchers"—shrimp fishermen on the U.S. Gulf Coast—were being overshadowed by high-tech others. Japan and Russia were taking well over a million tons a year of Pacific yellowfin sole and Alaska pollack. Fishing-buddy friends in White House and Congress passed laws giving fishers tax-credits to equip themselves for oceanic competition.

Thus by the 1980s, even once-local fishermen from U.S. ports—twice as many as in the 1970s—were at sea using million-dollar multipurpose boats, sonar-radar for moving in on the catch and kill, electric-shock harpoons for giant tuna, and forty-miles-around seine nets, plus endlessly aggressive tactics and techniques. Boat-builders constantly rebuilt fishing craft to fit new government rules which were aimed—often in futility—at limiting the catches. The resulting overkill and excess competition in the Atlantic FISHERY produced a glut of boats by 1990.

The old terms, fishing grounds, or spot, or hole, with a nice, local-sounding ring to them, no longer covered those vast seas making up THE FISHERY. The latter—once a statistical term used chiefly by federal agencies—had gone global. The open sea was becoming a realm of defined (though often-violated) twelve-mile and other national limits. The realm was codified into a Common Fisheries Policy for the European Community in 1983, which set up three-, five-, fifteen-, and two-hundred-mile exclusive economic zones—spelled out in international treaties to which the United States subscribed. U.S. coastal waters for two hundred miles out became an American monopoly, so that competition in international waters grew more intense.

Resulting shortages

Looming over all FISHERIES, domestic and global, is the possibility of being fished out. The Northeast Atlantic FISHERY, on which Europe depended, "declined by one-half between 1958 and 1968"—a similar decline having affected the California sardine FISHERY.[1] The great anchovy FISHERY off Peru failed in 1972. In the 1970s, the king crab FISHERY of Alaska boomed, with a peak catch of 195 million crabs in 1980, but in 1985 collapsed to 10 million. By 1921 the Upper Great Lakes FISHERIES had been invaded by sea lamprey and "were about to undergo one of the most drastic resource declines in history."[2]

Pollution and oil drilling

Meanwhile, oil spills (Valdez, Alaska, being the most notable) damaged portions of THE FISHERY, which was under many local pressures from urban and pesticide pollution, and ocean-bottom dumping. In 1990 the Bush administration canceled many offshore leases and banned drilling off 99 percent of the California and Florida coasts until the year 2000. But the Iraq-Kuwait crisis that year revived demands from U.S. oil promoters to relax offshore drilling rules. In a typical local compromise of another sort, rich fishing grounds twenty-six miles off San Francisco were approved for dumping Oakland harbor soils in 1988.[3]

Fish-farming

Fish farms—sometimes called "capture fisheries"— expanded along many coasts, especially the PACIFIC RIM where fish-raising in multilayered organic farms on tidal ponds is an ancient skill in Asian waters. By the 1980s about half the world's eels were being "produced" rather than caught. Responding to higher prices, catfish farms boomed in the 1980s across the U.S. South, as did crawfish farms in Louisiana, and a freshwater shrimp farm in Puerto Rico, shipping huge specimens to East Coast restaurants. In 1991, Americans' appetite for shrimp had contributed heavily to the U.S. trade deficit in seafood; we imported $1.7 billion worth of shrimp in 1980.[4] A whiskey distillery in Kentucky in 1990 experimented unsuccessfully with a two-acre enclosed fish farm to raise hybrid striped bass (in short supply within the Atlantic FISHERY) and tilapia, all fed on by-product distiller's mash. Man-made bait, rather than ocean-captured bait, was finding a larger niche in some markets. In response to widespread pressures to "save the environment" in the 1990s, recurring schemes for high-tech FISHERIES emerged in the concept of aquatic "energy farms." The intent was to transform urban sewage into a diet supplement for offshore farms producing huge tonnages of seaweed from which was (said to be) extracted uranium, methane, fertilizer, and hydrogen gas as fuel.[5] But, despite its potential, most urban sewage still carried toxic wastes downstream toward THE FISHERY, unwanted, unregulated, and unsafe.

Restoring migratory routes

No longer confined to deep water, THE FISHERY was increasingly identified as extending inland to upstream white waters where seagoing salmon once spawned. The lethal combination of midstream dams and aggressive logging of upstream watersheds combined to damage spawning grounds and cut into the valuable sockeye salmon catch in and off the Pacific Northwest. Under pressure from the salmon fishers, fish ladders were installed on some Pacific Coast river dams to boost salmon upstream to their old spawning grounds. But as 231 hydropower dams were licensed and installed on 105 rivers in twenty-four states, salmon migration declined. For the first time ever, it was proposed in 1990 by environmentalists and two U.S. federal agencies that the sixty-five-year-old Giles Canyon dam in Olympic National Park, Wash., be destroyed to restore the historic spawning run of Chinook salmon up the Elwha River.[6] The raising of trout, and of salmon, the latter to be released for return to oceanic feeding grounds, became a multi-million-dollar enterprise, some of it paid for by sports fishermen's fees.

The mobile fish, not being party to international treaties, make enforcement of regional quotas difficult. Some fish, like the bluefin tuna, have a seasonal range of six thousand miles; eels breed in the Sargasso Sea and their larvae ride the Gulf Stream to European waters for a six-year life (unless caught) before returning to spawn. Whales migrate between continents; salmon swim from ocean deeps, where they are classified as commodities, to shallow spawning beds many miles up coastal streams lined by fisherfolk who believe the fish to be tribal entitlements.

From offshore to upstream

Not only the fish, but THE FISHERY, as a place, has migrated inland. Here it is usually referred to as the sports FISHERY and is leased or owned in many guises: trout streams, fishing streams, spawning grounds/waters, or as tribal streams/waters. The percentage of fish raised in upstream hatcheries is on the rise—perhaps 80 to 90 percent of the Columbia River's "run" of salmon and steelhead trout in 1990.

The rise in recreational fishing has put greater pressure on the sports FISHERY. In 1988 Idaho began emphasizing "urban fisheries" to give city anglers closer-to-home fishing. The South Platte River in Colorado now has in Cheesman Canyon a "catch-and-release fishery that supports up to 700 pounds per surface acre of wild trout."[7]

Meanwhile, other local FISHERIES proliferate. Outside Atlanta, the once-muddy Chattahoochee River is now dammed, with cold bottom-waters released on schedule. This has cooled a twenty-mile reach of the river, creating a well-stocked suburban trout FISHERY.

Prospect

What is emerging is something new called "the unified world ocean habitat." As all FISHERIES—inland as well as oceanic—came to be identified, counted, and monitored, state fishing agencies busily kept track of "the catch" by sportsmen as well as by commercial fishers. Competition increased as the ranks of anglers expanded, and many a state agency found itself raising millions of fingerlings to keep up with sportsmen's pressure. In Alaska, the Pacific Northwest, and Wisconsin, local fishermen disputed with native Indians over the latter's treaty-endowed right to live and to fish—catching all they chose—from fishing subsistence villages. The assumed entitlement of fishermen to a dependable level of fishable water below federal dams has fueled many a debate over water levels, both above and below the dams.

Meanwhile, the goal of "maximum sustainable yield" has moved from timbering and farming to oceanic fishing. "Full ownership of property rights in the ocean pastures"—possibly exercised via a public corporation—was proposed in a seminal study in 1988.[8] Individual fishing quotas were scheduled to be in place in 1995 among some 7,500 long-line fishermen for halibut and black cod in the North Pacific. Once catch limits are set, quota-holders will know how much they can catch in 1995. Shares in such quotas were expected to be bought and sold in a new quotas market.[9]

All the foregoing represent faltering steps toward the conversion of still-ruthless hunting-at-sea and on land to the practice of 3-D husbandry on the moving waters. It's all come a long way from unclaimed waters, open seasons, and the still but-not-quite open seas.

See also DRAWDOWN, in this chapter, above.

GREENBELT

Variant: buffer zone.

Often short-lived, the U.S.-model GREENBELT was created with high hopes but inadequate legal protection. Its forebears lie in an English town-planning law of 1946 creating new "garden" or "greenbelt" cities. Each had formal, legally designated wide belts of forest, farms, swamps, and other OPEN SITES around new towns. GREENBELT'S deeper historical roots lie in the King's Forests which gradually became England's green lungs, open to some public pleasuring.

The British new-towns pioneer Frederic J. Osborn traces the GREENBELT concept back to roughly the thirteenth century B.C. "when The Lord said to Moses [that] the pasture lands of the cities . . . shall reach outward a thousand cubits all around."[1] It took a great leap forward in Britain with Ebenezer Howard's 1898 book, *Tomorrow: A Peaceful Path to Real Reform*—the holy text of the new-town-with-GREENBELT movement.

When GREENBELT occurs in the United States, it is usually bought or set aside in bits and pieces by public action to form a zone protected for its natural beauty, for food-raising, outdoor recreation, watershed value, or underground-water recharge, for air cleansing, urban forestry, or other city-serving purposes. GREENBELTS have two complementary purposes: preservation of nearby land in a natural, garden-like, or agrarian state; and the shaping and limitation of urban spread. True preservation by purchase happens infrequently. When it does occur, most purchases are carried out with great fanfare, high cost, and outcry from original landowners.

Following the British example, Canada created in 1970 the continent's largest GREENBELT, 64 square miles of it, 7.5 miles from the center of Ottawa, the national CAPITAL. Designed by famed French landscape architect Jacques Gréber as an element in the capital plan, its purpose was to "provide a visual and physical edge to the outward growth of the capital area."[2] The United States has nothing remotely resembling Ottawa's example.

Some U.S. GREENBELTS incorporate historic farms where some city kids may glimpse where food and Great-Granpa came from. However, a GREENBELT in the United States tends to last only so long as political pressures to sell it are muted, and where extraordinary sentiment and law build up locally to protect it.

"Belting" anything beyond a waistline or a machine is complicated business. It can unite against it landowners who believe their private good to be threatened by the public good they scorn. Once in place, the belt's purposes—including "green lungs" for the city—need constant protection from local capital's pressure for sites easy to develop.

The few U.S. public examples were products of the New Deal's attempts in the 1930s to solve unemployment by building new towns, surrounded by GREENBELTS. Best-known was Greenbelt, Md., followed by Green Hills, near Cincinnati, Ohio, and Greendale, Wis. After 1949, Greenbelt, Md., was sold to a veterans' cooperative, and most of its GREENBELT "preserved by gifts to federal and local public bodies for park and open space."[3]

Citizens of Boulder, Colo., continue to support a GREENBELT on the Rocky Mountains' Front Range overlooking their city—4,500 acres and growing in the 1990s. Meanwhile, speculators watch GREENBELT'S capacity to excite rising land values around it, and organize spasmodic drives to buy it up.

A major North American case of greenbelting is the Fraser River Valley's agricultural preserve to protect Vancouver's close-at-hand food sources. Other GREENBELTS enclose, in part, such privately developed new towns as Reston, Va.; Columbia, Md.; Valencia and Irvine, Cal.; as well as many large military bases—the latter's future uncertain.

A typical complex mix of public and private interest is EPCOT Center, Orlando, Fla., where Walt Disney Productions in 1962 assembled forty-two square miles for Disney World and associated developments. The firm "set aside no less than 35 square miles to serve as an undeveloped buffer [i.e., GREENBELT] to keep tacky outside developers away from the pristine Disney gates. But that was before Disney's earnings went into a tailspin"—and corporate raiders tried to take over the firm. To forestall takeover, in 1984 Disney joined Arvida Corporation in planning a "'comprehensive community' containing homes, resort hotels and high-tech industrial parks."[4]

Cast somewhat more in the British model is the so-called "Horse Farm GREENBELT," around Lexington, Ky., a legally defined and planned "urban county," with tight prohibitions against subdividing its famed Thoroughbred horse farms, deemed essential to the city's economy and way of life.[5]

In the late twentieth century, miniature so-called GREENBELTS and greenways spread widely in and around North American–type suburbs, chiefly to protect enclaves of middle- to upper-income residents seeking privacy and security. The terms descended into loose usage in real estate advertising.

See also SECURITY, in Chapter 3: Perks.

HAZARDOUS WASTE DUMP

DUMPS appear in many guises and disguises, as dump sites or disposal sites. But certain DUMPS are among the most politically dynamic places of the late twentieth century. They are either proposed for, or already committed to, the dumping, disposal, dispersal, or burial of waste materials judged—sometimes long after the fact, and occasionally without trial—to be hazardous to local, or to downstream, or to downwind life.

HAZARDOUS WASTE DUMP
A handy location for uranium mining waste—hurriedly dumped during World War II—turned out to be uncomfortably close to the town of Durango, Colo. It became one of the hundreds of toxic dumps de-toxed by removal during environmental clean-up beginning in the 1970s.

Such DUMPS are bitterly fought by most communities, regardless of the degree of danger (itself hotly disputed). By 1985, governors from "waste-producing states" were in a political struggle with governors of "waste-absorbing states" over proposed interstate shipments and proposed DUMPS. What was earlier said to be a technical debate became a political struggle.

Not until the disclosure of hundreds of once-concealed hazardous waste sites during the ecological scares of the 1970s was the 1976 Resources Conservation and Recovery Act passed by Congress. By 1979, from thirty-two thousand to fifty-one thousand "potentially dangerous sites" had been identified for the Environmental Protection Agency by Fred C. Hart Associates.[1] In spite of European examples, Congress refused to encourage in-plant disposal of hazardous wastes by burning and/or chemical-reaction disposal. Citizen protests were heightened after 1976 with the disclosure of Love Canal's pollution of ground surface and underground near Niagara Falls, N.Y., during the 1940s and early fifties.

Until the 1960s, toxic or HAZARDOUS WASTE DUMPS occupied—for the most part unknown to the general public—such locations as back lots, secluded valleys, remote locations, and other hideaways. Only after spectacular spills or kill-offs did aroused neighbors and legislators give the HAZARDOUS WASTE DUMP a new notoriety. When, however, it came time to haul away the waste to a toxic waste disposal facility, all hell broke loose along the proposed route, and at the proposed facility site. Unlike Germany, which discovered hundreds of moneymaking uses for such waste, the United States for some years chose to specialize in denigration rather than economic disposal and reuse.

HOLDING PATTERN

A casualty of the air traffic controllers' strike of 1981, when President Reagan dismissed 11,400 strikers, HOLDING PATTERN once consisted of giant circular zones designated in medium altitudes some fifteen minutes' flying time away from major airports. In bad weather, rush hours, or holiday jams, incoming planes could be directed to stack up at safe intervals, flying lazy circles while awaiting a controller's radioed OK for landing.

But the shortage of controllers following the Reagan dismissal changed all that. The Federal Aviation Agency stopped using AIRSPACE for holding, with its attendant risk of midair collisions. (In 1987, sixteen flights were scheduled to land in Denver simultaneously, at 8:10 A.M.) Instead FAA rules stashed planes at their last landing-place, sometimes waiting on runways in stately mile-long queues, until their destination airports were cleared for arrivals.

Such clearances were slow in coming, due to congestion on runways unfit for multiple landings in foul weather. This, in turn, led to airlines' pressuring local airports to expand runways, or to their beefing up service to airports with unused runway capacity. A consequence in the 1990s: a rash of new or continued airport expansion (Raleigh-Durham, Charlotte, Cincinnati, Louisville, Pittsburgh) and new airport construction (Denver); while Chicago considered a plan, but found it politically unwise, to thrust a new airport into Lake Calumet.

Formerly, use of the term HOLDING PATTERN was limited to a formally prescribed routine to be following by aircraft preparing for scheduled landing at a regulated airport. But it was generified in the 1980s to describe any situation in which a person or firm waits in an invisible HOLDING PATTERN to be notified of a contract award or other business decision. By extension, "my whole life has shifted into a HOLDING PATTERN."

See also FLIGHT PATH, in Chapter 8: Opportunity Sites.

IRRIGATED AREA/ LANDS

Until water scarcities moved into the Eastern United States in the 1960s, it was easy for Easterners—if they thought about it at all—to conclude that irrigation, as practiced Out West, was a form of creative vandalism whereby underground waters that can never be replenished are sucked out by irrigators who can never be satisfied, and distributed to farmers and ranchers convinced that they invented water in the first place. Now that water shortages have gone national, now that water flows in international commerce, now that irrigation has become a continental necessity, the tunes they are a-changing. Eastern waters grow scarce and expensive under the twin pressures of population and pollution. Easterners' understanding of Western water law and custom has barely begun to surface.

Wherever found, water itself is a high-energy system; a conductor, transporter, organizer. Compared to wind and sun energy, water's potential energy has already been naturally concentrated, and is available for immediate human uses. It has great "power density," which is a measure of its usability in the landscape.[1] Even where it comes from the sky in hundreds of thunderstorms per hour (forty-plus inches annually over much of the Eastern United States), there is still too little to satisfy everybody. These include city folk in 1991 averaging 168 gallons consumed per person per day[2] (350 gppd in Las Vegas),[3] as well as the farmer spreading thousands of acre-feet on crops of scarce lettuce, or on irrigated cotton to produce a surplus commodity.

Water has become not so much a gift of nature as a politicized commodity. By 1989, one-quarter of all Western farm acreage was IRRIGATED AREA. To millions of Western irrigators, water-as-usual with federal subsidy is an Entitlement. They confront the growing pollution of underground aquifers from "chemigators" who mix chemicals with their irrigation water; and accept, often reluctantly, the onset of new ground-water-control districts.

Irrigation techniques and water politics of the West are fast moving East. Congress extended the Western-oriented Water Facilities Act in 1954 to all states. Descendants of Western farmers who plowed up the Great Plains in the foolish belief that "rain follows the plow," now politick to dominate local flood-control dams and negotiable waters—in the face of growing demands from Western water-using cities. It was also believed that rain would follow the arrival of trains' smoke for moisture to collect upon; or the telegraph's induction of electricity into the air; or military battles because of gunshot and explosives. "There were even those who came right out and said what everyone else more or less hinted at: rain will follow settlement for no reason other than the presence of good people with a destiny to fulfill."[4] Water is becoming the manhandled fluid most widely and litigiously dispersed long-distance via canals, pipelines, and lawsuits.

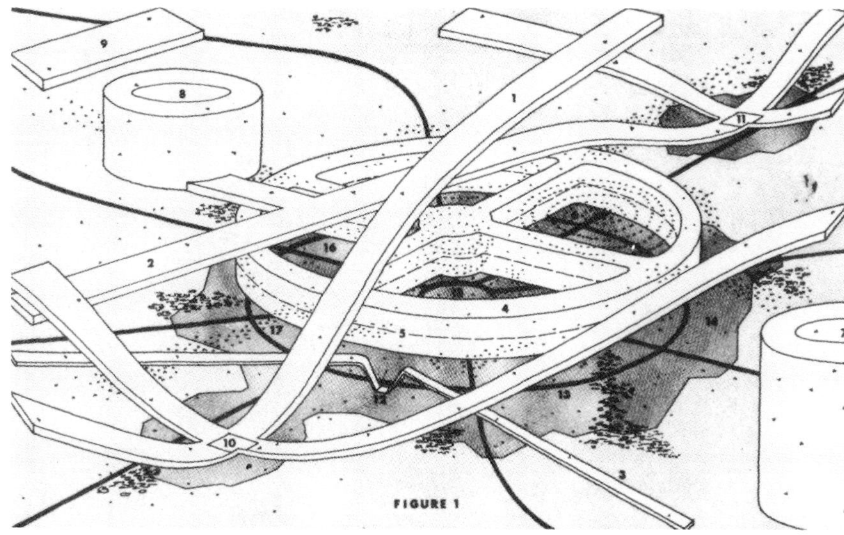

HOLDING PATTERN
Putting planes "on hold" required a global electronic network—and a careful carving-up of cubic miles of AIRSPACE, including far-out cylinders of unencumbered space, where planes, waiting to land, once circled in a HOLDING PATTERN. They were stacked up—for hours, so it seemed to impatient passengers. This ended in 1981, when planes were required to wait on the ground at their last stop until the AIRSPACE over their destination was clear. Map by Melville C. Branch, from "Airspace Jurisdiction, Environment, and Planning: Real Property and Transportation Utilization and Control," by Alvin G. Greenwald and Melville C. Branch, in *The Natural Resources Lawyer,* summer 1973.

IRRIGATED AREAS
Flooded front yards in Scottsdale, Ariz., are located in an old orange grove. The entire yard is surrounded by an earth berm, which holds water in place until it soaks into ground. The homeowner is entitled to a contracted share of water from the nearby irrigation ditch. Unneighborly aggression consists of running down the mother ditch and kicking your neighbors' watergates tight-shut to make sure they don't get "your" water.

Water has long been the man-managed liquid with the widest geographic high-volume spread. (Petroleum's tonnage is far less.) Historically its trans-mountain shipment and its manipulated flow between watersheds was a commonplace—to Egyptians, Syrians, Moslems, Romans, and to Spaniards and ancient Americans in the American Southwest.

The power-driven pump holds the key to this west-to-east migration. The first pumphouse on the landscape signals widespread change in land uses. Modern diesel or gas-driven pumps can suck out of the underground millions of gallons daily, thus "mining" waters that took millions of years to deeply accumulate. Pumps are tireless, mobile, quickly replaced. When powered by wind to irrigate small farms or fill stockers' tanks, they cost little. Once, run by giant windmills, they drained Holland's polders, England's mines and fens, and Italy's miasmic marshes. They owe much to their predecessors three thousand years ago in Egypt, where fieldworkers used footpedals on waterwheels that lifted Nile River water onto grain fields to help fill the breadbasket of the Egyptian and later the Roman Empire. Multiply that Egyptian footpower by millions of pumps and you see the means whereby the world's dry landscapes are being transformed, and wet landscapes dried out.

Not only landscapes but microclimates. Across many parts of the irrigated West, haze, morning fogs, and high pollen counts have replaced the once-crystalline atmosphere, thanks to moisture evaporated and pollen wafted from fields once semi-desert. Along with crops come weeds, so that no longer is moving to Phoenix or Tucson a guarantee against hay fever.

Because of pumps' efficiency, the long-distance transport of water occurs everywhere: from the Catskill Mountains down to New York City, across the Continental Divide to Denver, and hemstitched across Southern California. Flatlands get regraded under laser-beam control to move water by gravity's slow motivation. Only when the price of gasoline jumped after the Arab oil embargo of 1973/74 did irrigators suddenly confront new limits to how deeply they could dig and delve and pump with abandon. It marked the end of the unlimited spread of great green irrigated circles across the Great Plains.

The history of North American irrigation is filled with predictions, first of unlimited, inexhaustible supplies, and later of imminent disaster as overdrawing a limited resource continued. In 1988 Dan Goodgame reported that "high-stakes hydrobattles are brewing throughout the West as it runs out of new water sources." Over-pumping was lowering local portions of the Ogalala Aquifer (from Texas to Nebraska), "driving entire counties out of irrigated agriculture."[5]

The peak year for U.S. irrigation's expansion seems to have been 1989. Speculators' syndicates had thrown money from all over the nation into Nebraska CPI (center-pivot irrigation) deals in the 1970s. But "landscape irrigation"—of lawns, golf courses, resorts, etc.—fell off in 1993 some 30 percent, due to a drop in housing construction, among other causes; while farm irrigation stopped expanding when farm prices fell after '89. But the growing scarcities brought an overall shift from gravity flow-and-spread to the more parsimonious sprinklers and drip irrigation—an increase by some 2.5 million acres in 1992–93.[6]

The right to water from one's own land is hotly contested. This is especially so in the U.S. Southwest where ancient Spanish water law (first come, first served, also called first-in-time, first-in-use) is being modified by Eastern practices based on English common law that says you can't deprive your downstream neighbor of his rightful share. "Rightful share" is endlessly subject to debate.

It was evident by 1990 that a major shift in water policy and water rights was being impelled as "cities and farms are beginning to compete for available water; when supplies tighten, farmers typically lose out."[7] The Bureau of Reclamation has already begun brokering among competing claimants for water. It announced January 1, 1988, that it would become "the facilitator for water marketing proposals between willing buyers and sellers."[8] Even more

IRRIGATED AREA
Cheap, federally subsidized water keeps western cotton- and rice-farmers profitably raising price-supported crops. Cheap water also has made possible the greening of some eighty golf courses around Palm Springs, Cal., including the lush Marriott Desert Springs resort (below, 1988). By 1993, a new "Desert Marketplace" shopping center was being built (right) across Country Club Drive. (See also WATER RANCH.) Photo by John Barr for *Time* magazine; courtesy of Gamma Liaison Network, New York, N.Y.

drastic was the Omnibus Water Act of 1992 which explicitly encouraged farmers to sell their water to cities. The act "would reserve large amounts of water to repair environmental damage in California."[9]

Looming over these battles was what veteran reporter Gladwin Hill called a "greed machine" consisting of "the Iron Triangle," its elements sometimes in conflict, sometimes in concert. On one side was the Federal Bureau of Reclamation (generally prodevelopment); next, Congress on the lookout for pork-barrel projects; and finally the water-drawing beneficiaries; all together capable of swinging more big-project money for years to come.[10] Thus would continue the long downhill run of Western waters into the pockets of pioneers' descendants. Now they could pocket much the same sort of money by selling cheap water dear to thirsty cities. Taking a more modest tack, some Texas farmers began shifting to select dry-land crops, a trend noted in 1984.[11]

Beyond these regional struggles loom broader and not-so-level playing fields. There's the longstanding "Trans-Texas" plan to dam and canalize all Texas rivers to flow westward to water Texas deserts—reducing flow into the Mississippi and the international Rio Grande. There are later-generation "Reber Plans" to reorganize California freshwater flows. There's the perennial proposal for U.S. irrigators to reach northward via a "Water-and-Power Alliance" into the Columbia River's Canadian sources; and for Great Lakes states on the U.S. side to withdraw more Canadian water than now permitted under treaty. There's the tug-of-war between dry Spain of the south and wet Spain in the north over a twenty-year plan to redistribute water. In some ways, the U.S. West illustrates the much-disputed "hydraulic society" theory—from Egyptian and Chinese history—advanced by Karl A. Wittfogel, that a society based upon irrigation agriculture becomes dominated by measurement, rules, bureaucracy, and centralized government controls of a scarce resource.[12] "The luxury of our youth was taking water for granted. We are no longer young."[13] The tug-of-war continues.

See also WATER RANCH, in this chapter, below.

THE KUDZU

At first it was "getting out of hand." Then it was "covering whole woods." Eventually, it was accused of the more serious regional offense of "eating Southern real estate." By this time the voracious, rampageous legume known as kudzu vine has long since worn out its welcome as an exotic porch-vine plant. Within a century's aggressive residence in the Southern United States, it was recognized as having expanded far beyond its biological origins to become a geographic phenomenon, a generic place known as THE KUDZU.

With spectacular festoons, rendering them almost inaccessible, it had covered perhaps a million acres of Southern lands and forests. To get lost "out in THE KUDZU" was to be lost indeed. For many landowners it was still great cattle feed, but for others an uncontrollable, pestiferous, geographic presence.

Derived from one of Japan's most pervasive wild plants, kudzu fiber was made into cloth in the third and fourth century, and showed up on records of around A.D. 600 in a collection of Japanese poems, the *Manyoshu*. The plant is said to have originated in China millennia earlier, and migrated to Japan. By A.D. 1200 Japanese farmers were extracting the starch and preparing its chunky powder for home cooking and as a natural medicine. In the thirteenth century its tough stems were supplying fiber for cloth, and by 1600 a shop making kudzu powder opened in Nara Prefecture. Today it is widely used in Zen Temple cookery, and in traditional remedies and medicines. Its long stems and processed fibers go into making clothing, fabrics, paper, baskets, and other woven objects. By the 1970s it was widely cultivated in Korea. Its roots, used in the Orient, sometimes grow to four hundred pounds, the size of an overweight Japanese wrestler.[1]

Its first move to establish itself in North American soil was as a specimen exotic plant brought from Japan and shown at the Japanese Pavilion at the Philadelphia Centennial Exhibition, 1876, and the New Orleans Exposition, 1883. The noted plant explorer Henry Pratt Fairchild, after planting it around his Washington, D.C., home, found—and said in print in 1938—that it was "an awful tangled nuisance." But his warning came too late.

From 1902 in Chipley, Ga., C. E. Pleas became a plant propagandist, devoting thirty-five acres to kudzu pasture, and selling fodder by the wagonload. (Eventually, kudzu proved intractable for large-scale machine harvesting.) Pleas's pamphlet, *Kudzu—Coming Forage of the South,* was widely circulated. After his death in 1954, Pleas was memorialized by a bronze marker "Kudzu Was Developed Here." Little did Pleas foresee that his adopted broken-field runners would leap the garden wall and take all outdoors as their milieu.

Kudzu's geographic spread had been slowed by its early reputation for getting out of control. But New Deal farm agencies, especially the new Soil Conservation Service (SCS), after 1935 began promoting it for cattle feed and erosion control, growing it on thousands of demonstration plots. The Central of Georgia Railroad planted it alongside its tracks. (During the 1930s, the author of these commentaries was a teenage accomplice in the planting of kudzu on a family farm in middle Georgia. It did, in fact, flourish, cover the deep gullies in the raw clay soil, feed the cattle—and spread far and wide.) Slowly the erosion-control argument—together with SCS payments running as high as $8 an acre—carried the day. By 1940 SCS nurseries had produced over seventy-three million seedlings, some started with seeds imported from Japan. Plants from Japan were imported for use around Tennessee Valley Authority dams.

Plant ecologists point out that, unlike trees, the kudzu vine "directs its energies into stem elongation and leaf development instead of woody production." It "optimizes available sunlight for photosynthesis" by out-climbing and thus shading-out its competitors, even unto the top of one-hundred-foot-tall trees.[2]

Upward, as well as sideways, mobility is what kudzu, like most vines, requires. Some vines "send out exploratory shoots that trace circles in the air as they search sightlessly for something to climb on."[3]

The fervor spread backwards to Japan via a praiseful booklet, *Kudzu: The Plant to Save Japan.* Channing Cope, another plant propagandist writing in the *Atlanta Consti-*

tution, joined the crowd, forming the Kudzu Club of America and set a goal of one million acres for Georgia alone. Cope asserted that China's refusal to fully use its own native plant had caused the Gobi Desert. In a poverty-stricken region where all outdoors served as a dumping ground, THE KUDZU also served as useful camouflage for DUMPS. *Readers Digest* magazine in 1945 called it "Another Agricultural Miracle." Its extent had been mapped and appeared distinctly on high-level airviews.

But THE KUDZU as a generic place attracted Southern hyperbole, such as a claim that it "can cover an area the size of Rhode Island overnight without belching."[4] A Southern folklorist asserted it swallowed up a bird dog standing on point. It was known as "the mile-a-minute" plant. To plant it, "you drop it and run." But it also became the butt of national jokes on television, and turned into a regional put-down term for export, along with "Dog Patch."

By this time the miracle had expanded to cover some 580,000 acres. "Attitudes toward it approached reverence"; but "as kudzu overleaped its bounds and started marching through Dixie, some people began to panic."[5] By 1954 the U.S. Department of Agriculture removed kudzu from its list of acceptable cover crops. In 1976, on the hundredth anniversary of kudzu in America, a headline noted, "The South is fighting another war . . . and losing once again." Within a generation, kudzu acreage figures disappeared from popular media. In common with DEPRESSED AREA, THE KUDZU found itself virtually boycotted-by-neglect in the press.

As a transplanted foreigner, kudzu's first offense had been to disregard property

THE KUDZU

Since its introduction from Asia in the 1930s as a cheap forage crop, THE KUDZU segued in Southern talk from its status as a plant to a geographic place: "out in THE KUDZU" became a way to describe THE BOONDOCKS, 'way out, the outback, or other inaccessible locales. KUDZU specializes in aggressive takeover of ABANDONED AREAS and other untended tracts. Its coverage expanded to over five million acres by the 1990s. Photo by Byron Crawford; copyright 1988, *The Courier-Journal.* Reprinted with permission.

lines. Further, it climbed fences and grew a foot or more daily, up to one hundred feet a season. Cartoonists enjoyed playing with the same visual theme showing a rampageous plant reaching around the house. Shouts the anxious onlooker "Here it comes again." It also had become a fire hazard. The first killing frost could convert THE KUDZU "into a flash fuel, a fuse to nearby fields and timber. Eradication necessitates burning, plowing, and, salting the soil with herbicides."[6]

(Another pyrophyte, eucalyptus, was imported wholesale to California in 1856 as an ornamental curiosity. The eucalyptus-planting speculative bubble "began to burst in 1910 when it was discovered that the trees transplanted to California were virtually worthless [as timber], and that they constituted a serious fire hazard."[7] They still do. A freak frost in 1972 killed over a million eucalyptus from the root collar up, and only good fortune prevented another California holocaust.)

Gradual disenchantment set in as THE KUDZU'S nuisance value was recognized, especially by Northern tourists flabbergasted by natural growth of such magnitude. Continued research shows THE KUDZU has no competition for certain forms of erosion control. Further, say the authors of a popular book on kudzu, its "greatest future in America lies in the use of the root as a source of the remarkable kudzu powder and the medicinal kudzu root."[8]

But both its assets and liabilities THE KUDZU kept to the South. Its resistance to freezing fades as it passes "The Kudzu Line" somewhere around the thirty-eighth parallel through Missouri–Kentucky–southern Indiana. Unlike the former Confederate Army, THE KUDZU poses no threat to the North.

LULU

As an acronym for "locally unwanted land uses," LULU came into vogue in the 1980s, chiefly among planners, city officials, and attendees at zoning hearings. Its usage expanded as a by-product of legal environmental-impact hearings, and the rise of neighborhood protective (or watch) associations, coupled with suspicion of Big Government projects. Millions of commuters, on their daily runs, came face-to-face with sights, smells, and LULUS they'd never before encountered.

Aside from slaughterhouse (a nineteenth-century term still flourished chiefly at public hearings) and oil refinery, there are few universally agreed-upon LULUS. A LULU by definition is wholly derived from local conditions, attitudes, prejudices. The prevalence of LULUS goes hand-in-hand with migration: newcomers move into hitherto "pristine" territory and are outraged to find, already there, or proposed, some operation, activity, land-use, or enterprise they find undesirable. Anything proposed under the name "facility"—a favored circumlocution—is immediately suspect, headed for LULU-dom. Any suggestion that this would involve assembling, processing, recycling, modification, materials-handling, or heavy traffic could arouse opposition.

A partial list of locally branded undesirables assembled from local newspapers, includes ABANDONED AREA, auto sales lot, baseball field/park, city hall, crematorium, detention center, disposal area, DUMP, dumping ground, halfway house, industrial district, municipal center, police headquarters, prison, projects (especially as in public housing), sports and recreation center, waste processing facility, and workhouse (likely to be installed on a re-education campus). If all were outlawed, there would of course be no city left.

As the LULU list expanded, more than one city official would itch to find, somewhere out there, a jurisdiction that would welcome slaughterhouses or their equivalents. This led to a nationwide search for less-resistant communities, eager for another state's trash, or a new prison for its jobs potential—a backwater Siberia into which could be loaded the LULUS frozen out of more prideful purlieus. The search continues.

See also EARSHOT and IMPACT AREA, in Chapter 4: Ephemera; DOWNWIND, in this chapter, above.

NATIONAL BORDER

You can stand, or imagine yourself, at midnight on the Rio Grande, in company with one of the more frustrated U.S. Immigration Service employees you are likely to meet. From long exposure, he knows that the splashings in shallow distant waters are those of yet another wetback immigrant from Mexico, yet another illegal alien, yet another statistic in the future school population, yet another application for the nasty, dirty, stoop-labor, or domestic-labor jobs that Anglo-Americans turn down, yet another voter in some distant future for the first Spanish-surname U.S. President from a Southwestern American state.

One does not need the Bureau of Census's reminder that the present rate of immigration, legal and otherwise, from Latin and Central America, the Caribbean, and the Pacific, will substantially change the racial-ethnic-color composition of the United States. By the year 2020, or sooner, ethnic enclaves will have expanded to become neighborhoods; and towns and cities will have changed color, language, and voting habits. College courses in The History of Western (i.e., European) Civilization will have been discarded or drastically revised.

To reach their DESTINATIONS—whether it be the United States or any of hundreds of other countries—immigrants must cross a basically unstable line on a map. There being no such thing as a permanent balance of power among nations, there can be no such thing as a permanent NATIONAL BORDER or immigration policy. All are here today and gone in the long tomorrow. Footprints in the sands of time, each marks today's balance between a nation's ambitions and its power. In many places the NATIONAL BORDER is innocuous, invisible, or unenforceable. It may be passable, impassable, tangible, or impossible, depending on the temper of the times. In parts of Europe, protected borders are not merely national but regional, ethnic, religious, and/or dangerous mixtures. Some borders are still accompanied by that old-fashioned swathe of depopulated countryside called a "march," readily traversed by troops stationed a day's march from a porous border. Charlemagne protected his land frontier "by the creation of marches."[1]

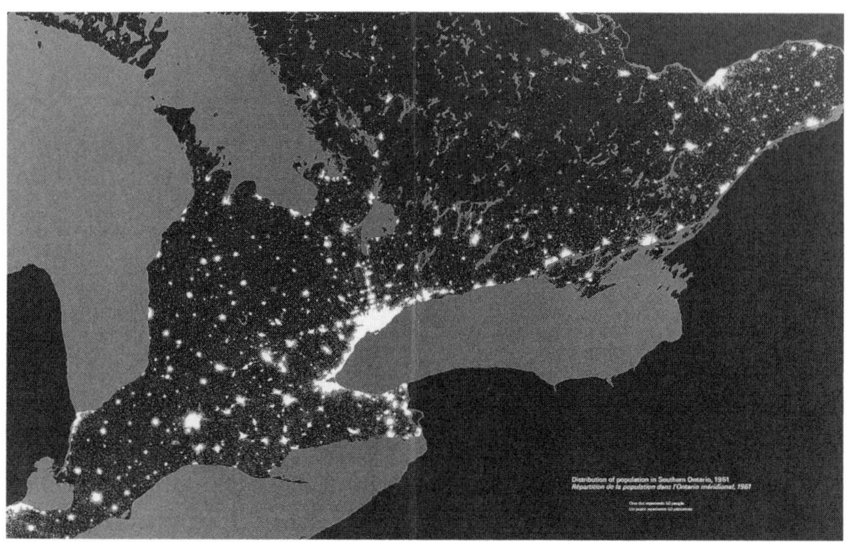

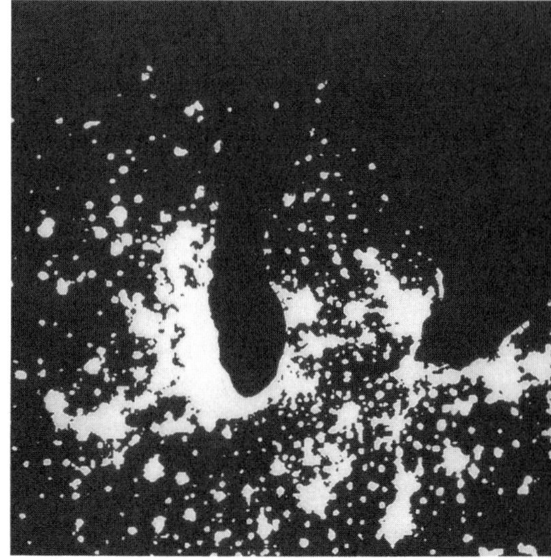

When border incidents heat up, and bordering nations react with hostility, and warlords and arms peddlers sniff the air, suddenly NATIONAL BORDERS appear on every media map. School children relearn map-reading, a skill normally abandoned between wars. Invisible borders become invested with national honor. Geopoliticians resurface with learned arguments in behalf of Our (as distinct from Their) national boundary. All agree "that the nationalism of other countries is absurd, but the nationalism of one's own country is noble and splendid."[2] Newspapers and TV move maps out of the morgue and into page one / prime time, and computermappers stay busy.

Nations believed to be peace-loving react vigorously to border-crossings or distant threats. The United States became adept, particularly (but not exclusively) during the Ronald Reagan administration, at finding reasons to cross other NATIONAL BORDERS by varied means. Over the decades, presidential moves included incursions, fly-overs, landings, penetrations, interferences, rescue missions, intimidations, bombing raids, and military ventures—as in Cuba's Bay of Pigs, Korea, Vietnam, Grenada, Nicaragua, El Salvador, Panama, Libya, and Iran.

But on January 1, 1993, BORDERS turned benign among the twelve European Economic Community nations, when customs and passport checks were no longer required. It was not, as the Associated Press prematurely reported January 2, 1993, "a Europe without borders."[3] But it was a notable reduction of ancient barriers, as was the U.S. Congress's approval of the North American Free Trade Agreement. This latter fused what was called the world's largest trading bloc.

Some NATIONAL BORDERS remain reinforced not only by the familiar borderlands of difficult-to-impassable terrain, but occasionally by minefields (an estimated one hundred million land mines still persisting in 1993), walls, fences, armed guards, tariffs, passport restrictions, etc. Here is where maps change color, guards change uniforms, and old maps bear witness to crusades, empires, treaties, and hostilities either long-gone or new-found. Satellite photos of the U.S.-Mexican border at Southern California show sharply different color patterns—reflecting historic differences in irrigation practices and water rights. These, in turn, reflect treaties that allocate Colorado River water in differing volumes to U.S. and Mexican irrigators.

In the nineteenth century, philosophers who spoke and wrote of "society . . . usually had in mind a human society transcending all state [i.e., national] frontiers."[4] But meanwhile, the NATIONAL BORDER was acquiring a growing burden of pride and prestige which reflected a limit to the geographic reach of a state's power, placed by royal marriage, war, or treaty, especially in the nineteenth century. BORDER became a symbol of national identity. True to its Anglophilic origins, the 1950 edition of *Encyclopedia Britannica,* in defining BORDER, narrowed its focus to the specific "territory on both sides of the boundary lines between England and Scotland, especially the Scottish side." Centuries of battles on both sides produced scores of frontier fortresses, a few of which remain as "sights, ruins, or grass-grown mounts." Another BORDER remnant is Gretna Green, a Scottish hamlet useful to English runaway couples. They could be married by a blacksmith, tollkeeper, or ferryman. The label "Gretna Green" is sometimes applied to marriage parlors across state lines in the United States.

Most boundaries act as screens and selectors rather than as outright blockades. The greater the disparity of power-potential, incomes, beliefs, history, and religious fervors on each side of a NATIONAL BORDER, the less likely is the BORDER to survive unbreached. At this writing, the huge expansion of trade between nations "has made every nation in the world more dependent on the outside world to maintain its 'good life.'"[5]

Yet many BORDERS still exert transforming power. Once across, an innocent agricultural product may become an illegal "controlled substance" (a.k.a. drug). At many a BORDER, tourists turn into foreigners, a draft dodger may be accepted as a "draft resister," travelers may become refugees, a runaway soldier turns deserter if not traitor, a language turns familiar or foreign, innocent trade goods may become contraband and their carrier a smuggler, and the possessor of innocuous research papers can be accused of stealing state secrets. "Ours" becomes "theirs" or vice versa. Crossing the BORDER, a deposed king or dictator seeks refuge; and a criminal hopes to resist extradition. Often freedom, real or fancied, lies on the other side. At many BORDERS it is believed that God or Mohammed is on this side, while the Devil occupies the other. Many BORDERS become DROP ZONES, where border-jumpers discard old clothes, incriminating papers, and identities.

BORDERS have been engaged in upward mobility between dozens of nation-states during the boom in new-nationhood encouraged by the United Nations since World War II. Those that follow geography, such as lofty mountain ridges, tend to survive; those less well marked may not get

NATIONAL BORDER

(Opposite, left) Many influences hover over the shoulders of mapmakers approaching a NATIONAL BORDER: patriotism, military security, budgets. Thus maps, as well as nighttime satellite photos, often blank-out the "Other Side" of the border. Here Ontario's *Economic Atlas* shows brightly lit towns and Canadian cities north of Lake Ontario, with Toronto giving off the biggest glow. The United States is blacked out. Photo, *Economic Atlas of Ontario,* 1969; reproduced with permission from the Queen's Printer for Ontario.

(Right) Companion satellite photo shows THE LIGHT only on the U.S. side of the Great Lakes—and in apparently greater luminosity than in the Canadian photo. Taken together, these two photos show how national interest, budgets, and electronic manipulation keep the "Other Side" of the border in THE DARK and, as in the U.S. photo, can also exaggerate THE LIGHT on "Our Side." Photo, EROS Data Center, Sioux Falls, S.D.

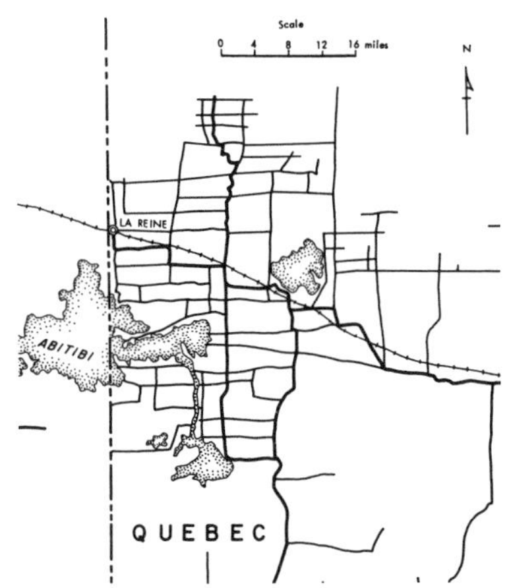

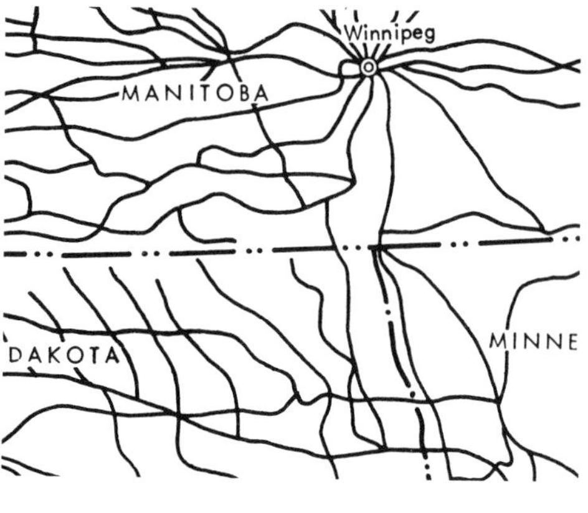

NATIONAL BORDER

"No traffic with THEM" appears to be the rule governing the extension of railroads and roads to, but seldom across, national and provincial borders, as shown by two maps. Right, American railroads from the south terminate just short of the U.S.-Canada border; and an extensive road network serves Quebec (left), but ends at the Ontario border. Original maps by Roy I. Wolfe, reprinted in *Locational Analysis in Human Geography,* by Peter Haggett (London: Edward Arnold, 1965) p. 69.

past an escalating "border incident." Between nations of peaceful coexistence (U.S.-Canada), incidents are few, checkpoints benign, and armed BORDER outposts long discontinued.

But until 1989–90, BORDER crossings between East and West Germany, between Soviet Bloc and Western Europe nations, between the United States and Mexico, and between many Middle Eastern nations were politicized, guarded, and routinely used as political devices. Border battles continued along the Jordan-Israeli border. But liberation movements in Eastern Europe, seizing on USSR "glasnost" policy, began in 1989 the removal or reduction of border restrictions that promised to open up relations between some but not all once-antagonistic European states. This newfound freedom of movement produced conflict when old ethnic/nationalist groups found or revived excuses to cross or disregard once-impassible BORDERS and to engage in "ethnic cleansing." By 1993 open warfare had broken out within former Yugoslavia and among former USSR satellite or client nations.

Post–World War II tensions between East and West Germany had focused attention on dramatized "Checkpoint Charlie," the lone passageway through the Berlin Wall that then separated Soviet and Allied (NATO-supported) zones. It was demolished beginning November 1989. Along the Rio Grande between the United States and Mexico, dozens of checkpoints were set up 1984–85 by U.S. officials to block illegal immigrants from Mexico. A major traffic block-up occurred February of 1985 when U.S. officers began searching all vehicles crossing the border, seeking clues to the kidnapping of a U.S. narcotics agent.[6]

Change a few words, and Osbert Lancaster's description of a dicey roadway in Greece could still apply to border crossings in troubled areas: "a point where suspicious gendarmerie stand forever poking the bundles of the indignant but not invariably innocent countryfolk in a routine search for tommy-guns, hand grenades, gold sovereigns and other rural produce."[7]

Maps of telephone traffic along the U.S.-Canadian border across the Great Plains have shown that the major flow of calls is east-west, rather than north-south across the NATIONAL BORDER. Like speaks to like.

Tourists who cross the border from Portugal into Spain in the vicinity of Ciudad Rodrigo will confront an amazing transition from inhabited countryside to open landscape—no settlements visible for the first five miles. Was this a leftover from the Franco regime's depopulation policy? Or, more likely, from centuries of warfare and fortressing which "during those centuries have led to a movement of population away from these areas."[8]

Two nations, measuring the length of

the same BORDER, can come up with quite different figures. The Portuguese atlas shows a border 20 percent longer than the Spanish atlas. Same border. Different measurement tactics.[9] An old French proverb observes of the mountains separating France and Spain, "Across the Pyrenees, truth is upside down."

Over time, according to geographer Richard Hartshorne, minorities in a border region—either leftover or recently migrated—tend to conform to their surroundings, if there is enough time and if both states undertake concerted policies of assimilation.[10] Thus, where tensions between two adjacent nations simmer down, border zones are the first place, outside nationalistic mass media, to show the effects.

Around every BORDER stretches a border zone of various size, serving as cushion as well as barrier. It functions somewhat like a no-man's-land in combat, absorbing force while resisting penetration. Many of these zones are hopeful places for the future—more so than NATIONAL BORDERS themselves—being zones alongside the BORDERS, where, many experts believe, future war tensions may deflate. Observed a noted geographer: "The chief cities of Texas lay within the blurred border zones of . . . [adjoining] regions and they not only displayed something of the elements of the cultures nearby but helped to draw them together and mitigate their differences."[11] And along the U.S.-Mexican border zone of San Diego–Tijuana lies one of the faster-expanding border regions of the continent. Commuting and shopping across this border was expanding in the 1990s. By 1990, a fifteen-mile stretch of the California-Mexican border was said to be "the most heavily traveled border in the world."[12] As the North American Free Trade Agreement of 1993 works its way into practice, both the Canadian and Mexican zones will reflect the changes.

Such BORDERS occupy a fuzzy zone where national identities go into neutral, and signs turn bilingual. Such ethnic ecotones, as well as the vast expansion in electronic traffic across BORDERS, tempt optimists to hope the world is not yet destined to split into a hundred warring nation-states.

NATIONAL SACRIFICE AREA/ZONE

Gradually in the 1980s the news began to spread: U.S. atomic and nuclear weapons plants were leaking and degrading. They were subject to accidents that put at risk millions of close and distant neighbors. Public debates earlier had suggested that some regions, with risky weapons plants, might have to be "sacrificed" in the interest of national security during the weapons race with the USSR. Especially during the Carter Administration (1977–81), the term NATIONAL SACRIFICE AREA came into use to mean "that for the greater good of the nation, parts of one region must be made unlivable." Recalled Ed Marston, editor of *High Country News,* Paonia, Colo., "Despite talk at the national level of making the West a 'national sacrifice area,' the leaders [of the West's rural establishment] either couldn't see the need to resist, wanted to cash in on the blood that would flow from the 'sacrifice,' or didn't know how to stand up to the city slickers that flooded the area."[1]

A major sacrifice zone was identified by a chemical union official as the eighty-five-mile stretch of petrochemical plants along the Mississippi River between Baton Rouge, La., and New Orleans. One-fifth of U.S. production is concentrated there in 135 chemical plants and seven oil refineries. Louisiana's attorney general "called this contaminated chemical corridor 'a modern form of barbarism,' and a chemical union leader now refers to it as 'the national sacrifice zone.'"[2]

In a 1986 study of the Savannah River reactor near Aiken, S.C., "we deliberately used the term 'national sacrifice area' to draw attention of the public" to the dangers, recalls Robert Alvarez, nuclear policy director, Environmental Policy Institute, Washington, D.C.[3] But the term migrated upward from the anti-nuke crowd into federal offices. On October 31, 1988, it appeared in Keith Schneider's report from Miamisburg, Ohio, in the *New York Times:* "Engineers at the Energy Department have privately begun calling such contaminated sites 'national sacrifice areas.'" Senator John Glenn of Ohio, where a nuclear feed materials center was located, observed, "We are poisoning our people in the name of national security."[4]

By 1988 all fourteen nuclear reactors in the Department of Energy's network were closed down, most of them permanently. A plutonium-processing building at Miamisburg, built in 1960 in five months, costing $1 million, would require $57 million and twenty-nine years to tear down. Such were typical costs of coping with domestic nuclear threats, after a 1942 cloak of wartime secrecy over nuclear plants began to be lifted. The huge task of cleaning up after forty years of frantic atomic-bomb-making had barely begun.

The term was applied to industry in general by William Ashworth in an examination of heavy-industry cities along the U.S.-Canadian border from Wisconsin to Ottawa, Ontario: "There is some sense of the region being a national sacrifice area for two nations."[5]

A new site category arose from past abuses: "Superfund Sites," where industrial or hazardous/toxic disposal had rendered a site unfit or unsafe for human use. These qualified to tap a federal "superfund" for cleanup. But the process appeared to have no end. The *U.S. News and World Report* raised the recurring issue in a headline March 27, 1989: "Will some [areas] be abandoned as 'national sacrifice zones'?"

As the flurry of atomic cleanup appeared to subside, the term NATIONAL SACRIFICE AREA lay little-used and quiescent, available—in some yet-to-be defined national emergency—for the next round of demands that a particular locale become a dumping-ground for more powerful others.

THE PUBLIC DOMAIN

This noplace-that-is-everyplace is the single great channel for bringing one's self and the larger world together. Here is where the law of the place and the place of the law are in constant tension. It includes street, gutter, curbing and sidewalk, the public square, the public markets. It embraces THE COURTHOUSE steps, public lobbies and reception halls, as well as places owned by governments. Shopping malls offer private-property space doubling as public space when it suits the owners. Television extends it to mean anyplace that is on-camera. Most Easterners think of it as Somewhere Out West.

Historically, THE PUBLIC DOMAIN was in possession of a ruler or any authority exercising rights and governance. It was a field or sphere of control; it was the Royal Domaine (with kinship to the French "demesne").

More recently, community rights entered the scene. THE PUBLIC DOMAIN came to signify the meeting ground between public and private interests, usually taking place on sites and properties belonging to the community at large—the town commons, or THE COURTHOUSE SQUARE. "The first recorded uses of the word 'public' in English identify the 'public' with the common good in society; in 1470 . . . Malory spoke of the emperor Lucyos . . . 'dictatour or procuratour of the publyke weal of Rome.'"[1] Sociologist Richard Sennett argues that only when we are situated "in-public" do we learn to be responsible citizens—or citizens in any reading of the term.

From subway to midway, from movie queue to traffic jam, in public places we come face to face with the hustle and heist, with newcomers and old-timers, with strangers and foreigners, with off-putting oddballs, off-beats, and deadbeats—and with unwritten rules about how we—and they—should act. Here, in open view, beyond the TV screen, is where "civility" does its work. Public places offer training-grounds for learning body English as public manners, for dealing with the world beyond self, family, kin, clan, tribe, locality, Us. Here in the public view, we learn, or may be forced to translate, private interest in terms of a larger value system beyond the self. Not everybody can handle it. Many public places are designed to prevent this kind of contact: no movable chairs, nothing loose, everything fixed in place and by ordinance to discourage human interaction.

The *New York Times,* defending its right to publish the ("secret") Pentagon Papers on the Vietnam War, argued that they "have long since belonged in the public domain," and therefore should be—as they later became—accessible to all.[2]

In the geographic sense, PUBLIC DOMAIN has also come to stand for the 1,131,353 square miles of land owned by the federal government, chiefly in Western states, much of it leased for ranching, mining, and logging at knockdown prices set in the nineteenth century. Which public is to be served is the contentious and continuing question.

See also THE COURTHOUSE DOOR/YARD/SQUARE and THE LANDING, in Chapter 1: Back There; and CURBSIDE, in Chapter 2: Patches.

SPEED TRAP/ZONE

This double-barreled place-name represents two separate worlds that may, in the dim past, have been separated by law and practice. But that was long ago. SPEED ZONE is generally taken to mean any street or highway stretch having a required speed limit formally announced by signs, and ended by a "Resume Speed" sign.

What unsuspecting motorists do not know is when and where a SPEED ZONE has been converted locally and with dubious legality into a SPEED TRAP. The latter is sometimes presaged but not revealed by an inconspicuous speed-limit sign that may be the tip-off to a concealed revenue-hungry local highway patrolman stationed nearby. Once the trap is sprung and the cover blown, the speeder is caught. But speeders are reluctant to go public with their complaints. Consequently, SPEED TRAPS survive until they happen to entrap a prominent citizen who finds a lawyer who finds a loophole which brings the whole edifice crumbling. But then again . . . .

WATER RANCH

"A new generation of entrepreneurs is selling a precious resource to the highest bidder, with potentially profound effects on development of the West," noted the *Los Angeles Times*.[1] The resource is irrigation water. The device for so doing in the 1990s is the WATER RANCH, a man-made oddity.

Not easily seen from the road, a WATER RANCH may have no livestock at all. It may cover a few acres or many square miles. Four facts set it apart: (1) it is under single ownership and management; (2) it has saleable quantities of water flowing through or past, or underground; (3) the owners are cropping their water as though it were hay, soybeans, or corn, sold to distant buyers, and (4) the water is scarce and under increasing demand in an active, speculative market. Some speculators build long pipelines to get their water to distant customers; they finance the cost by long-term borrowing. (But unlike ground crops, some underground waters are not replaceable, having been left by the melting of ancient, long-gone glaciers.)

Nineteenth-century political pressure on a farmer-dominated Congress kept Western irrigation water cheap—national taxpayers paid much of the real cost. In one California project, farmers-ranchers repaid only about 5 percent of cost over forty years. Thus did the humid East subsidize Western irrigators—especially large-scale corporations—when they converted semi-deserts into rangeland or irrigated cropland with a cheap, subsidized necessity, water.

Then came a population boom into Western states after World War II. By 1979 water rights were being briskly traded in Colorado, with over $100 million invested since then. The North Platte and Elkhorn River Valleys through Nebraska were scenes of a speculators' rush to set up CPI (center-pivot irrigation) systems around peak years 1956 and (all-time peak) 1976. This boom greatly expanded the spread of those round, green, irrigated fields and pastures across the Upper Midwest and Great Plains where water rights were negotiable. The underground water itself was of ancient origins and non-renewable. At the 1970s rate of withdrawal, someday after the year 2000 it would be gone.

Under continuing pressures, water ranching expanded in the 1990s. Around Reno, Nev., water rights rose from as little as $100 an acre in 1979 to as much as $3,000.[2]

All across the West, water tables were falling; some 50 percent of the West's rivers were already overused, and the political and other costs of "developing" new sources (via unpopular upstream dams, deeper wells) was getting prohibitive. Towns and cities, strapped for water, were turning to what they perceived as the only big untapped supply: irrigation water.

Meanwhile, farmers-ranchers in most dry states were claiming some 85 percent of water for crop and cattle consumption. "In Arizona, state law makes it difficult to buy rights to water independent of the land. So Tucson, Phoenix, and other rapidly expanding cities have taken to 'water ranching'—buying up land to get the associated water rights"—and acquiring in the process some 575,000 acres of ranch and farmland.[3] Under such pressures in Tucson's Pima County, irrigated agriculture is expected to disappear by A.D. 2020.

Meanwhile, the history of Owens River Valley, Cal., remains stark in Western memories—its water rights bought or stolen by Los Angeles for its own water (dramatized in the film *Chinatown*). The Owens Valley reverted to semi-desert, a fate more than one Western overdevelopment faces at some not-so-distant future.

The ultimate question hung heavy over the American West: how long and how far could drawing and overdrawing on underground water supplies continue? Water ranching would soon come to be viewed as merely the latest trading device for "rationalizing" water—for the final conversion of the last natural waters into a scarce commodity subject to tightening social controls. Western politicians would no doubt intensify their efforts to tap Canadian waters for a long overland, border-crossing trip. In such a scenario, water ranching would go international sooner, rather than later. Off in some not-so-distant future also lay the possibility that water-starved California cities could outvote the diminishing rancher-farmers and take further control of the state's waters for their own use. Thus might be prompted another historic migration—by thirsty farmers, moving back to the humid East.

WETLANDS

Variants: backwaters, bogs, bosques, bottoms and bottomland, crawfish lands, drainage areas and districts, duckshooting lands and preserves, fens, floodplains, FLOODWAYS, floodlands, flyways, inundation lands, mangroves, marshes/marshlands, miasma lands, the mire, mitigation lands, moist areas, moors, morass, mosquito lands, muskeg, overflow lands, peat land, pocosins, ponding areas, pothole country and prairie, playas, recharge areas, riparian lands, sloughs, hundred-year storm surge areas, submerged lands, swamps, shrubby swamps and swamplands, tidal bays and lands, tidewater, tidal flats and tidelands, tules, vernal pools, wastelands, "waters of the United States," water meadows, wet meadows, wet-weather ponds or wet prairie, the wet, and—along the way—"valuable croplands subject to occasional overflow."[1]

One man's slough is another's backwater. Necessity is the mother of lexigraphic invention. As the recent history of the term WETLANDS indicates, it has been swept up into international conventions and local politics via federal legislation, especially since the 1950s. As genuine WETLANDS disappeared, the names attached to them flourished.

If "endangered-species habitat" is added to the above list, it casts a still wider net. Scratch an Arkansas duck-hunter, a hydrol-

WETLANDS

Not many ditchers-and-drainers leave behind such evidence of their drying-out tactics. Here, in the midst of the former great swamp of the Kankakee River in northern Indiana, effigy pipe-toters serve to advertise pipe, an important facilitator in the drying-out process that has eliminated millions of wetland acres. Wide fields of corn and soybeans occupy the once-swampy background.

ogist, a marine biologist, a U.S. Corps of Engineers hearings officer, or an Iowa farmer and you may elicit five different definitions.

Until the 1960s, marshes and other WETLANDS components were still being kissed-off in print. Typical was a *New York Times* article in 1957, which followed accepted practice in describing a giant Delaware River marsh—about to be "reclaimed" by dumping—as a "marshy wasteland."[2] Even when New Jersey possessed vast tidal marshes opposite Manhattan, N.Y., it bragged of its "Jersey Meadows" while using them as a sports complex, regional DUMP, and urban outback.

The ability to wade through mire and picture it as a future meadow, or to see a swamp and envision a farm became a widely practiced visionary art. But not until the 1960s did the political utility of the term WETLANDS break into public consciousness.

Gradually WETLANDS emerged as the unifying term for conservationists, thus summoning up a contemporary form of order. Its modern usage "began only in the 1950s and its wide employment has only occurred since the 1970s." In 1954 the U.S. Fish and Wildlife Service had conducted "the first nationwide wetlands inventory," giving the term some prominence.[3] A typical more recent usage occurred when the *New Scientist* magazine on June 17, 1965, "defined [WETLANDS] to include marshes, bogs, swamps, and any still water less than six meters deep." As it turned out, this was a brilliant coinage suitable for wide circulation. The term WETLANDS offered a handy, all-embracing shorthand label during the burst of enthusiasm for environmental reforms since the 1960s.

Its re-interpretation remained a political morass, but the term helped focus national attention and Congressional debate on a vital, diminishing resource. The big change came slowly in the growing presumption that to substantially alter any WETLANDS would run counter to the public interest.

The disappearance of WETLAND components had long been forecast. Well before 1900, conservationists had noted how widespread ditching-and-draining had dried out parts of the Midwest. Swampland Reclamation Acts from before the Civil War had ditched and drained wide stretches of Virginia, Ohio, Indiana, Illinois for sale as "productive farmland." "Time and again it was found on examination that 75 percent of the land claimed was not in any sense swampy or subject to serious overflow."[4] By faking definitions, Illinois acquired from the federal government millions of acres of so-called swampland which it sold to speculators and settlers.

The federal Swamp Land Act of 1849 defined any place from slightly damp to supersaturated as "swamp or inundated lands within the meaning of the act." The federal government ceded to the states 64,433,870 acres of "swampland." The intent of the act was to grant only those lands marked on government plots as "swamp and unfit for cultivation." But the loopholed act itself was sidestepped, violated, and perverted by the states into a device for land speculation in high-and-dry land mislabeled "swampland."

When it was first settled, the United States had about 227 million acres in various forms and stages of permanent or temporary WETLANDS. For the next two centuries, states east of the 120th meridian treated water as an enemy. Just as "a good Indian was a dead Indian" in the view of aggressive nineteenth-century settlers, so "a good WETLANDS was a drained WETLANDS." The early history of WETLANDS management appears to derive from the belief that WETLANDS were wastelands. "Wetland drainage and destruction were simply the accepted norm."[5] In the West today water remains a basic commodity. Dried-out land is viewed as an Entitlement.

An American populace hellbent on conquering "the wet" has paved over its absorbent soils and sped up the downstream flow of rainfall, and of water supplies, sewage, and industrial wastes. They used levees and more efficient pumps and pipes to carry out the continent-wide drying-out process. New suburbs regularly short-circuit local natural drainage networks, preferring pipes. Together, pipe and pump have drastically shrunk the continent's power to absorb.

By this process, WETLANDS east and west were nearly squeezed off the map.

The Kankakee River swamplands of Indiana had been converted almost wholly to cropland for corn and soybeans. The meandering Kankakee itself, which once had some two thousand bends in 240 miles, was straightened to 88 miles with only a handful of bends. The Great Swamp south of Morristown, N.J., survived long enough to be proposed in the 1980s for New York's new jetport, which plan was eventually sidetracked by environmentalists.

Urbanization, as universally practiced, had turned out to be an effective form of desiccation. As the West became a giant ranchland, "every filet mignon is a spring dried up, every steak is another riparian zone degraded," observed a *New York Times* reporter.[6] "Irrigated farming, woodcutting, and recent population growth in the Southwest [by 1986 had] destroyed 90 percent of the region's original bosque."[7] These are wetwoods, pronounced "Boskee" in New Mexico.

During the 1960s—with the White House Conference on Natural Beauty in 1965, and Earth Day occurring in 1970—a new generation took over the battle against ditching and draining. President George Bush pledged in his 1988 campaign "no net loss" of the nation's WETLANDS. But federal controls continued to attract efforts to amend, which inspired media commentators to wax eloquent: the *Washington Post* headlined, "Wetland Agreement Watered Down." In other media, the new protections were said to be swamped, stuck, mired, and bogged-down in the halls of Congress. For years to come, states, localities, and Congress would be fighting over leftovers.

Amid all this, it is hard to imagine that backwoods swamps and borderland marshes once inspired a noted American geographer to call them "regions of survival of race and language." "The scattered islets of the Fens of England furnished an asylum to the early British Celts from Teutonic attack." Swamps, she wrote, "have always afforded a refuge for individuals and peoples. . . . [They] sustain and nourish the spirit of liberty."[8] No longer refuges, such places in the United States have become shrunken venues for seekers of wildlife and remnants of the great herds, flocks, and swarms which once thronged the landscape. It had become the U.S. Corps of Engineers' stage for exercising control over upper watersheds occupied by millions of American suburbanites.

See also SUBURBIA, in Chapter 6: Border Zones.

WHALE WATCHING SITE

So popular did whale watching become along the California coast, spurred by the environmental movement from the 1960s, that a multi-million-dollar touristic industry sprang up, operating from mid-December when some whale migrations begin. Whale-related spinoffs, such as whale artifacts, jewelry, and T-shirts, brought an estimated $50 million a year income to the whale watchers' territory during the 1980s. Most Southern California harbors became bases for whale watching cruises, half a dozen at San Diego alone.

The limited supply of WHALE WATCHING SITES became overcrowded. In 1984 some ninety-six thousand viewers traveled a windy route to Point Reyes Lighthouse thirty miles north of San Francisco's Golden Gate to watch for spouts of the migrating monsters. "Visitors need a good pair of binoculars, warm clothing and a sense of adventure because they must hike from a small parking lot to the top of a steep staircase leading to the lighthouse. Once there, 35-foot to 50-foot gray whales can be seen surfacing and diving in the shallow coastal water. . . . The spectacle begins in mid-December when the gray whales start moving through California waters on a 12,000-mile round trip from their Bering Sea feeding spots off Alaska to the warmer breeding grounds" in the waters of Baja California. "When they return north in March, the folks in the logging town of Fort Bragg, Cal., throw an annual whale festival to welcome them off the rugged Mendocino County shoreline. 'We've got 50- to 80-foot bluffs and you can look down on them,' said festival organizer Ron Smith. 'Most people come up here to get away. But five different kinds of whales is an attraction. You can see them spout, breach, and sky-hop when the weather's good.'" However, whale watchers and other environmentalists soon found themselves on a collision course with oil drillers along the California Coast and in the Bering Sea, the whales' summer feeding grounds. Addressing the possiblity that the Bering Sea could become a venue for seismic blasters and oil drillers, Birgit Winning of the Oceanic Society said, "We predict a dire impact if that occurs."[1]

EPILOGUE

Stand and Be Counted

The manufacture of man-made places in the United States is expanding beyond the wildest dream of its pioneers. It may well turn out that we have set a world's speed record in converting nature as we found it to habitat as we intend it. We have taken places once sacred to Native Americans and converted them to an expanded range of commodities in international trade.

Each of the 3,536,342 square miles of land and 351,083 square miles of waters in our fifty states has been surveyed, photographed from plane and satellite, mapped, identified, and converted to public or private domains of territory. Most of it can be classified as taxable real estate. By the year 2000, each acre, hectare, or square kilometer will have been identified in world data banks as an object of inspection and possible negotiation. Investors and tourists from Singapore to Berlin will be able electronically to summon up your place or mine for visual examination. Panning across continents, scanning across regions, and screening sites by the thousands provides an awesome spectacle, soon to be available for home computers at a competitive price.

We have overlaid upon the original mysterious terrain of American Indian possession fifty political states with 8,385,132 identifications on the list of Geographic Entities maintained by the U.S. Census. These designations include 310 Native/Indian areas and reservations; 3,141 counties; 49,961 census tracts and 6,961,150 census blocks by which we are counted; and 29,270 special DISTRICTS (irrigation, congressional, taxation) by which we are managed and sometimes self-governed. We get our mail at 115,391,000 addresses and mailbox numbers, and we identify our locations by 43,330 zip codes.

"Hundreds or even thousands of names [are being] added to each state file monthly."[1] More than 1.2 million of us change addresses every year. The total grows: some 600 new governmental districts added; and 1,311,959 building permits issued in a typical year, with uncounted other structures built without permits. And for every permit issued, two or more new parcels of land are carved out of the landscape.[2]

By 1993 the list of specific names for geographic places of human settlement—as distinguished from (but with some overlapping) the larger list of official subdivisions outlined above—totaled some 1.5 million, as enumerated by the U.S. Geological Survey Board of Geographic Names. The total is expected to rise to around 3.5 million by the year 2000.[3]

The number of man-made places we pass or traverse in a year—as shoppers, commuters, tourists, haulers, and fetch-ers—can hardly be estimated. We travel by car some 1.5 trillion miles per year,[4] and 432.4 billion miles by air.[5]

We make annual payments on millions of taxable parcels of real estate, including air, water, and mineral rights, and leaseholds. In the metropolitan county where this is written, taxes are levied on 256,000 parcels of real property, about one parcel per 2.6 persons.[6] Until the 1980s there were still thousands of untaxed and unlocatable parcels in the BOONDOCKS and outbacks of the land, but airphoto-mapping has picked up most of them for taxation, speculation, and other developmental purposes. The end of the Cold War is converting widespread military holdings to other uses yet to be identified.

Add those uncounted local plats, parcels, tracts, and minor subdivisions of land, and the list lengthens. Still uncounted (if not uncountable by current methods) are all our zoning districts, neighborhoods, easements, rights-of-way, leaseholds, mining and petroleum claims, and development rights—not to mention innumerable and often illegal conversions, extrusions, preemptions, and squatting. And far above these surface interventions loom designated FLIGHT PATHS and landing patterns for civilian aircraft, and restricted zones for the military. In another realm lie those millions of anchorages, assigned or limited FISHERIES, channels and landings recorded on navigation and fishing charts of inland and offshore waters. This never-ending flow of official and semi-official places continues to interplay with private ownerships, and to spread across once-natural landscape.

But even that is only the tip of the iceberg. Successful human intercourse required that our ancestors find or invent grunts and shouts, hand and body signals, and eventually, generic names for places they encountered or created. For each specific place-name, there had to be a generic name available for description and for trading purposes: VACANT LOT number four in the TEMPORARY HOUSING DEVELOPMENT adjacent to THE RUINS at the end of THE BEACH, 1.2 miles beyond THE EDGE OF TOWN. Those five generic descriptors, capitalized in the last sentence, are part of the man-made currency sampled in this book.

This ages-old naming was, and is, essential to the sorting-out process as we expand language to negotiate with, and to change, environments. When we add new terms, they further influence the way we think. The research leading to these pages has identified some four thousand generic man-made places, from ABANDONED AREA to Zoological Gardens. In their millions of specific examples, these have altered the rainfall, land use, and temperature patterns of the whole earth. From that great mountainside this book carves off only a small slice of 124 generic man-made places. Since this work began in 1984, the number of persons in the United States has increased by an estimated 18,420,000.[7] "The Greenhouse" has emerged from its speculative state to become a global presence. The DRUG SCENE, EVENT SITES, FLEA MARKETS, the HUB, the PACIFIC RIM, and SECURITY continue to become more noticeable and important to us.

By fastening attention and names to places, in all their generic and specific variations, we reveal the full extent of the historic transformation of North America. Little is left of our original national landscape, aside from designated "natural areas." By fastening our designation upon them, we change them inevitably. Thus we mass-produce a new reality.

Large portions remain in the PUBLIC DOMAIN as national parks and forests, reservations and preserves. But they undergo constant scrutiny as defensive occupiers and aggressive privateers try to maintain or to get control of public land and waters and access to them. Onto every scene, over every view, has been cast the transforming power of photos, satellite images, options and titles, land deeds and governmental or contractual rules. Their location and borders are described on census and other tapes.

Thus, Out There in the huge complex we call Earth, all transactions remain rooted at some point in identifiable places, even as their prices and ownership flicker around the world in electronic markets. New physical situations with names, numbers, and values are created 'round the clock. Each change of address is assembled into yet another data bank where it undergoes generic transformations. At one time or another, by many hands, each may be recorded as a taxable parcel, a construction site or an ABANDONED AREA. Tomorrow it may or may not be a CONVENIENT LOCATION, or within a newly designated ANNEXATION AREA. The next U.S. Census may find it in a CHANGING NEIGHBORHOOD. It could gradually become a TWILIGHT ZONE, located in a GHOST TOWN, within EARSHOT of noisy abutters, intruding into a neighbor's VIEW. Across every threshold we carry goods and matter that will end up, in one form or another, in a DUMP, DROP ZONE, FLEA MARKET, or DOWNWIND. Places we term beyond-the-pale will be rediscovered and marketed as GENTRIFYING NEIGHBORHOODS, or perhaps qualify as GROWTH AREAS. Few elements of THE SCENE will be wholly unmanaged; and all will be documented.

There was a time called "back then." That was back when the Rockies were a mere rumor among Easterners, back when there were blank spots on national maps. Even as late as 1929—back then—Bob Marshall, who co-founded The Wilderness Society, could be the first white man ever to explore a hidden valley in the Rockies' Brooks Range.[8]

Geographic discoveries at that scale are no more. "Back then" is no more. Millions of computer users now resort to personal electronic address files wherewith they call up thousands of addresses: clients, customers, prospects, contacts, friends, and related places. All can be "placed" and printed in quick order. One's addresses, past and present, at work or at play, are bought, sold, or pirated by electronic hustlers good and bad. They have thrown out of the electronic window the old notion that "information about me and my place is mine." Those of us who value personal privacy—a new minority?—are left to fight for it rather than expect it as a right.

All is up for grabs. Indian rights to tribal lands have been restored by court decisions; and tribal rights to develop those lands with gambling casinos have created new BOOMTOWNS. Water rights are in transition, and the fate of irrigated farms thrown into question as thirsty cities claim more of the Western waters that farmer-irrigators thought they owned. New forms or definitions of the PUBLIC DOMAIN are set up with every session of aldermen, and by county, state, and national legislatures.

As we tighten our grip on all these places, we leave bigger footprints. As the number of man-made places increases, the number and variety of living species diminishes. In what Karl Polanyi called "the great transformation," we have reorganized nature as a commodity, and all our relations to it.[9] Into every transaction is added the question of profit-or-loss.

What one learns from these places along our path is the inevitability of conflict between developers and sustainers; between local and "outside" interests; between those who got there first, and those who come later, wishing to change it. As the targeting of places for multipurposes expands over the face of the globe, what's new is its scope and scale. We will need all the flexible customs and institutions we can invent to contain the conflicts and to redefine, while protecting, the public's right of access.

All the more reason, then, for us to continue the sorting-out process which this book can only begin—digging out the functions, names, rights, and obligations that go with everyday places, and blowing them into proportion. Once we learn to look at the world this way, there is no chaos, nothing is wholly foreign, and we are never lost.

NOTES

Preface x

1. J. B. Jackson, "The Stranger's Path," *Landscapes: Selected Writings of J. B. Jackson,* ed. Ervin H. Zube (Amherst: Univ. of Massachusetts Press, 1970).

2. *The Exploding Metropolis* (New York: Doubleday, 1957), ed. Sam Bass Warner, Jr. (revised edition, Berkeley: Univ. of California Press, 1993).

3. Grady Clay, "Crossing the American Grain, or The Happiness of Pursuit," *Geographical Snapshots of North America,* ed. Donald G. Janelle (New York: Guilford, 1992).

4. *The National Road: Theater of American Culture* (projected title), ed. Karl Raitz (Baltimore: Johns Hopkins Press, scheduled publication Oct. 1995).

5. Grady Clay, *Closeup: How to Read the American City* (Chicago: Univ. of Chicago Press, 1980) 38.

6. O. B. Hardison, Jr., *Disappearing through the Skylight: Culture and Technology in the Twentieth Century* (New York: Viking, 1989) 49.

Introduction xviii

1. Michel Foucault, *The Order of Things: An Archaeology of the Human Sciences* (New York: Random House, 1973) 141.

2. William Cronon, *Nature's Metropolis: Chicago and the Great West* (New York: Norton, 1991).

Introduction to Section One: The Center 2

1. Richard C. Wade, *The Urban Frontier* (Chicago: Univ. of Chicago Press, 1959) 341.

2. Grady Clay, *Closeup: How to Read the American City* (Chicago: Univ. of Chicago Press, 1980) 34.

3. Peter Hall, *The World Cities* (New York: McGraw-Hill, 1977) 246.

4. David Everett, Providence, R.I., private communication, 13 Apr. 1992.

5. This gem of eloquence from Catherine Bauer found its way into my journals, with only the date 1962 to note its distinguished origin.

6. Sam Bass Warner, Jr., *Streetcar Suburbs: The Process of Growth in Boston (1870–1900)* (Cambridge: Harvard Univ. Press, 1978).

7. William Cronon, *Nature's Metropolis: Chicago and the Great West* (New York: Norton, 1991) 200.

8. Rudolf Arnheim, *The Power of the Center: A Study of Composition in the Visual Arts* (Berkeley: Univ. of California Press, 1982) 4.

9. Michael Abramowitz, "The Urban Boom: Who Benefits?" *Washington Post* 10 May 1992.

Chapter 1: Back There 10

THE CAPITAL 10

1. Vaughan Cornish, *The Great Capitals* (London: Methuen & Co., 1922).

2. "Austin Journal: In Texas Capital, A Growing Constituency for Bats," *New York Times* 18 Aug. 1992.

3. Susan Orlean, *Saturday Night* (New York: Knopf, 1990).

4. James T. Yenckel, "Tourist[s] Share the Secret of the Rich and Famous," *Washington Post* 13 Jan. 1985.

5. Amit Roy, "Countdown to the Big One," *Geographical Magazine* Dec. 1990: 44.

THE COURTHOUSE DOOR/YARD/SQUARE 13

1. Byron Crawford, "Loafers Abandon Benches as Leaves Start to Fall," *Courier-Journal* (Louisville, Ky.) 1 Oct. 1984.

2. John P. Callahan, "New Look Looms for Old Section," *New York Times* 12 Jan. 1958.

3. AP "Chicago Civic Center to Dominate Skyline," 31 Mar. 1962.

4. "Dearborn Civic Center," *Architectural Forum* Feb. 1959.

DECLINING AREA 17

1. Calvin L. Beale, "Quantitative Dimensions of Decline and Stability among Rural Communities," in Larry R. Wilding, ed., *Communities Left Behind* (Ames: Iowa State Univ. Press, 1974) 3.

2. Seth King, "Cape's Sales Continue Slide," *New York Times* 7 Jan. 1990.

THE GOOD ADDRESS 20

1. Geoffrey Hughes, *Words in Time* (London: Basil Blackwell, 1988) 162.

2. Sybil Baker, "Where Homes, Money Tend to Be Old," *Los Angeles Times* 27 May 1990.

3. Michael J. Weiss, *The Clustering of America* (New York: Harper, 1988) 62.

4. Vance Packard, *The Status Seekers* (New York: David McKay, 1959) 79.

HOLDOUT 21

1. Timothy Egan, "Big Discount Chains Joining Manhattan Toy Wars," *New York Times* 9 Dec. 1990.

2. Harold Henderson, "Yard Works," *Planning* July 1986.

3. Andrew Alpern and Seymour Durst, *Holdouts!* (New York: McGraw-Hill, 1984).

4. Tony Hiss, *The Experience of Place* (New York: Knopf, 1990) 103–105.

5. Paul Fussell, *Class* (New York: Ballantine, 1984) 215.

THE LANDING 23

1. James Malone, "Casino Boat Sets Sail; Town Hopes for Big Payoff," *Courier-Journal* (Louisville, Ky.) 24 Feb. 1993.

2. AP, "'Splash' Is Now Riverboat Boom to Struggling Town," *Courier-Journal* 8 Feb. 1993.

3. Kurt Eichenwald, "A Big Gamble on Mississippi Casinos," *New York Times* 31 Jan. 1993.

4. Francis X. Clines, "Gambling, Pariah No More, Is Booming across America," *New York Times* 5 Dec. 1993.

5. Todd Murphy, "Casinos Offer Only Fool's Gold, Experts Say," *Courier-Journal* 30 Oct. 1993.

6. Terence M. Green, "Offices with Shipshape Views," *Los Angeles Times* 25 Aug. 1986.

THE LIGHT 26

1. Jimmy Breslin, interview, National Public Radio, 19 Feb. 1988.

2. Daniel J. Boorstin, *The Discoverers: A History of Man's Search to Know His World and Himself* (New York: Vintage, 1983) 16.

3. Fernand Braudel, *Afterthoughts on Material Civilization and Capitalism* (Baltimore: Johns Hopkins, 1977) 26.

4. James R. Petersen, "Eyes Have They, But They See Not: A Conversation with Rudolf Arnheim," *Psychology Today* June 1972: 94.

5. Geoffrey Hughes, *Words in Time: A Social History of the English Vocabulary* (Cambridge: Basil Blackwell, 1988) 162.

6. Annie Dillard, "Sight into Insight," *Harpers* Feb. 1974: 45.

7. Sharon Begley and William J. Cook, "The SAD Days of Winter," *Newsweek* 14 Jan. 1985: 64.

8. Nancy Berla, "The Impact of Street Lighting on Crime and Traffic Accidents," Library of Congress, Legislative Reference Service, Education and Public Welfare Division, HV 6251 A, ED-108 (Washington 25, D.C.) 5 Oct. 1965: 1.

9. Philip Bess and Howard Decker, "The Chicago Architecture Police," *Inland Architect* Nov.–Dec. 1988: 26, 70–74.

10. "News," *Planning* Apr. 1985: 30.

11. "In a Daylight-Saving Fight, an Hour of Defeat," *New York Times* 28 Feb. 1993.

MEETING PLACE 29

1. "The Meeting Place," *Chicago Tribune* 2 Sept. 1984.

PHOTO OPPORTUNITY 30

1. Robert Guskind, "The Biggest Photo Opportunity of All," *Planning* July 1988: 4.

2. "Double Exposure," *Charrette* (Pittsburgh, Pa.) Nov. 1958.

PORNO DISTRICT 31

1. Harry F. Waters with Mark Starr, Richard Sandza, and Tony Clifton, "The Squeeze on Sleaze," *Newsweek* 1 Feb. 1988: 44.

THE ROW 32

1. "Book Row Is Gone, But Used Bookshops Aren't," *New York Times* 13 Mar. 1988.

2. Robert Lindsey, "Nevada County Squabbles over Brothels," *Atlanta Journal and Constitution* 24 Nov. 1977.

3. Dan Dimancescu and Gudrun Granholm, *Harvard Square: A New Pocket Guide* (Cambridge: Geneva Printing & Publishing, 1973).

4. Theodore James, Jr., *Fifth Avenue* (New York: Walker, 1971) 219.

5. Hope Cooke, "Ladies' Mile," *Seaport* spring 1987.

6. Alan S. Oser, "Historic-Area Plan Kindles a Debate," *New York Times* 23 Nov. 1986.

7. Martin Mayer, *Madison Avenue, U.S.A.* (New York: Harper, 1958) 6.

8. Theodore James, Jr., *Fifth Avenue* (New York: Walker & Co., 1971) 87.

9. Martha Groves, "Hot Streets," *Los Angeles Times* 21 August 1988.

10. Lita Solis-Cohen, "Antique Hunting in Bucks County," *New York Times* 15 Sept. 1985.

11. Carol Strickland, "New Orleans' Street of Antiques," *New York Times* 10 Sept. 1989.

12. "Eight Dealers Buy Sites in Ontario Auto Center," *Los Angeles Times* 13 July 1986.

13. Ruth Eckdish Knack, "Zipping Up the Strip," *Planning* July 1986: 22.

TOWN CREEK 40

1. Carol von Pressentin Wright, *New York: Atlas of Manhattan* (New York: Norton, 1983) 205.

2. Writers Program of Works Progress Administration, *WPA Guide to New York City* (New York: Oxford Univ. Press, 1940) 134.

Chapter 2: Patches 42

ARREST HOUSE 42

1. Josh Kurtz, "New Growth in a Captive Market," *New York Times* 31 Dec. 1989.

DISASTER AREA 46

1. Thomas J. Lueck, "Coping with the Next Calamity," *New York Times* 8 Feb. 1987.

2. Anders Wijkman and Lloyd Timberlake, *Natural Disasters: Acts of God or Acts of Man?* (Philadelphia: New Society, 1988) 23.

3. Daniel Goleman, "Doctors Say Key Assumptions Crumble for Disaster Victims," *New York Times* News Service 30 Nov. 1985.

4. Thomas Hine, "Don't Blame Mrs. O'Leary," review of *American Apocalypse: The Great Fire and the Myth of Chicago,* by Ross Miller, *New York Times Book Review* 15 July 1990.

DRUG SCENE 48

1. Terry Williams, *Crackhouse: Notes From the End of the Line* (New York: Addison-Wesley, 1992) 14.

2. Williams 15.

3. *CBS Evening News,* executive producer and director Lane Venardos, CBS, New York, 17 June 1989.

4. William Finnegan, "A Reporter at Large: Out There," *The New Yorker* 10 Sept. 1990: 63.

5. Nathan McCall, "Dispatches from a Dying Generation," *Washington Post* 13 Jan. 1991.

EMERGENCY CENTER 50

1. James L. Holton, "Planning for Chaos," *Quill* Nov. 1985: 21.

HUB 51

1. Christopher Burns, "Arrival of Airline Hub Brings Boon for Economy, Noise for Neighbors," *Courier-Journal* (Louisville, Ky.) 26 Apr. 1987.

2. Robert E. Dallos and Lee May, "Debate Still Rages over Deregulation," *Los Angeles Times* 2 Nov. 1986.

3. "The Art of Hubbing," *Condé Nast's Traveler* Jan. 1988: 108.

4. Alan Patureau, "Augusta Makes Culture Religious Experience," *Atlanta Journal / Atlanta Constitution* 1 Feb. 1987.

5. Mark McCain, "Growth Sought for Bustling Area," *New York Times* 1 Feb. 1987.

LIGHTING DISTRICT 53

1. Wolfgang Schivelbusch, *Enchanted Night: The Industrialization of Light in the Nineteenth Century* (Berkeley: Univ. of California Press, 1988) 97.

MIXED/MULTI-USE COMPLEX 54

1. Guy Brett, *Kinetic Art: The Language of Movement* (New York: Reinhold, 1968) 35.

2. Douglas Frantz, *From the Ground Up: The Business of Building in the Age of Money* (New York: Henry Holt & Co., 1991).

3. Steven Spalding, "A Unified Effort Is Needed to Attract National Conferences and Trade Shows," *Business First* (Louisville, Ky.) 6 Apr. 1987.

4. William E. Schmidt, "A Tara in Atlanta's Future?" *Courier-Journal* (Louisville, Ky.) 20 Jan. 1985.

5. Patricia Fuller, "Why Public Art?" *Miami Herald* 20 Jan. 1985.

6. Winston Williams, "Southland's Quest for Glamour," *New York Times* 16 Mar. 1986.

7. Sheldon Shafer, "Housing, Office Complex Planned by Texas Group," *Courier-Journal* 20 Jan. 1985.

8. William E. Schmidt, "Once-Rural Georgia County Now Has Fastest Growth in U.S.," *New York Times* 2 June 1985.

9. *A Glossary of Housing Terms,* compiled by a committee from five principal federal agencies concerned with housing, William C. Moore, Chairman (Washington, D.C.: Central Housing Committee, 1937).

10. John Rebchook, "Japan Firm Buys Aurora Complex for $13 Million," *Rocky Mountain News* (Denver, Colo.) 7 Nov. 1987.

11. Marshall Berman, *All That Is Solid Melts into Air* (New York: Penguin, 1982) 243.

12. Iver Peterson, "Dakota Water Project Is Cut by Reagan Panel," *New York Times* 16 Dec. 1984.

NO SMOKING AREA 59

1. Peter S. Greenberg, "Where There's Smoke, There's Ire," *Los Angeles Times* 7 Dec. 1986.

2. Carol McGraw, "Ban on Smoking in Beverly Hills Fails to Clear Air in Restaurants," *Los Angeles Times* 31 May 1987.

3. Mike Brown and Robert T. Garrett, "Ford Upholds Restrictions on Press-Gallery Smoking," *Courier-Journal* (Louisville, Ky.) 11 Dec. 1988.

4. Judith Egerton, "Anti-smoking Rules at Ford Draw Fire at Forum," *Courier-Journal* 9 June 1989.

5. William U. Chandler, "Smoking Epidemic Widens," *World Watch* Jan.–Feb. 1988: 39.

PARTY STREET 61

1. Martha Knight, letter to the editor, *Courier-Journal* (Louisville, Ky.) 6 June 1990.

2. Jim Adams, "Party Street Gives Police a Run for Their Money," *Courier-Journal* 6 May 1990.

3. Gabriel Escobar, "The Price of Popularity," *Washington Post* 26 Apr. 1992.

RIOTSVILLE/RIOT CITY, U.S.A./ RIOT SCENE 62

1. Abbie Hoffman, *Revolution for the Hell of It* (New York: Dial Press, 1968) 192.

SAFE HOUSE 63

1. Robert D. McFadden, "Fugitive in $1.6 Million Brinks Holdup Captured," *New York Times* 12 May 1985.

THE SCENE 64

1. Michael Jager, "Class Definition and the Esthetics of Gentrification: Victoriana in Melbourne," *Gentrification of the City,* ed. Neil Smith and Peter Williams (Boston: Allen, 1986) 86.

2. Erving Goffman, *Relations in Public* (New York: Basic Books, 1971) 19.

3. Philip Weiss, "Making the 'Downtown' Scene," *Columbia Journalism Review* Jan./Feb. 1988: 43.

SUPERBLOCK 66

1. Harvey S. Moskowitz and Carl G. Lindbloom, *The Illustrated Book of Development Definitions* (Piscataway, N.J.: Rutgers Univ. Press, 1981).

2. Ed Zotti, "Dreaming of Density," *Planning* Jan. 1986: 4.

Chapter 3: Perks 70

AIR RIGHTS AREA 70

1. Cleveland Rodgers and Rebecca B. Rankin, *New York: The World's Capital City* (New York: Harper, 1948) 220.

2. *New York City Guide ("The W.P.A. Guide")* ed. Lou Gody (New York: Random House, 1939) 234.

3. Jody Brott, "The Loop's Expanding Westward," *New York Times* 14 Feb. 1988.

4. Bob Weinstein, "Brooklyn's Fast-Emerging Office Market," *Crain's New York Business* 17 Oct. 1988: 33.

5. Julia Gilden, "Housing Built over a Store," *New York Times* 23 Oct. 1988.

6. "Ten Floors for Parking Planned on Top of a Chicago Church," *New York Times* 14 Jan. 1990.

CULTURAL ARTS DISTRICT 72

1. Sam Hall Kaplan, "Arts Park Concept Designed for Failure," *Los Angeles Times* 25 Jan. 1989.

2. Julie Collins (J.C.), "Urban Design Strategies for Dallas Arts District," *GSD News* Jan.–Feb. 1986: 10.

3. Calvin Tomkins, "The Art World: Dallas," *New Yorker* 13 June 1983: 97.

4. John R. Logan and Harvey L. Molotch, "The City as a Growth Machine," *Urban Fortunes: The Political Economy of Place* (Berkeley and Los Angeles: Univ. of California Press, 1987) 76.

5. Catherine C. Robbins, "Notebook," *New York Times* 11 Feb. 1990.

6. Grace Glueck, "East Village Gets on the Fast Track," *New York Times* 13 Jan. 1985.

7. Judith Weinraub, "Now Playing in Virginia," *Washington Post* 30 Sept. 1990.

8. Katherine Bishop, "Chinese-Americans Aim to Save Their History by Restoring Angel Island," *Courier-Journal* (Louisville, Ky.) 22 Nov. 1990.

DISTRICT 74

1. Mary Colby, "Invisible Governments Grow as Locals Face Fund Shortages," *City & State* 12 Mar. 1990: 11.

2. "Fifty Special Districts: Annual Financial Report," *City & State* 12 Mar. 1990: 23.

3. Stanley Crawford, *Mayordomo: Chronicle of an Acequia in Northern New Mexico* (Albuquerque: Univ. of New Mexico Press, 1988).

4. John Fisher, *Vital Signs, U.S.A.* (New York: Harper, 1975) 30.

5. Wynelle Wilson, "Analysis of the Special Disney World Legislation and the Economic Impact of Disney World," typescript, Florida Bureau of Economic Analysis, Florida Energy Office, Tallahassee, Fla., 1979: n.p.

6. Ernest R. Bartley, telephone interview, 30 Apr. 1992; Gail DeGeorge, "A Sweet Deal for Disney Is Souring Its Neighbors," *Business Week* 8 Aug. 1988: 48; Morton D. Winsberg, "Walt Disney World, Florida: The Creation of a Fantasy Landscape," typescript, Florida State Univ., 1992.

7. Gannett News Service and AP dispatches, "Lost That 'Magic' Feeling?" *Courier-Journal* (Louisville, Ky.) 29 Sept. 1991.

8. Lawrence J. Lebowitz, "State Rejects Disney Growth Plan," *Orlando Sentinel* 17 Jan. 1992.

9. Robert A. Caro, *The Power Broker: Robert*

Moses and the Fall of New York (New York: Knopf, 1974).

10. Patricia Nelson Limerick, *The Legacy of Conquest: The Unbroken Past of the American West* (New York: Norton, 1987) 51.

11. Kevin Lynch, *The Image of the City* (Cambridge: Technology Press, 1960).

EVENT/FESTIVAL SITE 78

1. Jerry Hopkins, Baron Wolman, and Jim Marshall, *Festival! The Book of American Music Celebrations* (New York: Macmillan, 1970) 5.

2. Hopkins, Wolman, and Marshall 114.

3. Hopkins, Wolman, and Marshall 5.

4. Grady Clay, "The Swarmers," *Courier-Journal Magazine* (Louisville, Ky.) 29 Aug. 1971: 13.

5. Grady Clay, "The Roving Eye: On Baltimore's Inner Harbor," *Landscape Architecture* Nov. 1982: 48–53.

PARADE ROUTE 79

1. Valerie Lagauskas, *Parades: How to Plan, Promote, and Stage Them* (New York: Sterling, 1982) 27–30.

2. Lagauskas 69–70.

3. "A Stately Procession to Miss Liberty," *New York Times* 29 June 1986.

4. "The Final Push," *Press Register* 17 Feb. 1985.

5. Carole Rifkind, *Main Street: The Face of Urban America* (New York: Harper & Row, 1977) 188.

6. Lewis L. Gould, *Lady Bird Johnson and the Environment* (Lawrence: Univ. Press of Kansas, 1988) 98–99.

PRESIDENTIAL SITE 83

1. Wilbur Zelinsky, "Nationalistic Pilgrimages in the United States," in *Pilgrimage in the United States*, ed. Gisbert Rinschede and Surinder M. Bhardwaj, Geographia Religionum, Band 5 (Berlin: Dietrich Reimer Verlag, 1990) 256.

2. Arthur M. Schlesinger, Jr., *The Imperial Presidency* (Boston: Houghton Mifflin, 1973, 1989) ix.

3. Dennis V. N. McCarthy with Philip W. Smith, *Protecting the President* (New York: William Morrow, 1985) 100.

4. Richard M. Pious, *The American Presidency* (New York: Basic Books, 1979) 243.

5. Bradley H. Patterson, Jr., *The Ring of Power* (New York: Basic Books, 1988) 339.

6. *Budget of the United States Government (Appendix)* (Washington, D.C.: U.S. Government Printing Office, Fiscal Year 1990) I-C1.

7. Daniel Patrick Moynihan, "The Peace Dividend," *New York Review of Books* 28 June 1990: 5.

8. Bruce Sievers, "Putting a President's Library in Its Proper Place," *Los Angeles Times* 24 May 1987.

9. Sam Allis, "A Small Town Goes Prime-Time," *Time* 9 Jan. 1989.

10. Zelinsky 262.

SECURITY 87

1. Geoffrey Hughes, *Words in Time* (London: Basil Blackwell, 1988) 222.

2. George Steiner, *The Death of Tragedy* (London: Faber, 1951) 56–57.

3. David Gelman and Rich Thomas, "Banality and Terror" *Newsweek* 6 Jan. 1986: 60.

4. Grady Clay, *Closeup: How to Read the American City* (Chicago: Univ. of Chicago Press, 1973, rpt. 1980).

SKYLINE 90

1. Florence King, review of *Texasville*, by Larry McMurtry, *Los Angeles Times* 12 Apr. 1987.

2. Robin Garr, "What's Doing in Louisville," *New York Times* 9 Apr. 1989.

3. Evelyn DeWolfe, "Irvine Tower Reflects Rising-Skyline Trend," *Los Angeles Times* 10 Aug. 1986.

4. Brendan Gill, "The Sky Line: Above/Below," *New Yorker* 8 Feb. 1988: 90.

5. Tim Loughran, "Looking at a Dixie Cup Skyline," *Crain's New York Business* 17 Oct. 1988: 47.

6. Denise Goodman, "Heavens," *Boston Globe* 30 Oct. 1988.

Chapter 4: Ephemera 102

ARRIVAL ZONE 102

1. Wilbur Zelinsky, "Where Every Town Is Above Average: Parsing Those Welcoming Signs along America's Highways," *Landscape* vol. 30, no. 1 (1988): 1–10.

2. Donald Appleyard, Kevin Lynch, and John R. Myer, *The View from the Road* (Cambridge, Mass.: MIT Press, 1964).

3. William Attoe, *Skylines: Understanding and Moulding Urban Silhouettes* (Chichester: John Wiley & Sons, 1981).

4. Paul Daniel Marriott, office of planning, Rochester, N.Y., letter to the author, 13 Sept. 1991.

BOOMTOWN 104

1. Edward K. Spann, *The New Metropolis: New York City, 1840–1857* (New York: Columbia Univ. Press, 1981) 24.

2. Spann 158.

3. Stephen J. Pyne, *Fire in America: A Cultural History of Wildland and Rural Fire* (Princeton: Princeton Univ. Press, 1982) 165.

4. Michael VerMeulen, "Boomtown!" *Parade* 17 Feb. 1985: 8.

5. Charles Hillinger, "Washington Apple Area Yields Juicy Gold Mine," *Courier-Journal* (Louisville, Ky.) 1 Dec. 1985.

6. Mary Williams Walsh, "The Great Canadian Diamond Rush," *Los Angeles Times* 4 Oct. 1992.

7. Scott Thybony, "Rocky Mountain Refuge," *Outside* Jan. 1986: 69.

8. Art Seidenbaum, "Booming Houston: A City under Glass Lids," *Los Angeles Times* 6 Apr. 1975.

9. Geoffrey Hughes, *Words in Time: A Social History of the English Vocabulary* (Oxford: Basil Blackwell, 1989) 86.

CHANGING NEIGHBORHOOD 107

1. Samuel G. Freedman, "Metropolis of the Mind," *New York Times Magazine* 4 Nov. 1984: 68.

COMMUNITY BONFIRE 109

1. Mary Ann Sternberg, "Putting up the Christmas Lights Louisiana-Style," *Smithsonian* Dec. 1989: 147.

2. *Oxford English Dictionary,* 1971 ed.

3. American Heritage editors, *American Album* (n.p.: American Heritage Publishing Co., 1968) 329.

DESTINATION 109

1. U.S. Federal Highway Administration, *Highway Statistics* (Washington, D.C.: Government Printing Office, 1989); U.S. Federal Highway Administration, Motor Carrier Division, phone call, July 1992; U.S. Department of Transportation, *Air Carrier Traffic Statistics Monthly* 1991.

DROP ZONE 112

1. Raymond E. Murphy, *The American City* (New York: McGraw, 1966) 368.

2. William Warntz, *Macrogeography and Income Fronts,* Monograph Series No. 3 (Philadelphia: Regional Science Research Institute, 1965).

EARSHOT 114

1. David White, "Noise Is the Sound That Other People Make," *New Society* 28 May 1981: 343.

FLOODWAY 115

1. William D. Moore, Waldo L. Born, and Judon Fambrough, "The Truth about Flooding," *Tierra Grande No. 27* (Austin: Texas Real Estate Research Center, 1985) 20.

2. Stephen A. Thompson and Gilbert F. White, "A National Floodplain Map," *Journal of Soil and Water Conservation* Sept.–Oct. 1985: 417.

3. Associated Press, "New Law to Help Flood Victims to Relocate," *New York Times* 5 Dec. 1993.

4. David Holmstrom, "US Rethinks Flood-Relief Policy," *Christian Science Monitor* 1 Dec. 1993.

GENTRIFYING NEIGHBORHOOD 117

1. Briavel Holcomb, review of *Gentrification, Displacement and Neighborhood Revitalization,* ed. J. John Palen and Bruce London, *Professional Geographer* 1986: 441.

2. Ruth Glass, *London: Aspects of Change* (London: MacGibbon & Kee, 1964) xviii.

3. Richard T. LeGates and Chester Hartman, "The Anatomy of Displacement in the United States," *The Gentrification of the City,* ed. Neil Smith and Peter Williams (Boston: Allen & Unwin, 1986) 198.

4. Jeffrey R. Henig, "Neighborhood Response to Gentrification," *Urban Affairs Quarterly* Mar. 1982: 353.

5. Neil Smith, "New City, New Frontier," in *Variations on a Theme Park,* ed. Michael Sorkin (New York: Hill & Wang, Noonday Press, 1992) 85.

GUNSHOT 119

1. David Freed, "Proliferation of Guns May Be Bloody Legacy of Riots," *Los Angeles Times* 17 May 1992, A22.

2. David Freed, "L.A. County Found Armed, Dangerous," *Los Angeles Times* 17 May 1992, 1.

HANGOUT 120

1. H. L. Mencken, *The American Language: Supplement I* (New York: Knopf, 1945) 444; *The American Language* (New York: Knopf, 1937) 582.

2. *Webster's Third New International Dictionary,* 1967 ed.

3. Susan Orlean, *Saturday Night* (New York: Knopf, 1990).

4. John Voskuhl, "Jackson," *Courier-Journal* (Louisville, Ky.) 5 Jan. 1992.

5. Proprietor of The Hilltop, author's interview, 30–31 May 1980.

6. Marcida Dodson, "Hidden Hot Spots," *Los Angeles Times* 14 Apr. 1991.

HURRICANE PATH 122

1. Wallace Kaufman and Orrin Pilkey, *The Beaches Are Moving* (New York: Anchor-Doubleday, 1979) 129.

2. E. B. White, "The Eye of Edna," *The Essays of E. B. White* (New York: Harper, 1979) 29–33.

3. William Booth, "South Florida Left Looking Like the End of the World," *Courier-Journal* (Louisville, Ky.) 30 Aug. 1992.

4. Mark Seibel, "Hurricane Andrew—An Old Lesson," *Nieman Reports* winter 1992: 19, 21.

5. Mike Clary, "Hurricane Andrew Not Over," *Los Angeles Times* 23 Aug. 1993.

6. Peter Kerr, "How Big a Disaster Can

Insurers Survive?" *New York Times* 13 Sept. 1992.

7. James Barth, professor of finance, Auburn (Ala.) University, conversation with author, 17 Nov. 1993.

THE SETTING 126

1. Real estate ad, *Robb Report* Aug. 1984: 146.

2. Author's interview with an editor of the former English-language edition of *Geo* in New York, N.Y., c. 1982.

THE SETUP 126

1. Bonnie S. Schwartz, "Setting the Scene for the Stones," *Metropolis* Mar. 1990, 22–23.

2. Schwartz 22–23.

3. Roger Fristoe, "When 'Starlight Express' Rolls into Town, Technicians Become the Stars of the Show," *Courier-Journal* (Louisville, Ky.) 3 June 1990.

4. Leo Sandon, "Religion in America," *Tallahassee Democrat* 2 Dec. 1989.

TOADS 129

1. Michael R. Greenberg, Frank J. Popper, and Bernadette M. West, "The TOADS: A New American Urban Epidemic," *Urban Affairs Quarterly* vol. 25, no. 3 (Mar. 1990): 435.

2. Greenberg, Popper, and West 435.

3. Greenberg, Popper, and West 449.

4. Greenberg, Popper, and West.

VACANT LOT 130

1. John H. Niedercorn and Edward F. R. Hearle, *Recent Land Use Trends in Forty-Eight Large American Cities,* table 8 (Santa Monica, Cal.: Rand Corporation Memorandum RM-3664-FF, June 1963); see also National Commission on Urban Problems, "Land Use in 106 Large Cities," *Three Land Research Studies,* Study no. 2, Research Report no. 12 (Washington, D.C.: Government Printing Office, 1968).

2. Charles E. Winter, *Four Hundred Million Acres* (Casper, Wyo.: Overland Publishing Co., 1932) 42.

VIEWSHED/THE VIEW 131

1. R. Burton Litton, Jr., "Descriptive Approaches to Landscape Analysis," *Our National Landscape* (Berkeley: USDA Forest Service-Pacific SW Forest & Range Expt. Station, 1979) 77–87.

2. Hardesty Associates, "Guidelines for Land Management, Portola Valley Ranch," report, Palo Alto, Cal., 1982.

3. *Journal of the American Planning Association* summer 1987: 363.

4. Michael A. Hiltzik, "Skyscraper Prompts an Environmental Uproar," *Los Angeles Times* 17 Dec. 1989.

5. Hiltzik.

WRECK SITE 133

1. Bruce Keppel, "New Private Eyes Focus on Disasters," *Los Angeles Times* 13 Jan. 1985.

Chapter 5: Testing Grounds 136

ACTIVE ZONE 136

1. Nehl Horton, "Police Pact with Neighborhood Credited with Decline in Crimes," *Intown Extra* (Atlanta, Ga.) 31 Oct. 1985.

ANNEXATION AREA 136

1. Tom Moore, "Town Delays Decision on Annexations," *Chapel Hill Newspaper* 12 Apr. 1988.

2. Douglas Vaughan, "Colorado Springs Annexes Ranch, Growing Nearly 50 Percent," *New York Times* 19 Feb. 1989.

CONVENIENT LOCATION 138

1. Josh Getlin, "Planners Seek Solution to Suburban Gridlock," *Los Angeles Times* 8 Jan. 1989.

GROWTH AREA 139

1. Marshall Berman, *All That Is Solid Melts into Air: The Experience of Modernity* (New York: Penguin, 1982) 61.

2. Anthony DePalma, "Newark Airport Border Lures Developers," *New York Times* 9 Dec. 1984.

3. Anthony DePalma, "A 10,000 Acre City Is Growing in the Everglades," *New York Times* 2 Oct. 1988.

4. "Drivers in Virginia Shun New Car Pool Lanes," *New York Times* 7 Dec. 1986.

5. Tom Gorman, "Desert Bloom: Dizzying Pace of Growth Takes Its Toll on Temecula, Murrieta," *Los Angeles Times* 24 Nov. 1991.

6. Paul Goldberger, "Orange County: Tomorrowland—Wall to Wall," *New York Times* 11 Dec. 1988.

GROWTH CONTROL DISTRICT 141

1. George Will, "'Slow Growth' Is the Liberalism of the Privileged," *Los Angeles Times* Aug. 1987.

2. Robert Lindsey, "Sunny San Diego Acts to Slow Its Rapid Growth," *New York Times* 28 June 1987.

3. David Dubbink, "I'll Have My Town Medium-Rural, Please," *American Planning Association Journal* fall 1984.

SHORTCUT 144

1. Robert Lloyd, review of *Route Choice: Wayfinding in Transport Networks,* by Piet H. L. Bovy and Eliahu Stern, *Annals of the American Association of Geographers* June 1992: 323.

Chapter 6: Border Zones 150

BROWSERS' VILLAGE 150

1. William Trombley, "Westwood Village Looks to Its Past for Its Future," *Los Angeles Times* 5 July 1987.

2. *Oxford English Dictionary.*

3. Eric Garland, "Building a Cozy 'Village' for Suburban Shoppers," *New York Times* 13 Sept. 1987.

THE BURN 151

1. Stephen J. Pyne, *Fire in America: A Cultural History of Wildland and Rural Fire* (Princeton: Princeton Univ. Press, 1982) 18.

2. Christine M. Rosen, *The Limits of Power: Great Fires and the Process of City Growth in America* (Cambridge: Cambridge Univ. Press, 1986).

3. Exhibit, Cape Cod Museum, 1986.

4. Pyne 252.

5. Pyne 330.

6. U.S. Department of Commerce, *Historical Statistics of the United States, Part 1* (U.S. Dept. of Commerce, 1975).

7. Judith Cummings, "Thousands of Firefighters Struggle against Infernos in Western States," *Courier-Journal* (Louisville, Ky.) 6 Sept. 1987.

8. Ron Hodgson, California landscape architect and firefighter, conversation with author, 7 May 1988.

9. James B. Davis, "Use of the 1990 Census to Define Wildland Urban Interface Problems," *Fire and the Environment: Ecological and Cultural Perspectives,* ed. Stephen C. Nodvin and Thomas A. Waldrop (Asheville, N.C.: U.S. Department of Agriculture, Forest Service, Southeastern Forest Experiment Station) 384.

10. Pyne 390–91.

THE EDGE 154

1. Barry Schiff, "Summer on the Icy Edge of the World," *Los Angeles Times* 5 Jan. 1986.

2. S. D. Joardar and J. W. Neill, "The Subtle Differences in Configuration of Small Public Spaces," *Landscape Architecture* Nov. 1978: 487.

3. William H. Whyte, Jr., *City: Rediscovering the Center* (New York: Doubleday, 1988) 107–8.

4. Marshall McLuhan, *War and Peace in the Global Village* (New York: McGraw-Hill, 1968) 91.

5. Patt Morrison, "Western Avenue," *Los Angeles Times* 28 July 1985.

6. Ray Raphael, *Edges: Human Ecology of the Backcountry* (Lincoln, Neb.: Univ. of Nebraska Press, 1986) 112.

7. Rachel Carson, *The Edge of the Sea* (Boston: Houghton, 1955) 7.

8. Israel Rosenfield, review of *Vision: A Computational Investigation into the Human Representation and Processing of Visual Information,* by David Marr, *New York Review of Books* 11 Oct. 1984: 55.

9. Grady Clay, "This Must Be the Place, But Where Is Everybody? Observations in Cultural Geography," lecture text, Louisville Collegiate School, Louisville, Ky., 13 May 1992.

10. Battery Park City Authority, map, "Filling in the Edges of Manhattan," *New York Times* 22 Mar. 1987.

THE EDGE OF TOWN 157

1. Thomas E. Jacobson, "Sins of the Neotraditionalists," letter to the editor, *Planning* Nov. 1989: 29.

LOVERS' LEAP 160

1. "Noccalula Falls Park and Campground," brochure, Noccalula Falls Campground, Gadsden, Ala., 1989.

2. Jeffery Ray Jones, *Noccalula: Legend, Fact and Function* (Collinsville, Ala.: Jeffery & Jones Gang, 1989).

SUBURBIA 163

1. Robert C. Wood, *Suburbia: Its People and Their Politics* (Boston: Houghton, 1959) 4.

2. Camilo Jose Vergara, "Detroit Waits for the Millennium," *Nation* 18 May 1992.

3. Ed Ayres, "Breaking Away," *World Watch* Jan./Feb. 1993: 13.

4. Jim Newton and Mark Landsbaum, "Developers Lead Campaign Funding in Orange County," *Los Angeles Times* 3 Nov. 1991.

5. Marshall Kaplan, *The Dream Deferred: People, Politics, and Planning in Suburbia* (New York: Vintage, 1977) 14.

6. This account also benefits from the author's membership in The President's Task Force on Suburban Problems, 1967–1968, chaired by Charles M. Haar. O. Harold Folk, in accordance with guidelines supplied by the Task Force, prepared *Final Report: Urban Development Bank,* December 2, 1968, one of seven volumes of the *Final Report of The President's Task Force on Suburban Problems.*

TWILIGHT ZONE 165

1. A. O. Lovejoy, *The Great Chain of Being* (Cambridge: Harvard Univ. Press, 1964) 56.

2. E. A. Gutkind, *The Twilight of Cities* (New York: Free Press of Glencoe, 1982) 197.

3. *Dictionary of Sociology,* ed. Henry Pratt Fairchild (New York: Philosophical Library, 1944) 342.

THE VILLAGE 167

1. Richard Lingeman, *Small Town America: A Narrative History 1620–The Present* (Boston: Houghton Mifflin, 1980) 259.

2. John McMullan, "An Urbanized Nation Still Needs Its Villages," Knight-Ridder News Service 6 Aug. 1982.

3. Denise Cabrera, "California Community Rated Tops in Survey of 'Micropolitan' Area," AP *Indianapolis Star* 22 Apr. 1990.

4. Donald S. Connery, *One American Town* (New York: Simon, 1972) 39.

5. Lawrence O. Houstoun, Jr., "Living Villages: Thoughts on the Future of the Village Form," *Small Town* Nov/Dec. 1988: 14.

6. Rob Brofman and Sue Allison, "Village Bashing," *Life* vol. 12, Apr. 1989: 10.

7. Sheila Rule, "Critics Wonder If Ethiopia Is Creating Villages or Collectivism," *New York Times* 22 June 1986.

8. Herbert Gans, *The Urban Villagers* (New York: Free Press, 1962).

9. Ruth Knack, review of *Village Planning Handbook from Bucks Co., Pa.,* in *Planning* Mar. 1990: 6–7.

10. Susan Diesenhouse, "Subsidies Aid 'Farm Village,'" *New York Times* 25 Mar. 1990.

11. Sam Hall Kaplan, "Urban Village: A Grab Bag of Goals," *Los Angeles Times* 14 Jan. 1990.

12. Diana Scott, "A New Idea in New Haven," *Metropolis* Apr. 1990: 18.

13. Laura Thomas, "Can a New Suburb Be like a Small Town?" *U.S. News & World Report* 5 March 1990: 32.

14. Beth Mollard, "Static from Radio Church," *New York Times* 28 May 1989.

15. Fox Butterfield, "The Perfect New England Village," *New York Times* 14 May 1989.

16. Cited by Richard Critchfield, *Villages* (New York: Anchor/Doubleday, 1983) 207.

Introduction to Section 3: Out There 170

1. Anthony DePalma, "Office Development Surges in Suburbs," *New York Times* 25 May 1986.

2. Barbara Ehrenreich, *Fear of Falling* (New York: Pantheon Books, 1989) 206.

3. William F. Buckley, Jr., "I Am Lapidary but Not Eristic When I Use Big Words," *New York Times Book Review* 30 Nov. 1986: 3.

4. Edward K. Spann, *The New Metropolis: New York City, 1840–1857* (New York: Columbia Univ. Press, 1981) 417.

5. Paul Goldberger, "When Suburban Sprawl Meets Upward Mobility," *New York Times* 27 July 1987.

6. William K. Stevens, "Beyond the Mall: Suburbs Evolving Into 'Outer Cities,'" *New York Times* 8 Nov. 1987.

7. Joel Garreau, *Edge City* (New York: Doubleday, 1991).

8. Marjie Lundstrom, "Sandra Atchison: Pioneer in Her Own Right," *Denver Post* 7 Aug. 1985.

9. Timothy Egan, "Outrage Grows in West over City Slickers as Ranching Neighbors," *New York Times* 22 Dec. 1991.

10. Anne Matthews, "An Academic Couple Brings an Unwelcome Message to the People of the Great Plains," *High Country News* (Paonia, Colo.) 16 Dec. 1991.

11. Mary Beth Regan, "EPA Gives Up on Cleaning Nation's Worst Toxic Sites," *Orlando Sentinel* 23 Aug. 1992.

12. James Warner Bellah, "Maryland," *American Panorama,* A Holiday Magazine Book (Garden City, N.Y.: Doubleday, 1960) 286.

13. AP, "Teamster Aide Held in Contempt for Balking at Outside Supervision," *New York Times* 10 Dec. 1989.

Chapter 7: Power Vacuum 178

ABANDONED FARM/AREA/TOWN 178

1. Steven A. Channing, *Kentucky: A Bicentennial History* (New York: Norton, 1977) 87.

2. John Stilgoe, *Common Landscape of America 1580–1845* (New Haven: Yale Univ. Press, 1983) 186.

3. James Howard Kunstler, "Schuylerville Stands Still," *New York Times Magazine* 25 Mar. 1990: 50.

4. Jack Doherty, "A Camping Trip during the Great Depression," *Los Angeles Times* 2 Mar. 1986.

5. Deborah Epstein Popper and Frank J. Popper, "The Fate of the Plains," *High Country News* (Paonia, Colo.) 26 Sept. 1988.

THE BOONDOCKS 180

1. William Least Heat Moon, *Blue Highways* (New York: Ballantine, 1982) 7.

2. Anthony DePalma, "It's Boom Time in What Once Was the Boonies," *New York Times* 10 Feb. 1985.

3. Anna Quindlen, "'Five O'Clock Dads,'" *New York Times* News Service 20 Feb. 1990.

4. Berkeley Rice, "Keeping a Country Inn," *US Air* magazine Sept. 1986: 49.

5. National Public Radio discussion, 18 Oct. 1988.

6. John R. Logan and Harvey L. Molotch, *Urban Fortunes: The Political Economy of Place* (Berkeley: Univ. of California Press, 1987) 264.

THE DARK 181

1. Ralph Vartabedian, "'Black World' of Defense Cloaks Firms in Secrecy," *Los Angeles Times* 10 Apr. 1988: 7.

2. M. Mitchell Waldrop, "Taking Back the Night," *Science* 9 Sept. 1988: 1288.

3. "News: Darkness on the Edge of Town," *Planning* Apr. 1985: 30.

4. AP, "San Diego Panel Turns Off Lights for Astronomers," *Courier-Journal* (Louisville, Ky.) 15 Dec. 1984.

5. Waldrop 1289.

DEPRESSED AREA/REGION 182

1. Lettice Stuart, "In Houston, Bottom-Fishing Brings Up Bargains: Out-of-Towners Pursue Deals in Depressed Area," *New York Times* 28 Jan. 1990.

2. Robert Burroughs, "What to Do (or Not Do) When the Local Economy Goes Bust," *Publisher's Weekly* 7 July 1989.

3. Susan Diesenhouse, "Bank Failure Adds to Small Town's Housing Woes," *New York Times* 3 Dec. 1989.

4. Ronald J. Ostrow and William J. Eaton, "Areas Most in Need May Be Real HUD Scandal Victims," *Los Angeles Times* 2 July 1989.

5. Laurie Thomas, letter, *Planning* Apr. 1988.

6. Kevin Lynch, "On Wasting," manuscript circulated for review, 1986, 25.

7. Raad Cawthon, "Houston's Lack of Zoning Turns Porn into Neighborhood Affair," *Atlanta Journal-Constitution* 9 Nov. 1985.

8. Clifford Pugh, "Street People," *Houston Post* 2 May 1982.

9. Nancy Pieretti, "Summer Stirs Hope around New Hampshire Lakes," *New York Times* 8 July 1990.

10. J. Brandt Hummel, "Two Developers Bucking Slump," *New York Times* 8 July 1990.

DUMP 185

1. Michael Thompson, *Rubbish Theory: The Creation and Destruction of Value* (New York: Oxford Univ. Press, 1979).

2. Martin V. Melosi, *Garbage in the Cities:*

Refuse, Reform, and the Environment, 1880–1980 (College Station, Tex.: Texas A&M Univ. Press, 1981) 22.

3. "The Recycling of 'Dump,'" *New York Times* 16 Sept. 1990.

4. "Municipal Waste," editorial, *Science* 12 June 1987: 1409.

5. "Mount Trashmore: Dealing with Garbage While Protecting Groundwater," *The Futurist* Dec. 1984: 68.

6. Brendan Gill, "The Sky Line," *The New Yorker* 20 Aug. 1990: 69.

7. William Ashworth, *The Late, Great Lakes: An Environmental History* (New York: Knopf, 1986) 61.

8. Bob Hill, "Your Garbage," *Courier-Journal* (Louisville, Ky.) 15 Mar. 1992.

9. Peter T. White, "The Fascinating World of Trash," *National Geographic* Apr. 1983: 450.

10. White 432.

11. "The Big Squeeze" *Courier-Journal* 9 Mar. 1992.

12. Commentator, *Market Place,* National Public Radio, 22 Apr. 1992.

13. Timothy Egan, "A Garbage Success Story," *New York Times* 19 Mar. 1989.

14. Michelle Faul, "Recycled Trash and Junk Is One of Zimbabwe's Major Natural Resources," *Louisville Times* 27 Dec. 1983.

15. White 432.

GHOST TOWN 187

1. Lambert Florin, *Ghost Towns of the West* (n.p.: Promontory Press, 1970) 8–14.

2. Florin 111.

3. AP, "Ten Ghost Towns in County," 28 May 1947.

4. B. Drummond Ayres, Jr., "Delmarva: The Island on a Peninsula," *New York Times* 12 Sept. 1982.

5. Charles Colcock Jones, *The Dead Towns of Georgia* (Savannah Morning News Steam Printing House, 1878), cited by Calvin L. Beale in "Quantitative Dimensions of Decline and Stability among Rural Communities," in *Communities Left Behind: Alternatives for Development,* ed. Larry R. Whiting (Ames, Iowa: Iowa Univ. Press, 1971) 3.

6. Lambert Florin, *Ghost Town Album* (New York: Bonanza Books, 1962);Donald Harington, *Let Us Build Us a City: Eleven Lost Towns* (San Diego: Harcourt, 1986);J. R. Humphreys, *The Lost Towns and Roads of America* (New York: Harper & Row, 1967).

7. J. B. Jackson, *The Necessity for Ruins* (Amherst: Univ. Mass. Press, 1980).

SINK 189

1. John Chandler, "Pumping Threatens to Sink High Desert's Future," *Los Angeles Times* 17 Mar. 1991.

2. Chandler.

3. Grady Clay, *Close-Up* (Chicago: Univ. of Chicago Press, 1980) 143.

SUICIDE SPOT 190

1. Constance Holden, "A New Discipline Probes Suicide's Multiple Causes," *Science* 26 June 1992: 1761.

2. David H. Rosen, "Suicide Survivors: A Follow-up Study of Persons Who Survived Jumping from the Golden Gate and San Francisco–Oakland Bay Bridges," *Western Journal of Medicine* Apr. 1975.

3. Allen Brown, *Golden Gate: Biography of a Bridge* (Garden City, N.Y.: Doubleday, 1965) 199.

4. Jenifer Warren, "City Trying to Stop Golden Gate Jumpers," *Los Angeles Times* 8 Dec. 1993.

5. Jenifer Warren, "Dimming the Fatal Allure of Golden Gate Bridge," *Los Angeles Times* 7 Nov. 1993.

6. Tom McNichol, "Choosing Death," *Los Angeles Times Magazine* 13 Oct. 1991: 25.

7. Lisa Petrillo, "Bridge of No Return," *San Diego Union* 6 Apr. 1987.

8. Rosen 289–91.

9. Petrillo.

10. John F. Bonfatti, "Niagara Falls a Favored and Lethal Spot for Suicide Tries," Associated Press 11 Nov. 1987.

11. McNichol, "Choosing Death."

12. Warren, "Dimming the Fatal Allure."

TENT CITY 191

1. Eric Schmitt, "The Name Game," *Courier-Journal* (Louisville, Ky.) 4 Jan. 1987.

2. Katherine Bishop, "Shanty Power," *Courier-Journal* 3 Sept. 1989.

3. National Affairs section, *Newsweek* 20 Aug. 1984: 38.

4. Photo caption, *New York Times* 19 Aug. 1984.

5. Katherine Bishop.

6. Bill Boyarsky, "Can Downtown L.A. Survive Its Own Heady Prosperity?" *Los Angeles Times* 6 Jan. 1985.

7. Estes Thompson, "Army Tests Tent City That Offers the Troops the Comforts of Home," *Courier-Journal* 26 Dec. 1992.

Chapter 8: Opportunity Sites 196

AIRSPACE 196

1. *National Ocean Survey, 1989,* Public Information Office, Air Transport Association, Washington, D.C.

BLAST SITE 200

1. Bryan Di Salvatore, "Vehement Fire," *New Yorker* 27 April 1987: 72, quoting from Hudson Maxim, *Dynamite Stories* (New York: F. A. Stokes, 1916).

2. Douglas Pike, "Texas," *The Orlando Sentinel* 19 April 1987.

3. Jay Lawrence, "Sewer Blasts Leave City to Pick Up the Pieces," *Courier-Journal* (Louisville, Ky.) 13 Feb. 1981.

BYPASS 201

1. D. W. Meinig, *Southwest* (New York: Oxford Univ. Press, 1971) 76.

2. See sketches on pp. 92–93 of Grady Clay, *Closeup: How to Read the American City* (Chicago: Univ. of Chicago Press, 1973).

3. Joel Garreau, *Edge City: Life on the New Frontier* (New York: Doubleday, 1991).

CAMP 202

1. Ann Cornelisen, review of *The City and the House,* by Natalia Ginzburg, *New York Times Book Review* 13 Sept. 1987: 30.

2. William Cronon, *Changes in the Land* (New York: Hill & Wang, 1983) 46.

FLIGHT PATH 206

1. Joy Aschenbach, "FAA Control Centers Keep Skyway Rush Hour Traffic Flowing," *National Geographic* News Service 6 Apr. 1987.

2. Peter Ward, "Position of Strength Needed for Arms Talks," *Calgary Herald* 16 Oct. 1987.

THE FURROW 207

1. Robert Trow-Smith, *Life from the Land: The Growth of Farming in Western Europe* (London: Longmans Green, 1967) 14–15.

2. Fernand Braudel, *Afterthoughts on Material Civilization and Capitalism* (Baltimore: Johns Hopkins Univ. Press, 1977) 26.

3. Philip E. L. Smith, *Food Production and Its Consequences* (Menlo Park, Cal.: Cummings Publishing, 1976) 46–48.

4. John W. Oliver, *History of American Technology* (New York: Ronald Press, 1956) 224.

THE ICE 209

1. Dean MacCannell, *The Tourist: A New Theory of the Leisure Class* (New York: Schocken, 1976) 115.

THE MAILBOXES 210

1. William H. Whyte, Jr., *The Organization Man* (New York: Simon, 1956).

PACIFIC RIM 212

1. Clayton Jones, "Contest over Asia," *Christian Science Monitor* 17 Nov. 1993: 12.

2. Robert Strausz-Hupe, *Geopolitics: The Struggle for Space and Power* (New York: Putnam's, 1942) 68.

3. New York Times News Service, "As Asian Influence Rises, West Coast Struggling for Sense of Equilibrium," *Courier-Journal* (Louisville, Ky.) 10 May 1989.

4. Peter Grier, "U.S. Defense Goals Reflect Rising Importance of Asia," *Christian Science Monitor* 17 Nov. 1993: 13.

5. Simon Winchester, "Millard Fillmore Was Right," review of *The Pacific Century*, by Frank Gibney, *New York Times Book Review* 11 Oct. 1992.

6. John A. Alwin, "North American Geographers and the Pacific Rim: Leaders or Laggards," *Professional Geographer* vol. 44, no. 4 (Nov. 1992): 369.

7. John Lippman, "In Global Village, TV Is Lingua Franca," *International Herald Tribune* 23 Oct. 1992.

8. Jock O'Connell, "The Pacific Rim May Be Little but a Bill of Goods," *Los Angeles Times* 10 July 1988.

PRESENCE 214

1. Michael Bociurkiw, "U.S. Aid to a Troubled Philippines Refocuses Debate on Military Presence," *Los Angeles Times* 12 Aug. 1990.

2. Lee Hockstader, "Contras Maintaining Presence in Nicaragua," *Washington Post* 22 Sept. 1989.

3. Elizabeth Drew, "Letter from Washington," *New Yorker* 1 Apr. 1991.

4. Wilbur Zelinsky, "The Imprint of Central Authority," *The Making of the American Landscape*, ed. Michael P. Conzen (Boston: Unwin Hyman, 1990) 329.

5. Kathy Scruggs, "Federal Presence in Alabama Town Helps Cut Crime, Police Chief Says," *Atlanta Journal* Feb. 1990.

6. Zelinsky 311.

7. Kenneth L. Warriner, Jr., "Interpreting Presence," *Avant Garde* summer 1990: 50–51.

THE SAND CASTLES 216

1. Ted Siebert, *The Art of Sandcastling* (Seattle: Romar Books, 1990) 125.

2. Siebert 9.

3. Siebert 14–15.

4. Siebert 140.

SOLAR FARM 218

1. Robert H. Annan, Photovoltaic Energy Technology Division, Conservation and Renewable Energy, Department of Energy, Washington, D.C. 20585, letter, *Science* 4 Jan. 1985: 8.

WINDFARM 219

1. Frank R. Eldridge, *Wind Machines* (New York: Van Nostrand Reinhold, 1980) 77.

2. James A. Throgmorton, "Community Energy Planning: Winds of Change from the San Gorgonio Pass," *APA Journal* summer 1987: 359.

3. Devin Odell, "Wind Energy Makes Waves," *High Country News* (Paonia, Colo.) 29 June 1992.

4. Jerry Belcher, "Whirligigs, California's Forest of Windmills Is Generating Power,

Controversy," *Courier-Journal* (Louisville, Ky.) 9 June 1984.

5. Odell.

6. Throgmorton 360.

7. Throgmorton 364.

8. James R. Chiles, "Tomorrow's Energy Today," *Audubon* Jan. 1990: 62.

9. Chiles 61.

10. Jenifer Warren, "A Second Wind for Energy Industry," *Los Angeles Times* 24 Dec. 1989.

11. Eugene Linden, "Megacities," *Time* 11 Jan. 1993: 36.

12. John Saintcross, "Windmills on Top of New England's Horizon," *Christian Science Monitor* 8 Feb. 1993.

Chapter 9: The Limits 224

AVALANCHE ZONE 224

1. Roderick Peattie, *Mountain Geography* (Cambridge: Harvard Univ. Press, 1936) 57.

2. Peattie 55.

THE BEACH 225

1. "1694 Papers Show Fire Island Shift," *New York Times* 11 May 1958.

2. Wallace Kaufman and Orrin Pilkey, *The Beaches Are Moving: The Drowning of America's Shoreline* (New York: Anchor-Doubleday, 1979) 105.

3. Kaufman and Pilkey 117.

4. Author's interview with Per Brunn, former chairman of the Department of Coastal and Oceanographic Engineering, University of Florida.

BIOREGION 228

1. Alan AtKisson, "The Eco^3 Solution: An Interview with Donald Conroy," *In Context: A Quarterly of Humane Sustainable Culture* no. 24 (late winter 1990): 49.

DOWNWIND 231

1. Bryan DiSalvatore, "Vehement Fire," *New Yorker* 27 Apr. 1987: 49.

2. Jeffrey L. Rabin and George Stein, "Safety Fears in Torrance Strain Ties with Mobil," *Los Angeles Times* 16 Apr. 1989.

3. William Ashworth, *The Late, Great Lakes: An Environmental History* (New York: Knopf, 1986) 232.

4. Samuel G. Freedman, "Downwind from the Bomb," *New York Times Magazine* 9 Feb. 1986: 34.

5. John R. Logan and Harvey L. Molotch, *Urban Fortunes: The Political Economy of Place* (Berkeley and Los Angeles: Univ. of California Press, 1987) 225.

DROUGHT AREA 236

1. Albert J. Fritsch, "Communities at Risk: Environmental Dangers in Rural America," report of Renew America, Washington, D.C., 1989.

THE FISHERY 237

1. Raymond F. Dasmann, *The Conservation Alternative* (New York: Wiley & Sons, 1975) 67.

2. William Ashworth, *The Late, Great Lakes: An Environmental History* (New York: Knopf, 1986) 113.

3. Katherine M. Griffin, "Fishermen Assail Dumping of Silt in Ocean," *Los Angeles Times* 22 May 1988.

4. Linda Garman Weimer, "About Fish: Fish Farming," *New York Times* 20 Jan. 1991.

5. Elisabeth Mann Borgese, *Seafarm: The Story of Aquaculture* (New York: Abrams, 1977) 108.

6. Timothy Egan, "Dams May Fall to Give Rivers Back to Salmon," *New York Times* News Service 15 July 1990.

7. David S. Taylor, "The Battle for the South Platte," *Orvis News* Jan. 1989: 7.

8. Elmer A. Keen, *Ownership and Productivity of Marine Fishery Resources* (Blacksburg, Va.: McDonald & Woodward, 1988).

9. Krys Holmes, "IFQ Plans Inch Ahead in the Pacific and North Pacific," *National Fisherman* Nov 1993: 15.

GREENBELT 241

1. Mark Correll, Jane H. Lillydahl, and Larry D. Singell, "The Effects of Greenbelts on Residential Property Values: Some Findings on the Political Economy of Open Space," *Land Economics* May 1978: 207 (referring to Numbers 35:4).

2. Robert M. Leary, "Capital on the Ottawa," *Town Planning Review* Jan. 1970: 8.

3. K. C. Parsons, "Clarence Stein and the Green Belt Towns: Settling for Less," *APA Journal* spring 1990: 178.

4. "Cinderella's Condos?" *Newsweek* 4 Feb. 1985: 64.

5. Robert Kaiser, "Greenbelt Aimed at Controlling Growth, Keeping Bluegrass Green," *Herald-Leader* (Lexington, Ky.) 25 June 1989.

HAZARDOUS WASTE DUMP 243

1. Samuel S. Epstein, M.D., Lester O. Brown, and Carl Pope, *Hazardous Waste in America* (San Francisco: Sierra Club Books, 1982) 203.

IRRIGATED AREA/LANDS 244

1. M. T. Brown and Howard T. Odum, *A Basic Science of the System for Humanity and Nature* (Gainesville: Univ. of Florida Center for Wetlands, 1981).

2. Mark S. Hoffman, ed., *The World Almanac and Book of Facts, 1992* (New York: Pharos Books, 1991).

3. M. E. Anderson, "A Low-Flow Way to Quench Las Vegas's Thirst," *High Country News* (Paonia, Colo.) 6 Apr. 1992.

4. George Sibley, "The Desert Empire," *Harper's* Oct. 1977: 53.

5. Dan Goodgame, "Just Enough to Fight Over," *Time* 4 July 1988: 16.

6. Charles Putnam, executive vice president, Irrigation Association, Washington, D.C., personal interview, 2 Mar. 1993.

7. William Robbins, "Irrigation's Water Use Found to Be Mounting Problem," *New York Times* 10 Dec. 1989.

8. Philip Shabecoff, "Builder of West's Dams Now Sees to the Water," *New York Times* 1 Jan 1988.

9. Robert Reinhold, "New Age for Western Water Policy: Less for the Farm, More for the City," *New York Times* 11 Oct. 1992.

10. Gladwin Hill, "When the Bill for the Marvels Falls Due," review of *Cadillac Desert*, by Marc Reisner, *New York Times Book Review* 14 Sept. 1986: 40.

11. "Less Irrigation in the Future," *The Futurist* Nov.–Dec. 1986: 47.

12. Karl A. Wittfogel, "The Hydraulic Civilizations," in *Man's Role in Changing the Face of the Earth*, ed. William L. Thomas, Jr. (Chicago: Univ. of Chicago Press, 1956).

13. "Dying of Thirst," writer/director Susan Winslow, Public Broadcast Service Special, KET, Lexington, Ky., 14 Dec. 1993.

THE KUDZU 248

1. William Shurtleff and Akiko Aoyagi, *The Book of Kudzu: A Culinary and Healing Guide* (Brookline, Mass.: Autumn Press, 1977).

2. C. Furqueron, "Kudzu in the Southern

Landscape," *Georgia Landscape* spring 1990: 17.

3. Carol Kaesuk Yoon, "Vines: Upwardly Mobile Marvels," *Courier-Journal* (Louisville, Ky.) 19 Feb. 1993.

4. Frances Cawthon, "Is Kudzu Really a Secret Weapon?" *Atlanta Constitution* 20 Mar. 1983.

5. Shurtleff and Aoyagi 15.

6. Stephen J. Pyne, *Fire in America: A Cultural History of Wildland and Rural Fires* (Princeton: Princeton Univ. Press, 1982) 187.

7. Pyne 188.

8. Shurtleff and Aoyagi 17.

NATIONAL BORDER 253

1. "Charlemagne," *New Columbia Encyclopedia,* 1975 ed.

2. Bertrand Russell, *New Hopes for a Changing World* (London: n.p., 1951) 69; reprinted New York: Minerva Press, 1968.

3. Terril Jones, "Crossing into New Era," *Courier-Journal* (Louisville, Ky.) 2 Jan. 1993.

4. Norbert Elias, *The Civilizing Process* (New York: Urizen Books, 1978) 240.

5. Jean Gottmann, *The Significance of Territory* (Charlottesville: Univ. Press of Virginia, 1973) 138–39.

6. UPI dispatch, Eagle Pass, Tex., 18 Feb. 1985.

7. Osbert Lancaster, *Classical Landscape with Figures* (London: John Murray, 1947) 96.

8. *Baedeker's Portugal,* 2d edition (New York: Prentice Hall, 1983) 12.

9. O. B. Hardison, Jr., *Disappearing through the Skylight* (New York: Viking, 1989) 63.

10. Julian V. Minghi, "Boundary Studies in Political Geography," *Annals of the American Association of Geographers* vol. 38, no. 3 (Sept. 1963): 415.

11. Donald W. Meinig, *Imperial Texas: An Interpretive Essay in Cultural Geography* (Austin: Univ. of Texas Press, 1975) 56.

12. Bill Barich, "La Frontera," *New Yorker* 17 Dec. 1990: 72.

NATIONAL SACRIFICE AREA/ZONE 257

1. Ed Marston, "The West Lacks Social Glue," *High Country News* (Paonia, Colo.) 26 Sept. 1988.

2. David Maraniss and Michael Weisskopf, "Jobs and Illness in the Petrochemical Corridor," *Washington Post* 22 Dec. 1987.

3. Robert Alvarez, research director, Environmental Policy institute, Washington, D.C., personal interview, 9 Nov. 1991.

4. Ed Magnuson, "'They Lied to Us,'" *Time* 31 Oct. 1988.

5. William Ashworth, *The Late, Great Lakes: An Environmental History* (New York: Knopf, 1986) 65.

THE PUBLIC DOMAIN 258

1. Richard Sennett, *The Fall of Public Man* (New York: Vintage, 1978) 16.

2. "The Vietnam Documents," editorial, *New York Times* 16 June 1971.

WATER RANCH 260

1. Michael Parrish, "Speculating in Water," *Los Angeles Times* 10 Sept. 1989.

2. Parrish.

3. Sandra Postel, "The Selling of Western Water," *New York Times* 18 Feb. 1990.

WETLANDS 261

1. See, for example, William J. Mitsch and James G. Gosselink, *Wetlands* (New York: Van Nostrand Reinhold, 1986).

2. "Delaware's Silt Reclaims Marsh," *New York Times* 8 Dec. 1957.

3. U.S. Dept. of Interior, *Wetlands of the*

That universal man-made place called "There" survives innumerable uses and abuses, notably in Oakland, Cal.: after that city was quotably maligned by its former resident Gertrude Stein—who said, "When you get there, there's no there there"—the city of Oakland commissioned this downtown sculpture, called "There." The Oakland *Tribune* headlined this photo and its dedication story "So There!"

United States: Current Status and Recent Trends Mar. 1984: 1.

4. Benjamin Horace Hibbard, *History of Public Land Policies* (New York: Peter Smith, 1939) 278.

5. Mitsch and Gosselink 22.

6. Elizabeth Royte, "Showdown in Cattle Country," *New York Times Magazine* 16 Dec. 1990: 70.

7. "In Desert, Land's Fate by Rivers Is Unclear," *New York Times* 7 Dec. 1986.

8. Ellen Churchill Semple, *Influences of Geographic Environment* (New York: Henry Holt, 1911) 370–72.

WHALE WATCHING SITE 264

1. John M. Leighty, "Whale Watching Anchors New Industry to California Coast," *Houston Chronicle* 20 Jan. 1985.

Epilogue 266

1. Roger L. Payne, chief, Branch of Geographic Names, U.S. Geological Survey, Washington, D.C., letter to the author, 4 June 1992.

2. John Antenucci, Plangraphics, Inc., Frankfort, Ky., private communication, 1992.

3. Geographic Names Information System (database), Geographic Names Board, U.S. Geological Survey, Dec. 1993.

4. U.S. Federal Highway Administration, *Highway Statistics* (Washington, D.C.: U.S. Government Printing Office, 1989).

5. U.S. Department of Transportation, *Air Carrier Traffic Statistics Monthly*, 1991.

6. Information from Property Evaluation Office, Jefferson County, Louisville, Ky., July 1992.

7. Information from U.S. Bureau of Census, Jeffersonville, Ind., July 1992.

8. Bill McKibben, "The End of Nature," *New Yorker* 11 Sept. 1989.

9. Karl Polanyi, *The Great Transformation: The Political and Economic Origins of Our Time* (New York: Farrar & Rinehart, 1944).

GARAGE
PALM COURT
BART
CENTER WALK
J.B. WILLIAMS
PLAZA

INDEX

D

E